Plants in 16th and 17th Century

Medical Traditions

Edited by
Alain Touwaide

Scientific Committee
Michael Friedrich, Jost Gippert, Marilena Maniaci,
Paolo Odorico, Steve M. Oberhelman,
Dominik Wujastyk

Volume 8

Plants in 16th and 17th Century

Botany between Medicine and Science

Edited by
Fabrizio Baldassarri

DE GRUYTER

ISBN 978-3-11-221384-1
e-ISBN (PDF) 978-3-11-073993-0
e-ISBN (EPUB) 978-3-11-074000-4
ISSN 2567-6938

Library of Congress Control Number: 2023935148

Bibliographic information published by the Deutsche Nationalbibliothek
The Deutsche Nationalbibliothek lists this publication in the Deutsche Nationalbibliografie;
detailed bibliographic data are available on the internet at http://dnb.dnb.de.

© 2025 Walter de Gruyter GmbH, Berlin/Boston
This volume is text- and page-identical with the hardback published in 2023.
Typesetting: Integra Software Services Pvt. Ltd.
Printing and binding: CPI books GmbH, Leck

www.degruyter.com

Acknowledgments

The present collection of 8 essays stems from exchanges started at the annual meeting of the History of Science Society (HSS) at Utrecht in July 2019. Together with Alain Touwaide and the Institute for the Preservation of Medical Traditions I co-organized a panel entitled "Herbs, Plants, and Vegetal Bodies: Botanical Knowledge in Medical, Naturalistic, and Philosophical Contexts." The lovely encounters and rich discussions we had at that time furthered some serious discussion on the topic of plants in the pre-modern period. As we started planning to publish the contributions in a volume, Alain Touwaide offered to consider publication in the series *Medical Traditions* and graciously invited me to be taken the lead of this possible publication. A few months after the meeting the world changed: the pandemic broke out and everything took a halt. The project of the volume slowed down and took a different shape: whereas some authors readily accepted to take part in this programme, some had other opportunities to have their research published sooner. From the discussion in Utrecht, the volume developed into presentations in the seminar series "Plants in Early Modern Knowledge: History, Philosophy, and Medicine" I have coordinated at Ca' Foscari University of Venice and History and Philosophy Department, Indiana University, Bloomington, in addition to several informal exchanges either online or in person in Venice, Rome, Wolfenbüttel, Bloomington, Chicago, Ames, and Leiden. In spite of these changes, the original aim and meaning of the Utrecth meeting remained substantially identical.

In particular, I would like to thank all authors who have taken part in the volume for their patience and very important and proficient work, as well as all attendees to panels and seminars for their questions and availability in discussing plants. Special thanks go to Alain Touwaide for having discussed the project of the panel first, and of the volume later, and having significantly helped me at different stages of its preparation in an abundant exchange of emails and phone calls, and to Emanuela Appetiti for her support. I would also like to thank Sarah Kyle, for her help at several editorial stages, Florike Egmond, for the botanical discussions in Venice and Rome, and Iolanda Ventura, who supported me in the rough seas of the preparation of the volume and for the meetings in Bologna and Wolfenbüttel. I also thank Karen M. Reeds for having kindly accepted to read the whole manuscript, and then for having discussed it with me, and written a foreword: her knowledge of Medieval and Renaissance botany has been crucial in the production of the volume.

Last but not least, I would like to thank my supervisors, Marco Sgarbi and Domenico Bertoloni Meli for their support through the difficult tasks of a Marie Curie project. Financial support for this work has been provided by the European Union's Horizon 2020 research and innovation programme under the Marie Skłodovska-Curie Grant Agreement n.890770, "VegSciLif."

Contents

Acknowledgments —— V

List of Illustrations —— IX

List of Contributors —— XIII

Karen Reeds
Foreword —— 1

Fabrizio Baldassarri
Introduction: The World of Plants in Premodern Medical Knowledge —— 3

Sarah R. Kyle
A More Modern Order: Virtual Collaboration in the *Roccabonella Herbal* —— 19

Barbara Di Gennaro Splendore
Mediterranean Botany. Making Cross-Cultural Knowledge about Materia Medica in the Sixteenth Century —— 53

Tassanee Alleau
A Bridge to the Underworld? An Explanation of the Act of Digging up Plant Roots in Early Modern Medical Fictions —— 71

Aleida Offerhaus, Anastasia Stefanaki, and Tinde van Andel
Not just a Garden of Simples: Arranging the Growing Floristic Diversity in the Leiden Botanical Garden (1594–1740) —— 99

Fabrizio Baldassarri
From the Analogy with Animals to the Anatomy of Plants in Medicine: The Physiology of Living Processes from Harvey to Malpighi —— 121

Edoardo Pierini
Opium Taking: Blurring Experimentation and Pharmaceutical Theories —— 145

Federica Rotelli
The Accommodation of New World Plants in Early Modern Pharmacology: The Case of Cinchona Bark and the Challenges to Seventeenth-Century Galenism —— 169

Bettina Dietz
**Knots in a Web: Botany, Materia Medica, and South Asian Languages
in the Publication of Paul Hermann's** *Ceylon-Herbaria* **(ca. 1690–1770)** —— 197

Bibliography —— 211

Index —— 253

List of Illustrations

Fabrizio Baldassarri, The World of Plants

Figure 1   *Preface*, in Otto Brunfels, *Herbarum vivae eicones*. Strasbourg: Schottum, 1530–1532: p. 1. Source: www.gallica.fr /BNF —— **7**

Figure 2   *Musa cum fructu*, in Pietro Andrea Mattioli, *Commentarii, in libros sex Pedacii Dioscoridis Anazarbei, De materia medica* . . . Venezia: Valgrisi, 1573: p. 244. Source: www.gallica.fr /BNF —— **9**

Figure 3   *De exercitio eius: Analysis Herbariae*, in Adam Zalužanský y Zaluzian, *Methodi herbariae libri tres*. Prague: Draczieeni, 1592: f. Ee2 verso. Source: HAB Wolfenbüttel, Germany —— **13**

Sarah R. Kyle, A More Modern Order

Figure 1   *De Meliloto* (*Lotus corniculatus* L., bird's foot trefoil), by Andrea Amadio, in *Liber de simplicibus* (Roccabonella Herbal). Venice, Biblioteca Nazionale Marciana, *Lat*. VI, 59 (coll. 2548): f. 1 recto. 28.5 x 20.5 cm, watercolour on paper, Venice, ca. 1430–1459. Su concessione del Ministero della Cultura – Biblioteca Nazionale Marciana. Divieto di riproduzione —— **25**

Figure 2   *Del Meliloto* (*Lotus corniculatus* L., bird's foot trefoil), in *Carrara Herbal*. London, British Library, Egerton 2020: f. 15 recto. 35 x 24 cm, gouache on vellum, Padua, ca. 1390–1405. © British Library Board, Egerton 2020 —— **27**

Figure 3   *De popolo* (*Populus nigra* L., black poplar), by Andrea Amadio, in *Liber de simplicibus* (Roccabonella Herbal). Venice, Biblioteca Nazionale Marciana, *Lat*. VI, 59 (coll. 2548): f. 156 recto. 28.5 x 20.5 cm, watercolour on paper, Venice, ca. 1430–1459. Su concessione del Ministero della Cultura – Biblioteca Nazionale Marciana. Divieto di riproduzione —— **29**

Figure 4   *De Cicorea* (*Cichorium intybus* L., chicory), by Andrea Amadio, in *Liber de simplicibus* (Roccabonella Herbal). Venice, Biblioteca Nazionale Marciana, *Lat*. VI, 59 (coll. 2548): f. 207 recto. 28.5 x 20.5 cm, watercolour on paper, Venice, ca. 1430–1459. Su concessione del Ministero della Cultura – Biblioteca Nazionale Marciana. Divieto di riproduzione —— **30**

Figure 5   *De Cicorea* (*Cichorium intybus* L., chicory), by Nicolò Roccabonella, in *Liber de simplicibus* (Roccabonella Herbal). Venice, Biblioteca Nazionale Marciana, *Lat*. VI, 59 (coll. 2548): f. 207 verso. 28.5 x 20.5 cm, ink on paper, Venice, ca. 1430–1459. Su concessione del Ministero della Cultura – Biblioteca Nazionale Marciana. Divieto di riproduzione —— **32**

Figure 6   *Portraits of Hippocrates and Johannitius (above) and Hippocrates and Galen (below)*, by Lippo Vanni or Roberto d'Oderisio, in *Tractatus de herbis et plantis* (Herbal of Manfredus de Monte Imperiale). Paris, Bibliothèque nationale de France, Lat. 6823: f. 1 verso. 34.5 x 24.7 cm, Naples, ca. 1330–1340. Source: Bibliothèque nationale de France, Département des manuscrits, Latin 6823 —— **48**

Figure 7   *Portraits of Mesue and Bartolomeo of Salerno (above) and Averroes and Porphyry (below)*, by Lippo Vanni or Roberto d'Oderisio, in *Tractatus de herbis et plantis* (Herbal of Manfredus de Monte Imperiale): f. 2 recto. 34.5 x 24.7 cm, Naples, ca. 1330–1340. Source: Bibliothèque nationale de France, Département des manuscrits, Latin 6823 —— **49**

Tassanee Alleau, A Bridge to the Underworld?

Figure 1   *Man as an inverted tree*, in Laurent Van Haecht Goidtsenhoven, Μικροκόσμοσ *Parvus mundus*. Antwerp: Plantin, 1579. French translation: *Le Microcosme contenant divers tableaux de la vie humaine*. Amsterdam: chez Théodore Pierre, 1613: p. 35. Bibliothèque nationale de France, département Arsenal, 4-BL-3164. Source: www.gallica.fr /BNF —— **75**

Figure 2   *The metaphor of the underground, cave or grotto*, by Théodore de Bry (engraver), in [Grasshoff (ed.)], *Dyas Chymica Tripartita*. Frankfurt: Luca Jennis, 1625: frontispiece. Courtesy of Science History Institute. Source: https://digital.sciencehistory.org/works/5t34sk63f —— **77**

Figure 3   *Nature's correspondences in the Microcosm and macrocosm: "Integrae Naturae speculum Artisque imago"*, in Robert Fludd, *Utriusque cosmi maioris*. Oppenheim: Hieronymi Galleri, 1617: *planche hors texte*. Source: London, Wellcome Library —— **80**

Figure 4   *The female mandrake*, in [Johannes von Cuba], *Herbarius*. Mencz [Mainz]: [Peter Schöffer], 1485: f. 208 verso. Courtesy of Wellcome Library, Latin. n 93074319 Source: https://wellcomecollection.org/works/sr2sujvm —— **86**

Figure 5   *Uprooting the mandrake with a dog*, in Pierre Boaistuau, *Histoires prodigieuses*. London, Wellcome Library, MS. 137: f. 138 verso. Public Domain. Source: https://wellcomecollection.org/works/zx3es3zp —— **93**

Aleida Offerhaus, Anastasia Stefanaki, and Tinde van Andel, Not just a Garden of Simples

Figure 1A   *Yarrow (Achillea millefolium L.)*, in Bartholomeus Mini de Senis, *Tractatus de Herbis*. London British Library, Egerton MS 747: f. 66 recto. © British Library Board —— **100**

Figure 1B   *Yarrow (Achillea millefolium L.)*, in [Apuleius], *Herbarium*. Oxford, Bodleian Library, MS. Ashmole 1431: f. 23 verso. Digital Bodleian, licensed under CC-BY-NC 4.0 —— **100**

Figure 2A   *Specimen of Salvia canariensis L., described as 'Sclarea'*, in *Clifford Herbarium*. London, BM000557604, now kept at London, Natural History Museum. Right granted by the Board of trustees of the Natural History Museum, license CC-BY —— **114**

Figure 2B   *Specimen of Salvia canariensis L., described as 'Horminum'*, in *Boerhaave specimens*. Leiden, Naturalis Biodiversity Center, L 0142243. © Naturalis Biodiversity Center —— **114**

Figure 2C   *Specimen of Salvia canariensis L., described as 'Salvia'*, in *D'Oignies Herbarium*. Leiden, Naturalis Biodiversity Center, Book 6: f. 25. © Naturalis Biodiversity Center —— **114**

Figure 3A   *Specimen of Clutia pulchella L.*, in *Zierikzee Herbarium*. Zierikzee, the Stadhuismuseum: No 327 —— **115**

Figure 3B   *Specimen of Clutia pulchella L.*, in *D'Oignies Herbarium*. Leiden, Naturalis Biodiversity Center, L.3961055 and L.3961056, Book 2: f. 56. © Naturalis Biodiversity Center —— **115**

Figure 3C   *Specimen of Clutia pulchella L.*, in *Clifford Herbarium*. London, British Museum, BM000647328, now kept at London, Natural History Museum. Right granted by the Board of trustees of the Natural History Museum, license CC-BY —— **115**

Figure 3D   *Specimen of Clutia pulchella L.*, in *D'Oignies Herbarium*. Leiden, Naturalis Biodiversity Centre, Book 5: f. 47. © Naturalis Biodiversity Center —— **115**

Figure 4   *Asplenium scolopendrium L.*, in *D'Oignies Herbarium*. Leiden, Naturalis Biodiversity Centre, Book 4: f. 68 (accompanied by an inexpertly copied text of Boerhaave's catalogue [1720, I.24]: "Lingua Cervina qua pilltris Major: in uno pediculo quadroqua bifida"). © Naturalis Biodiversity Center —— **117**

Figure 5   *Gloriosa superba L., described as 'Metonica malbarorum'*, in *D'Oignies Herbarium*. Leiden, Naturalis Biodiversity Center, Book 5: f. 16. © Naturalis Biodiversity Center —— **117**

List of Illustrations — **XI**

Fabrizio Baldassarri, From the Analogy with Animals to the Anatomy of Plants in Medicine

Figure 1   *Branch of verbena* (as a whole, *Figura iiii*, and cut off, *Figura iii*), in Hieronymus Fabricius ab Aquapendente, *Tractatus Quatuor, IV. De venarum ostiolis*. Frankfurt: Harman Palthenij, 1624: plate II, pp. 152ff. Source: www.gallica.bnf.fr / BNF —— **125**

Figure 2   *Stinging nettle*, in Robert Hooke, *Micrographia*. London: John Martyn and James Allestry, 1665: scheme XV. Source: www.gallica.fr /BNF —— **129**

Figure 3   *Representation of channels and threads within woods*, in Nehemiah Grew, *The Anatomy of Plants*. London: W. Rawlings, 1682: tab. xxxix. Courtesy of Leiden University Library, 661A18 —— **134**

Figure 4   *Trachea*, in Marcello Malpighi, *Anatome plantarum*. London: John Martyn, 1675–1679, vol. 1: tab. vii, figura 29. Courtesy of Bibliothèque municipale de Lyon – Numelyo, cote doc. 22678 —— **136**

Figure 5   *The growth of the leaves of citrus*, in Marcello Malpighi, *Anatome plantarum*. London: John Martyn, 1675–1679, vol. 1: tab. xiii, figura 63. Courtesy of Bibliothèque municipale de Lyon – Numelyo, cote doc. 22678 —— **137**

Figure 6   *The stalk of lemon*, in Marcello Malpighi, *Anatome plantarum*. London: John Martyn, 1675–1679, vol. 1: tab. xv, figura 77. Courtesy of Bibliothèque municipale de Lyon – Numelyo, cote doc. 22678 —— **137**

Figure 7   *The surface of a leaf with utricles*, in Marcello Malpighi, *Anatome plantarum*. London: John Martyn, 1675–1679, vol. 1: tab. xvii, figura 88. Courtesy of Bibliothèque municipale de Lyon – Numelyo, cote doc. 22678 —— **138**

Figure 8   *A seed of opium in the plant*, in Marcello Malpighi, *Anatome plantarum*. London: John Martyn, 1675–1679, vol. 1: tab. xlvi, figura 262. Courtesy of Bibliothèque municipale de Lyon – Numelyo, cote doc. 22678 —— **139**

Figure 9   *Uterine vessels*, in Marcello Malpighi, *Opera posthuma*. London: A.&J. Churchill, 1697: tab. xi, figura v. Source: HAB Wolfenbüttel, Germany —— **140**

Edoardo Pierini, Opium Taking

Figure 1   *Domestic poppy*, in Pietro Andrea Mattioli, *I discorsi di M. Pietro Andrea Matthioli sanese . . . nelli sei libri di Pedacio Dioscoride Anazarbeo della materia medicinale*. Venezia: Appresso gli Heredi di Vincenzo Valgrisi, 1573: p. 675. Source: Bibliothèques d'Université Paris Cité, Histoire de la santé —— **149**

Figure 2   *Injection into crural vein of a dog*, in Johann Sigismund Elsholtz, *Clysmatica nova sive ratio qua in venam rectam medicamenta*. Coloniae Brandenburgicae: Reichelius, 1665: p. 13. Source: Wellcome Collection. Public Domain Mark —— **156**

Federica Rotelli, The Accommodation of New World Plants in Early Modern Pharmacology

Figure 1   *Planisphere, called "Mappa del Cantino"*, 1502, in Biblioteca Estense Universitaria di Modena, C.G.A.2. Courtesy of Biblioteca Estense Universitaria of Modena. Photo credits of the Ministero della Cultura-Gallerie Estensi, Biblioteca Estense Universitaria —— **171**

Figure 2   *Hyacum et lues venerea*, in Jan Van Der Straet, *Nova reperta*, Antwerp: Philip Galle, c. 1600, Plate 7. Courtesy of Civica Raccolta Stampe Achille Bertarelli, Milan, Castello Sforzesco, Albo.E.293 —— **177**

Figure 3   *Sassafras*, in Nicolás Monardes, *Segunda parte del libro, de las cosas que se traen de nuestras Indias Occidentales, que sirven al uso de medicina*. Sevilla: en casa Alonso Escrivano, 1571, f. 27 recto. Courtesy of Biblioteca Nazionale Centrale of Rome, 8.36. H.17 —— **183**

Figure 4   *Cinchona tree*, in Charles-Marie de La Condamine, "Sur l'arbre du quinquina", *Histoire de l'Académie Royale des Sciences avec les Mémoires de mathématique et de physiques, Année 1738*. Amsterdam: P. de Coup, 1745?: plate 6, p. 346. Courtesy of Wellcome Collection.
Source: www.wellcomecollection/org/works/fxmtmy9x —— **188**

Figure 5   *Cinchona tree*, in Charles-Marie de La Condamine, "Sur l'arbre du quinquina", *Histoire de l'Académie Royale des Sciences avec les Mémoires de mathématique et de physiques, Année 1738*. Amsterdam: P. de Coup, 1745?: plate 7, p. 346. Courtesy of Wellcome Collection. Source: www.wellcomecollection/org/works/fxmtmy9x —— **189**

Figure 6   *Title page*, in Francis Bacon, *Instauratio Magna*. London: Ioannem Billium, 1620. Courtesy of Biblioteca Statale of Cremona, CIV.A.EE.5.30 —— **194**

Bettina Dietz, Knots in a Web

Figure 1   *Herbarium sheet of the Paris Herbarium*, in Paul Hermann, *Collectio plantarum Ceylanensium quas olim peritissimus Botanicus Paulus Hermannus in ipsa Ceylona observavit atque colligit [. . .] Thesaurum meum Zeylanicum conscripsi & edidi anno MDCCXXXVII*. Paris, Bibliothèque de l'Institut de France, Collection Benjamin Delessert, Ms 3912 Réserve: f. 66. Bibliothèque de l'Institut de France, Collection Benjamin Delessert, Ms 3912 Réserve. Reproduced with kind permission of the library —— **200**

Figure 2   *Urinaria Indica*, in Johannes Burman, *Thesaurus Zeylanicus*. Amsterdam: Janssonio-Waesbergios & Salomonem Schouten, 1737: pp. 231–232. Digitized by Bayerische Staatsbibliothek München. Source: Google books —— **202**

Figure 3   *Cinnamon sive Canella Zeylanica*, in Paul Hermann, *Herbarium*. Ceylon 1670s, now London, Natural History Museum: ID 480. Source: https://www.nhm.ac.uk/research-curation/scientific-resources/collections/botanical-collections/hermann-herbarium/. Accessed 11 July 2021 —— **205**

Figure 4   *Laurus*, in Carolus Linnaeus, *Flora Zeylanica sistens plantas indicas Zeylonae insulae . . .* Stockholm: Salvius, 1747: p. 61. Digitized by the Bayerische Staatsbibliothek München Source: https://www.digitale-sammlungen.de/de/view/bsb10302269. Accessed 11 July 2021 —— **208**

Figure 5   *Laurus*, in Carolus Linnaeus, *Materia medica, liber 1 de plantis*. Stockholm: Laurentius Salvius, 1749: p. 64. Digitized by the Bayerische Staatsbibliothek München. Source: https://www.digitale-sammlungen.de/de/view/bsb10302269?page=,1. Accessed 11 July 2021 —— **209**

List of Contributors

Tassanee Alleau is a Ph.D. candidate in history at the Centre d'études supérieures de la Renaissance in Tours (France) and teaches early modern history at the University of Tours. Her research focuses on the natural histories of plants' roots to unravel the structures of botanical knowledge, vegetal symbols and metaphors, rituals, practices, and cultural beliefs in early modern European societies and their colonial territories. She is working under the supervision of Pascal Brioist and Concetta Pennuto. She has published several articles on botany, the materiality of plants, vegetal remedies, herbs and roots and has co-published a book entitled *Sciences et société France et Angleterre 1680–1789* with Pascal and J.J. Brioist (Atlande éditions, 2020).

Fabrizio Baldassarri is a Marie Skłodowska Curie fellow at Ca' Foscari University of Venice and Indiana University Bloomington. He is running a project on plants in early modern natural philosophy, ranging from plant physiology to medicine. He co-edited several special issues, as well as volumes such as *Vegetative Powers* (2021) and *Scientiae in the History of Medicine* (2021), and has widely published on Descartes's naturalistic studies and philosophy, and on the early modern natural philosophy of plants.

Barbara Di Gennaro Splendore is a fellow of the Research Institute of the University of Bucharest. She earned a PhD in 2021 at Yale University in the History Department and in the History of Science and Medicine Program. Her dissertation focuses on the medicines culture and market in early modern Italy, using theriac—the most famous drug in the Western world up to the nineteenth century—as a case study. In 2015, she was awarded the Jerry Stannard Memorial Award for the History of Pharmacy and the Annals of Science Essay Prize for "Craft, money and mercy: an apothecary's self-portrait in sixteenth-century Bologna". She also published her work in *Nuncius* and in several edited volumes. Before going to Yale, she worked as an editor for the two major educational publishers in Italy—Zanichelli and Mondadori Education—and helped publish some of the best-selling history textbooks in Italy.

Bettina Dietz, associate professor at Hong Kong Baptist University, is a historian of early modern science and knowledge with a focus on seventeenth- and eighteenth-century botany. She has been working on the collaborative knowledge culture of eighteenth-century botany, the botanical correspondence as an information system, botany and its use of books, the networked practices of botanical nomenclature, the history of scientific authorship and publishing, and translations and language skills in the history of science.

Sarah R. Kyle is Professor and Chair of the Department of Art and Visual Culture at Iowa State University, USA. An historian of art and ideas, her research focuses on illustrated manuscripts as sites of intersection for pan-Mediterranean medical traditions, humanist enterprises, and artistic currents, particularly in the courts of northern Italy and in Venice during the fourteenth and fifteenth centuries. Her publications include *Medicine and Humanism in Late Medieval Italy: The Carrara Herbal in Padua* (2017).

Aleida Offerhaus (1964) studied theology and went on to write a (sadly unfinished) thesis on the reconstruction of an early Aramaic gospel. In essence, this is what she still does in trying to reconstruct the history of the herbaria discussed in this chapter. In between, she worked as librarian, raised her children and fell in love with botany. As guest researcher at Naturalis she published on the Zierikzee herbarium in 2021, decided to enrol as a PhD student at Leiden University and went on to publish on the subject of the d'Oignies herbarium and the Boerhaave collection within the van Royen herbarium. Both publications are forthcoming in 2022. As part of her PhD thesis she will also research another herbarium attributed to Boerhaave. Her promotion is scheduled in 2025.

Edoardo Pierini is a PhD candidate in history of medicine at the University of Geneva (IEH2), working on a project on the opiates in early modern European medicine. He especially focuses on the cultural and social aspects related to the consumption of psychoactive drugs, particularly dealing with a comparative view on different civilizations. He has studied at Roma Tre. He is also interested in the role of images and scientific illustration in the transmission of medical knowledge. He has recently published an article on the *Canadian Bulletin of Medical History*, entitled "Different Peoples, Different Inebriations: The Recognition of Different Cultures of Intoxication in Early Modern Medicine".

Federica Rotelli obtained a PhD in Bioeconomics at the University of Verona in 2007. She has deepened her researche on the historical and economic importance of Plant Biodiversity. She is currently preparing a monograph on the transformation of Mediterranean environments through the introduction of new foreign plants of Asian and American origin from the Middle Ages to the Seventeenth-century. In recent years she has also focused her research interests on Premodern Medicine and its transformation after the introduction of new exotic plants in the European culture of the early modern era.

Anastasia Stefanaki is a botanist specialized on sixteenth-century botany and book herbaria. She is affiliated with Wageningen University and Naturalis Biodiversity Center in The Netherlands. Besides history of botany, she also does research on medicinal and aromatic plants, their essential oils and uses, rare and threatened plants, plant conservation and ecology.

Tinde van Andel (1967) was trained as a tropical ecologist and defended her PhD in ethnobotany in 2000, after which she continued her research on useful plants in the Guianas and western Africa. In 2015, she became special professor in Ethnobotany at Wageningen University and Clusius professor in History of Botany and Gardens at Leiden University. She is employed by Naturalis Biodiversity Center as senior researcher. Van Andel and her research team focus on historical herbaria and botanical drawings (16th-18th century), kept in the treasure rooms in Dutch museums and libraries.

Karen Reeds
Foreword

> *Modernitätzwang amputiert permanent die Geschichte der Wissenschaft.*
> Heinz Schneider,
> "200 Jahre « Tentamen Floræ Basileensis » von C.F. Hagenbach"
> https://tube.switch.ch/videos/OoGBIa1WnY (36:59 /1:12:16)
> Botanische Abendkolloquien, Basler Botanische Gesellschaft
> Universität Basel, 18 December 2021

When did botany become modern? 1753, the publication year of Carl Linnaeus's *Species Plantarum*, and the date that the taxonomists at the Paris International Botanical Congress (1867) fixed as the official starting point for scientific names of plants. Linnaeus's success at imposing order on plant nomenclature had, the taxonomists agreed, allowed botanists to sweep aside the huge tangle of plant synonyms that had hampered progress in their discipline from antiquity onward.

However, the Congress's decision had the unintended consequence of dividing botany's history into two very unequal periods. The vast human experience with plants that had accumulated before that cut-off date of 1753 was relegated to the pre-Linnaean era—and then increasingly ignored by professional botanists, pharmacists, physicians, naturalists, and historians of science and medicine.

I must count myself among the many historians who have looked back at early botany primarily through the Linnaean lens. I have spent plenty of time looking at Linnaeus's forerunners and their quests for a consistent system of plant names and classification, but I have paid far less attention to the other ways that plants had spurred imaginations in the centuries before Linnaeus. So, for me, *Medical Botany in Pre-Modern Times: Tradition and Innovation* comes as an eye-opener.

It is a truism that in early modern Europe the main reason for studying plants was medical. Indeed, Linnaeus (1707–1778) himself had grown up taking it for granted that the practice of medicine and the knowledge of plants were inseparable. His own handbook for medical students, *Materia Medica: Liber I. De Plantis* (1749)—written while he was still a "pre-Linnaean"—underscored the ancient assumption that learning the names, properties, and uses of medicinal plants was a necessity for any aspiring physician.

Fabrizio Baldassarri and his contributors, however, point to new kinds of medical motives and opportunities for thinking about the plant world. The volume examines striking novelties that captured physicians' curiosity, innovative technologies that enlarged their botanical tool-chests, and puzzling medical results that challenged their long-held medical theories. Thanks to the construction of botanical gardens, the observations recorded in herbals and herbaria, and the influx of plants previously unknown to Europeans, powerful exotic drugs could now be grown, exchanged, tested, and studied more closely than ever before. The invention of the microscope revealed minute plant structures that invited philosophically inclined investigators to grapple with continuities between vegetal and animal life.

The opening essay here—Sarah R. Kyle on the *Roccabonella Herbal*—characterizes that extraordinary manuscript as "a virtual conversation" carried on by its fifteenth-century author with his ancient sources, his artist, and his fellow physicians, pharmacists, and herbalists. I see the present volume as a conversation in the same spirit, bringing the discussion forward into the centuries between Nicolò Roccabonella and Linnaeus. Reader, I invite you to join in the conversation!

Karen Reeds, FLS
Princeton Research Forum

Fabrizio Baldassarri
Introduction: The World of Plants in Premodern Medical Knowledge

The study of plants has traditionally developed as an important complement to medicine, especially as it pertains to the uses of medicinal plants in the preparation of recipes and therapies. This is what scholars generally call medical botany: its emphasis on the knowledge of plants as the basis for treatments, hygiene, therapies, and body balance has thrived since the ancient times,[1] and it is sometimes differentiated from the study of plants in their own right. While the latter largely fell out of favour from late antiquity to the Renaissance, medical botany overshadowed the study of plants. Notwithstanding this differentiation, the extent to which plants and plant products were used to prepare remedies can serve as a benchmark to appreciate the changes in the history of botany, as Anne Stobart and Susan Francia have recently outlined. Indeed, analyzing the wide variety of connections between medicine and plant studies provides a way to see how the still-fragmented world of plants was understood in the pre-modern period.[2] Through this volume's case studies, we aim to delineate how much, in the transformations from the Renaissance world of plants to a modern science of plants, medical botany played a relevant role, combining traditions with innovative approaches.

Historians of botany have generally recognized a clear divide between (a) the Renaissance period, generally acknowledged as a moment of renewal in the approach to plants, but still related to the classical connection between plants and medicine, and (b) the late seventeenth century, when plant studies acquired autonomy, and botany became a science in the modern sense. As Claudia Swan has noted, botanical studies became "emancipate[d] from practical exigencies and acquired disciplinary status through the study of similarities and differences between appearances and internal structures",[3] and the physiological studies of plant functions, such as reproduction and how they transmit fluids gained momentum. In *Outlines of the History of Botany*, Robert John Harvey-Gibson described "the gradual evolution of the different aspects or departments of botanical knowledge. At first, [. . .] botanical treatises were purely utilitarian—horticultural, agricultural, or medical [. . .]. When plants began to be

---

1 Guthrie 1961.
2 Stobart and Francia 2014: 5–8.
3 Swan 2008: 64.

Note: This project has received funding from the European Union's Horizon 2020 research and innovation programme under the Marie Skłodowska-Curie grant agreement No 890770, "VegSciLif". This introduction was written during a visiting period at the Herzog August Bibliothek Wolfenbüttel. I thank Brenton M. Wells for his comments and discussion of this text, and Karen M. Reeds for her comments and edits.

regarded as subjects of study for their own sakes and not merely as sources of drugs or as articles of food, an instinctive feeling was awakened that they must be related in some way to each other, in other words that they could be classified".[4] In the more recent *History of Botanical Science*, Alan G. Morton repeated a similar claim, suggesting that botany as a "field of enquiry was broadened [by the outburst of scientific activity in the second half of the seventeenth century] and [botany] was brought for the first time into contact with other sciences besides medicine, whence experimental and technical methods began to penetrate into botany; thus the way was opened to the formation of new theoretical concepts".[5] Similarly, Hilda Leyel has claimed that botany and medicine, after "[coming] down the ages hand in hand", in the seventeenth century "their ways parted".[6] While these authors grasped a significant point, the history of the study of plants is much more nuanced—scholars in the Middle Ages and Renaissance also investigated the nature of plants per se, or tried to travel different roads from the study of simples, while traditions and medicine still played a crucial role throughout the seventeenth-century world of plants. However, as Agnes Arber has brilliantly pointed out, "workers [in the] field of medicine [. . .] lay the foundations of the copious and exact knowledge of plants",[7] making the connections between medicine and plants a crucial perspective from which to understand botanical science, as a few scholars have more recently explored.[8] Despite historians' assertions of a growing divide between medical botany and a science of botany in the modern sense, the process of understanding the world of plants over the long terms does not (to adopt a botanical simile) look like two saplings leaning in opposite directions although they started from the same taproot; instead, it is much more like a shrub of intertwining branches.

In this sense, the assumption of clear-cut divisions between the "dark ages of botany", the renewal of a botany grounded on materia medica in the Renaissance, and the early modern constitution of a new science based on observation—that is, a linear progress from one period to another—has kept historians from recognising the complexity of earlier knowledge of the the world of plants. Recently, Florike Egmond has challenged the "logical loop in defining sixteenth-century activities in terms of (an absence of) seventeenth century characteristics";[9] she argues that accepting a single hierarchical progress from the sixteenth to the seventeenth century underestimates the links uniting the scientific enterprises of the pre-modern times.

---

[4] Harvey-Gibson 1919: 22.
[5] Morton 1981: 166; Gerber 1927. See also, the more recent Findlen 1994; Ogilvie 2006; Touwaide 2008a and 2008b; Anagnostou et al. 2011; Bellorini 2016.
[6] Leyel 1959: xiii. I too accepted this claim until I started working on this volume. After a closer look on the 16th- and 17th-century studies of plants I realized that the connection between medicine and botany deserved more attention and a diverse reading.
[7] Arber 1912/1970: 6.
[8] Boutroue 2002; Ventura 2013 and 2017.
[9] Egmond 2017: 40, and 2022.

This volume takes Egmond's objections to heart: its aim is to investigate those continuities seriously. By closely examining the ways that Renaissance and early modern scholars engaged with medical botany—through the production of herbals, the introduction of new plants with therapeutic properties, the comparison with diverse authorities, the experimentation in botanical gardens,[10] the medical trials with parts of plants, the confrontation with traditional pharmacology, and the medical analogy between plants and animals—and by measuring their distances and connections with traditional systems, the present volume sheds new light on the ways medical knowledge favoured the shaping a science of botany through a combination of innovation and tradition.

In the Middle Ages, the study of plants intersected the field of medicine, as "plants themselves were usually just called by their medical uses: *simplices medicinae* or *simplices*, 'simple drugs" or, most often, 'simples'",[11] but it also crossed into symbolism, magic, folklore, and horticulture, as well as natural philosophy.[12] As Iolanda Ventura has brilliantly specified, "during the Middle Ages, botany [. . .] emerged from a mélange of several branches of knowledge [. . .]. Moreover, medieval botany was a theoretical science that with few exceptions did not involve any practical experience".[13] Scholars with medical training benefited from Galen's (A.D. 129-after [?] 216) *De simplicium medicamentorum facultatibus* or from Dioscorides' (*fl.* 1st cent. A.D.), *De materia medica*, the two main pharmacological collections written during Antiquity and the most influential during the Middle Ages.[14] As this tradition expanded, the study of plants began to combine the description of pharmacological virtues with the attention to plants classification, ultimately highlighting a theoretical aim that gave a philosophical perspective to the study of plants.[15] Nevertheless, the combination of herbal knowledge, pharmacology aims, and classificatory goals uncovered problems and lacunae, disclosing a clear gap between a theoretical and practical plant knowledge. Even before the Renaissance humanistic attempt to bridge such gaps—notably Nicolao Leoniceno's (1428–1524) enterprise to renovate scientific knowledge, amend errors, and develop a more certain pharmacology through the humanist and philological approach to classical texts[16]—some late medieval herbals serve as significant case studies in their conscious presentations of a new approach to botany. Besides the famous case of the Carrara herbal (a late-fifteenth century vernacular Italian translation of the *Liber aggregatus in medicinis simplicibus*, that clearly combined Arab and medieval tradition, close observation of specimens,

---

[10] Wijnands 1988.
[11] Reeds 1991: 3.
[12] Paravicini Bagliani 2009.
[13] Ventura 2013: 100–101; 2017, and 2018.
[14] Touwaide 1997; Touwaide and Dendle 2008; Reeds 2012.
[15] For an investigation of the natural philosophical study of plants in Christian Middle Ages, see Panarelli 2020 and 2021; Cerrito 2021.
[16] Stannard 1969/1999a; Van Arsdall 2014; Touwaide 2022b.

and pharmacological knowledge), another fifteenth-century illustrated herbal demands special attention: the *Roccabonella Herbal*, deliberately created by the Venetian physician Nicolò Roccabonella (1386–1459) and the artist Andrea Amadio (*fl.* 15th cent.) to prevent errors by apothecaries (see Sarah R. Kyle in this volume).[17] Roccabonella claims that his manuscript, while based on ancient knowledge of simples, adopts a more modern order in the world of plants, sparking new light on the pharmacological and natural historical knowledge of the plants in the mid-Quattrocento.[18]

More or less at the same time, German illustrated, early printed herbals such as the *Herbarius latinus* (1484), the *Hortus sanitatis* (*Gart des Gesundheit*, 1485) and the *Herbarius in Dyetsch* (published in 1483, 1484, 1500) incorporated a connection between botanical knowledge and pharmacology that paved the transition to a closer attention to plants and represented a step from the medieval herbal to the modern flora.[19] Despite the contradictions and limitations in their visual depictions of plants (as recent scholars have noted),[20] these late medieval books played a central role in transmitting the knowledge of plants inherited from the ancient times, with a special focus on medicinal qualities.

The study of plants gained momentum with the publication in 1530 of the landmark printed herbal *Herbarum vivae eicones*, written by Otto Brunfels (ca. 1488–1534) with naturalistic images by Hans Weiditz (ca. 1495–1537) (see Figure 1).[21] Its innovative visual approach—although still coupled with traditional descriptive system—set a new standard for the study of plants. Indeed, highly accurate and detailed illustrations of plants quickly superseded the earlier schematic representations, in part because they served as a shortcut to learning about individual plants, cancelling the distance between written knowledge and the object of such knowledge.

The realistic representation of plants became a central aspect of medical botany of the period, primarily as an aid to plant identification, but also as a stimulus to readers to go out and make their own direct observations of plants. In the famous herbal of Leonhart Fuchs (1501–1566), *De historia stirpium* (1542), the visual depiction of nature (though conceptually remodelled from Brunfels') took on a new relevance in the text; Fuchs had supervised his artists to ensure that illustrations measured up to his ideals of clarity and accuracy.[22] The books by German contemporaries of Fuchs and Brunfels recognised the value of pictures even if they could not always include them. The first edition of the first regional herbal, *Kreuter Buch* (1537) by Hieronymus Bock (1498–1554)—a Lutheran pastor, not a physician—lacked pictures, but had very clear descriptions instead; later editions in German and Latin included illustrations, mostly copied from

---

17 Smith 2008: 16–18; Kyle 2017.
18 See also Ineichen 1959.
19 Stannard 1974/1999b: 30. On flora from the fifteenth century onwards, see Tongiorgi Tomasi 1997.
20 Swan 2006; Givens 2006; Olariu 2014; Moran 2017; Van Leerdam 2021.
21 Bertoloni Meli 2022.
22 Kusukawa 2012.

Figure 1: *Preface*, in Otto Brunfels, *Herbarum vivae eicones*. Strasbourg: Schottum, 1530–1532: p. 1.
Source: www.gallica.fr /BNF.

Brunfels and Fuchs. The posthumously published annotations on Dioscorides (1549) and the herbal (1561) by Valerius Cordus (1515–1544) incorporated systematic, precise, detailed descriptions of plants and firsthand observations that the short-lived botanist had made on his extensive European journeys. His father, Euricius Cordus (1486–1535), had also engaged in plant studies: his *Botanologicon* (1534) is an unillustrated treatise in dialogue form on the medical necessity of accurate plant identifications.[23] The Swiss physician-botanist, Conrad Gessner (1516–1565), who had edited both Bock's Latin herbal and Valerius Cordus' herbal, was unable to bring his own hufe *Historia plantarum* to fruition; it was published posthumously with illustrations based on his detailed watercolours.[24]

Fuchs and Pietro Andrea Mattioli (1501–1578) established the foundations for a botanical science that would become more and more autonomous from medicine.[25] In 1544, Mattioli published the Italian translation of antiquity's foremost text on medicinal plants, *De materia medica* by Dioscorides' (*fl.* 1st cent. A.D.). A decade later Mattioli followed up his Italian text (which had followed Jean Ruel's Latin version of 1516) with his own elegantly illustrated Latin edition and extensive commentary: *Commentarii in libros sex Pedacii Dioscoridis Anazarbei, de Materia Medica . . .* (1554, and later expanded editions). Both in Latin and several vernaculars, these commentaries proved to be immensely successful in Renaissance pharmacology and medicine, thanks to the ways Mattioli aligned contemporary botanical knowledge with the classical tradition,[26] supplemented Dioscorides' text with data from the field, and introduced new plants, such as the banana tree (see Figure 2).[27]

Two contemporaries of Mattioli commented on classical authorities in ways that similarly introduced evidence from first-hand experience with plants in the field. Andrés Laguna de Segovia (1499–1559) published a Castilian translation, with commentary, of Dioscorides in 1555: *Pedacio Dioscorides Anazarbeo. A cerca de la materia medicinal . . .* (1555). It brought a direct knowledge of plants and specimens, acquired through encounters among scholars and unschooled healers, to bear on the medical and pharmacological content of Dioscorides' *Materia medica*. The Ferrara physician and botanist Antonio Musa Brasavola (1500–1555) published several *examens* on the pharmacological virtues of plants and specimens. His *Examen omnium simplicium medicamentorum* (1536) set its long dialogue on the nature and virtues of simples against the backdrop of an Alpine plant-collecting trip. As Leoniceno's student, Brasavola was well-grounded in his teacher's pioneering critique of Pliny's (A.D. 23/24–79) botany and developed both a fine-grained sense of the natural world and a clear-eyed recognition of the limitations of ancient sources.

---

23 See Swan 2006: 245–249.
24 On herbals in Germany, see Arber 1912/1970: 66.
25 Crisciani 1999; Repici 2003.
26 One should note that an Italian version was published in 1544. On Mattioli, see Ferri 1997; Fausti 2004; Ciancio 2015; Touwaide 2022a.
27 Mattioli, *Commentarii* 1.126: 219–220.

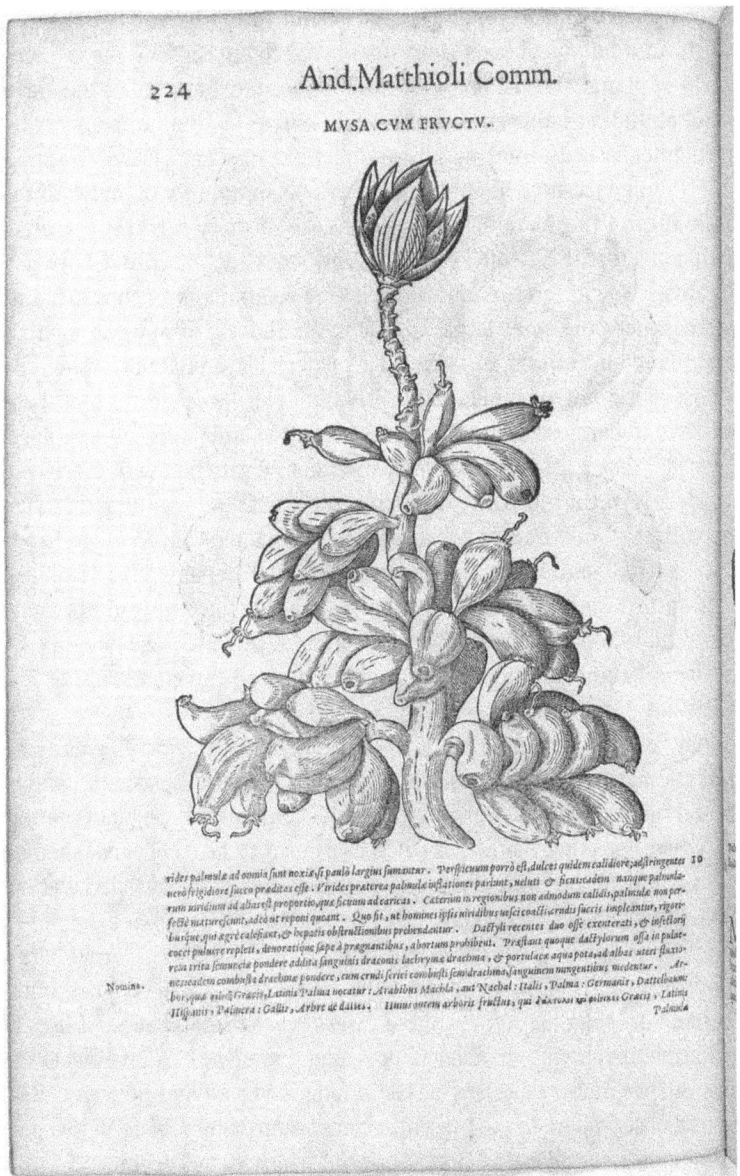

Figure 2: *Musa cum fructu*, in Pietro Andrea Mattioli, *Commentarii, in libros sex Pedacii Dioscoridis Anazarbei, De materia medica* . . . Venezia: Valgrisi, 1573: p. 244.
Source: www.gallica.fr /BNF.

The short list compiled by Lucia Tongiorgi Tomasi and Tony Willis of Renaissance botanists, naturalists, and botanical texts is a valuable aid to envisioning the growing interest in the world of plants.[28] Yet, it does not claim to be an exhaustive catalogue of the participants and abundant publications in the field. When Mattioli urged his readers to send him specimens, seeds, information, and pictures of plants, he set the stage for botany's role in the Renaissance Republic of Letters. A community of lovers of the natural world, in and out of universities—physicians, apothecaries, travellers, amateurs and virtuosi, collectors, horticulturalists—found ways to share their curiosity and passion for plants.[29] The difference among their experiential approaches to plants characterized Renaissance studies at large. In effect, the linking of materia medica, botany, and efforts to amend traditional knowledge (especially the Galenic system) by means of direct observation of specimens took advantage of the eagerness of these diverse *hybrid* experts to connect their practical knowledge and skills with explorations of natural world.[30] The growing audience would not be satisfied with mere copies of traditional, stereotypical illustrations. Early in the sixteenth century they had already begun to collect plants and to preserve their specimens in *herbaria* (*horti sicci*) to compare to classical and contemporary accounts. Their goal was to "establish a new catalogue of nature, starting from the premises that the ancients did not know everything [. . .] and that their modern successors could best follow their example by surpassing them",[31] as Brian Ogilvie puts it. In going beyond the classics, the new study of plants grew out of them.

Herbals gradually fostered studies of the natural history of plants, emerging as a crucial combination of attempts to know plants more directly, to observe specimens, and to describe qualities and medical virtues—an enterprise that continued throughout the seventeenth century, when books on materia medica continued to be published, shaping medical botany.[32] In line with the work of Laguna, Mattioli, and Brasavola, special attention began to be paid to exotic vegetal bodies. Leonhard Rauwolf's (ca. 1535–1596) *Aigentliche Beschreibung der Raiß [. . .] inn die Morgenländer* (1583) and Prospero Alpini's (1553–1617) *De plantis Aegypti* (1592), to name two influential works, reveal the great interest in plants from non-European nature. Yet, these cases reveal a combination of uses of ancient sources, intermediation with diverse cultures, and the attempt to investigate the curative properties of specimens, especially re-combining medicine and botany, classic sources and first-hand observation. In the present volume, contributors focus on two particular cases: balsam and mandrake. The first is balsam, which gained momentum as it figured among the ingredients of theriac. In Alpini, the botanical knowledge and the therapeutic virtues of this specimen intertwined with politics and with the encounter with

---

28 Tongiorgi Tomasi and Willis 2009. See also Arber 1912/1970.
29 Olmi 1991. For the epistolary exchanges in a later period, see Egmond 2007a.
30 Klein 2008: 780; Egmond 2017.
31 Ogilvie 2006: 139.
32 Barreiros and Fontes da Costa 2021.

indigenous cultures, as this unusual plant surfaces as a model to delve on the Renaissance intersections between botany and medicine at the cross between cultures (see Barbara Di Gennaro in this volume). The second is the case of Mandrake, generally considered a wondrous plant with various therapeutic uses. Scholars engaged with the mandrake at different levels, ultimately trying to frame it within medical fictions to make it available to a large audience. This has two consequences. First, encompassing the knowledge of this plant within allegorical and magic interpretations of nature (especially connected to the Doctrine of Signatures and to the mythology of the underground, or *mundus subterraneus*) raised challenges to the occult and irrational strands of medicine. Second, scholars portrayed mandrake by means of these allegories to provide medical evidence for the uses of plants, ultimately shaping medical botany in a more modern fashion (see Tassanee Alleau in this volume on the medical narratives on mandrake).

A more direct investigation of plants was favored by the practices of didactic *herborisations* and field trips to collect plants (both introduced by Fuchs at the University of Tübingen) and by the construction of botanical gardens. Over the course of the sixteenth century, scholars funneled their need to observe living plants into a new conception of garden architecture: Renaissance universities started building botanical gardens as spaces for the cultivation of simples and the investigation of specimens.[33] In 1543, Luca Ghini (1490–1556) founded the Botanical Garden in Pisa,[34] and the Botanical Garden in Padua was founded in 1545.[35] In 1568, Ulysses Aldrovandi (1522–1605) founded the Botanical Garden in Bologna. In 1590, the request to establish a *hortus academicus* in Leiden was granted, and in 1593 Carolus Clusius (1526–1609) became prefect of the botanical garden.[36] In 1593, Pierre Richer de Belleval (1564–1632) founded the botanical garden in Montpellier.[37] The original purposes of these gardens was the study of simples: they were built to give medical students, pharmacists, physicians, and scholars direct experience with the plants used for producing remedies. But the gardens soon played an outstanding role in focusing attention to the study of plants in their own right.

Andrea Cesalpino (1524–1603) was at once a physician, botanist, and philosopher.[38] He brought his first-hand experience with plants as the director of the botanical garden in Pisa for several years and as a collector of plants (the herbarium he constructed in 1563 is now collected in Florence) to bear on his theoretical study of plants, *De plantis libri XVI* (1583). This unillustrated treatise contains both descriptions of individual plants and a natural philosophical section—rooted in the Aristotelian

---

33 On medieval gardens, see Landsberg 1996.
34 On Ghini, see Findlen 2017. On the botanical garden of Pisa, see Garbari, Tongiorgi Tomasi and Tosi 1991.
35 Minelli 1995.
36 Stearn 1962.
37 Rioux 2004.
38 On Cesalpino, see Baldassarri and Martin 2023.

science of the sixteenth century—that encompassed the general features of plants and ultimately attempts to classify plants from their forms.[39]

Works by Cesalpino's contemporaries—Giovanni Costeo (Costeus, 1528–1603), *De Universali stirpium natura libri duo* (1578), Franz Tidike (Tidicaeus, 1554–1617), *Phytologia generalis* (1582), and Adam Zalužanský y Zaluzian (1558–1613), *Methodi herbariae libri tres* (1592)—reveal other efforts to provide botany with a natural philosophical framework. In particular, Zalužanský's *Methodi herbariae* is a notable attempt to apply Petrus Ramus' (1515–1572) principles of logical division to the world of plants. Zalužanský sought to analyse both the causes and processes of plants (book 1) and their history (book 2), and finally to offer an exercise on understanding botany through the senses (book 3) (see Figure 3). In book 2, he applied the theoretical definitions presented in book 1 to specific cases, where the variety and diversity of plants surfaces in relation to their pharmacological virtues. Despite the important differences between these texts, they share a similar structure: first, a more theoretical attempt to systematise plants by knowing their processes and activities; second, a description of particular bodies, varieties, qualities, and virtues, consistent with the theory.

This philosophical and methodological addition to the world of plants appears therefore central in the botanical studies in the Renaissance. In some ways, it developed from the naturalistic approach to specimens and from the efforts to reduce the abundance of nature to classifications and catalogues. The debate between the Renaissance scholars Girolamo Cardano (1501–1576) and Julius Caesar Scaliger (1484–1558) about the natural functions of plants certainly spurred on the theoretical enterprise by the end of the sixteenth century.[40] Ultimately, a combination of activities ranging from the commentaries on ancient texts, the naturalistic investigations of plants through *herbaria* and cultivation in botanical gardens, the development of therapies, to use of theoretical frameworks for classifying both old and new plants, came to shape the sixteenth-century world of plants at large.[41]

Ogilvie has reduced these features to four moments or scholarly generations in the Renaissance study of plants. The first phase of sixteenth century botanical studies testified to a philological approach to plants and, specifically, to plants used in medical botany. The second phase (from Fuchs to Mattioli) revealed a naturalistic approach to plants, in which the primary motive for studying nature as a collection of bodies, or histories, was its medical value. The third phase saw the development of exchanges of specimens and the constructions of *herbaria*, setting the ground for modern *flora*, and the founding of botanical gardens, where lectures in materia medica had a major role. And the fourth phase underscored the need for a theoretical framework to explain plants and their medical uses in terms of classifications and

---

[39] Jensen 2000.
[40] On Cardano, see Siraisi 1997; Ventura 2013: 133–140.
[41] Baldassarri 2020.

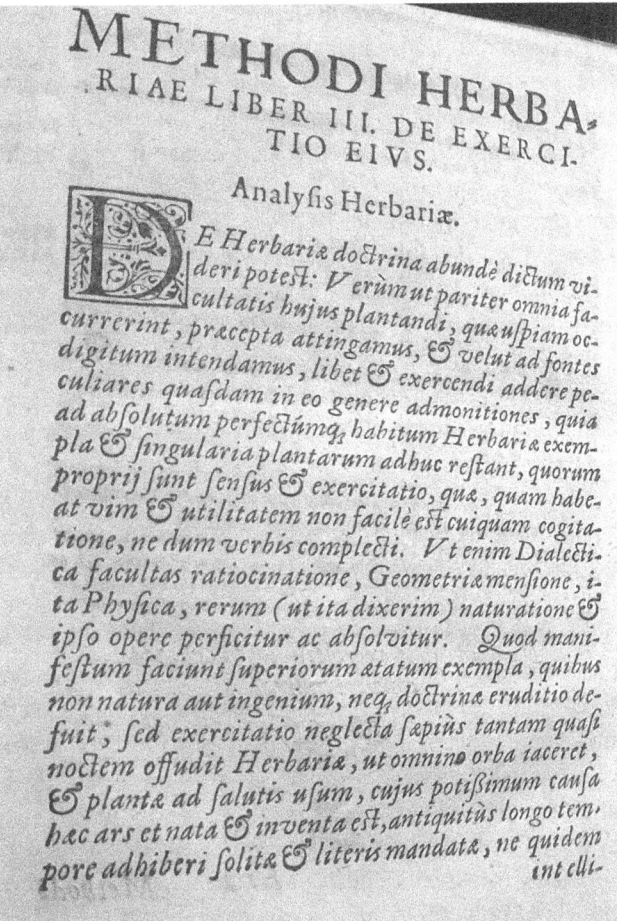

Figure 3: *De exercitio eius: Analysis Herbariae*, in Adam Zalužanský y Zaluzian, *Methodi herbariae libri tres*. Prague: Draczleeni, 1592: f. Ee2 verso.
Source: HAB Wolfenbüttel, Germany.

taxonomy.[42] In acknowledging the soundness of Ogilvie's chronology, it is essential to emphasise that medical botany was the root of all these various branches of the Renaissance interest in plants.

These threads persisted in the seventeenth-century world of plants. The science of botany that gradually took shape in the seventeenth century combined (1) the theoretical investigation of the processes and functions (and forms) of vegetal bodies; (2) the study through catalogues and *flora*, of the differences among plants; (3) the direct observation of physical bodies through *herbals* and *herbaria* and at botanical gardens;

---

42 Ogilvie 2006: 29.

and (4) the knowledge of plants' medical virtues and therapeutic powers. Although these aspects already characterized much of Renaissance botany, some significant innovations began to emerge: (a) new technologies for cultivating plants and for observing plants' previously unseen structures through microscopes, and (b) a philosophical and scientific framework that incorporated the new technological approach to the world of plants.

Continuities could be detected in all these fields. While a distinctively new philosophical framework characterized the seventeenth-century studies, the Renaissance natural philosophical investigations of "plant descent and propagation, spontaneous generation, metamorphosis, the movement of plants, and the distinctions between plants and animals"[43] anticipated the early modern observations of the functions of plants. The connections between Giovanni Battista Della Porta's (1535–1615) and Francis Bacon's (1561–1626) experimentation have been thoroughly investigated.[44] In particular, Della Porta offered an alternative approach to the field: on the one hand, he examined ecology and some of the doctrine of signatures; while on the other hand he discussed the ways to make 'new' plants through magical processes. Somehow similarly, French Libertine and Paracelsian physician and botanist Guy de La Brosse (1586–1641), who founded the Jardin royal in Paris in the late 1620s, promoted a chymical experimentation with plants as a way to investigate their forms and processes, a method whose roots are in Renaissance natural history and alchemy. In his philosophical text, *De la nature, vertu et utilité des plantes* (1628), his theory embedded experimentation.[45] Yet, at the same time, La Brosse was also attentive to the attributes of individual specimens—his *Description du Jardin Royal des plantes medecinales* (1636) contains a catalogue of plants cultivated in the garden.

A similar approach (but within a utterly different framework) surfaced in *Isagoge in rem herbariam libri duo* (1633), in which Flemish physician Adrianus Spigelius (1578–1625) dealt with the forms of plants, including particular plants (especially those collected at Padua) as well as the uses and faculties of plants. A combination of the study of vegetal processes and forms, botanical experimentation, representation of singular plants, and pharmacological knowledge characterized the early seventeenth-century studies of plants. This text was re-published in Leiden with a list of plants collected at the Dutch *hortus botanicus* in 1633. Indeed, the three-pronged programme of the Leiden Botanical Garden included teaching physicians about medicinal botany, cultivating exotic and indigenous plants, and promoting experimentation with plants (see Aleida Offerhaus, Anastasia Stefanaki and Tinde van Andel in this volume on the diverse approaches to the study of plants at the Leiden *hortus botanicus*). Among Spigelius' students, two scholars played a particularly important role in developing a

---

43 Egmond 2017: 39.
44 Rusu 2017 and 2020.
45 Baldassarri 2021.

methodology to study plants. The first was Joachim Jungius (1587–1657), whose posthumously published text on plants, *Isagoge phytoscopica* (1679), offered a methodological attempt to use the nature, structure, and processes of plants—that is a morphology of vegetal bodies—as a way to classify plants (for which he followed Caspar Bauhin's [1560–1624] binomial system, whose in its turn was an extension of Matthias L'Obel's [or Mathias de Lobel, 1538–1616]). The second scholar was the Cartesian physician Henricus Regius (1598–1679), who took a mechanistic approach to the vegetal nature. On the one hand, he included several chapters on materia medica in his *Fundamenta medica* (1647), listing the therapeutic virtues of plants and parts of plants, and published both a *Hortus Academicus Ultrajectinus* (1650), i.e., a plan for the *hortus botanicus* in Utrecht, and a *Pharmacopea ultrajectina* (1657) for Utrecht's medical community. On the other hand, Regius claimed physics or natural philosophy should precede the knowledge of plants in the field and the medical treatment one could derive from them. That premise was explained in Regius' *Fundamenta physices* (1647, and its later re-elaborations); its natural philosophical explanation of the general nature, form, and functioning of plants paved the way to the knowledge of singular bodies and to medical botany.

Roughly contemporary with these Germand and Dutch attempts to frame the study of plants, natural philosophers and naturalists within the Hartlib circle and the Royal Society in England promoted experimentation with vegetation. Following Bacon's project for the advancement of learning, scholars in these circles produced natural histories of plants that collected observations and discussed them within a natural philosophical framework. Despite the difference among investigations by scholars such as Thomas Browne (1605–1682), John Evelyn (1620–1706), and Nehemiah Grew (1641–1712), to name just a few, a unified picture seems possible.[46] However, the standpoint that ultimately won out was based on the foundational work of Grew and Marcello Malpighi (1628–1694), who undertook parallel projects during the 1670s. Both presented a similar anatomical investigation of plants in front of the Royal Society, making clear how far (microscopic) observation had become relevant to botany. While both scholars were interested in specifying the processes and functions of vegetal bodies, their aims diverged. On the one hand, Malpighi generally concentrated on the anatomy of plants, that is, a field strictly connected to medicine. Grew, on the other hand, proposed a wider focus that included an idea of philosophical and phytological study of vegetation. Malpighi and Grew brought a new vitality to the field, as both their theoretical frameworks and microscopic observations of previously unknown plant structures made new insights into the growth and classification of plants possible. From their anatomical exploration of plants, both Grew and Malpighi argued for the use of plants as models to describe animal physiology, which in turn opened

---

[46] Recent work on this period has tried to present a unified picture, despite the differences between authors. See Clericuzio 2018; Jalobeanu and Matei 2020 and 2022.

new prospects for late seventeenth-century medical knowledge (see Fabrizio Baldassarri in this volume).

Parallel to the observations of plants and the attempt to provide a theoretical approach to vegetal bodies, seventeenth-century investigations on the world of plants benefited from the compositions of flora and materia medica, which were especially helpful to understand exotic *naturalia*. From the Renaissance into the early modern period and beyond, the description of specimens in herbaria, herbals and floras comprised both botanical and pharmaceutical information, and these two elements became increasingly intertwined over time. The compilation of these works was, however, complex, as the creation of a flora was tied to finding a new systematization of nature, as Alette Fleischer has pointed out.[47] Ever since the Renaissance, naturalists had constantly compared plants, generally confronting the specimens they had in hand with relevant literature and proposing new interpretations of specimens and a new understanding of plants in themselves. In this sense, as scholars interlinked the information contained in materia medica and existing literature with the material available to them, their deeper understanding of the divergences and similarities among herbals shaped botanical studies at large in important ways (see Bettina Dietz in this volume). More specifically, this approach intersects with the uses of spices and plant remedies, especially with those from exotic places. Indeed, as plants and plant products from the West or East Indies entered the European medical system, they challenged not only systems of botanical classification, but also the Galenic humoral theory, i.e., the very basis of European pharmacology. As highlighted by Saskia Klerk, the introduction of exotic specimens proved to be a central case study for understanding the foundations of modern European medicine, as scholars felt compelled to move beyond Galen's pharmacology and a medical system still connected to the classics.[48] In investigating and testing the properties of drugs and the virtues of plant remedies, early modern scholars slowly rejected the traditional pharmacopeia and developed an alternative epistemological approach. In doing so, they grounded their knowledge on direct observation and experimentation, through which they re-elaborated the understanding of the properties of specimens into a new system. This was the case of opium (see Edoardo Pierini in this volume) and the case of cinchona bark (see Federica Rotelli in this volume). These two vegetal substances, among many other cases, specifically challenged Galenic medicine by *not* acting on the human body as Galenic humoral theory and pharmacology seemed to predict. To overcome the limitations of traditional systems, scholars tried experimental studies of plant bodies upon human bodies, chemical experimentation, and other innovative attempts to test the properties of exotic drugs. For all their novelty, however, as the authors show, these attempts grow out of the sixteenth-century investigations of medicinal plants.

---

47 Fleischer 2017.
48 Klerk 2015.

In dealing with the composition of herbals and materia medica, the experimentation with drugs and spices, the acquisition of exotic plant products, the microscopical investigation of plant anatomy, and the cultivation and observation of plants in botanical gardens from the Renaissance to the late seventeenth century, the contributions collected in this volume nuance the role of medical botany in the transformations of the study of plants. Indeed, while botany acquired more autonomy, as it evidently surfaces in the work of John Ray (1627–1705) and Joseph Pitton de Tournefort (1656–1708), the importance of medicine for the world of plants persisted beyond the sixteenth century, shaping the study of plants at large. At the same time, the contributions collected in this volume offer the reader with a privileged perspective from which to rearrange the fragmented world of plants in the early modern period, uncovering in the medical study of botany the origins of a science focused on the morphology of plants. During the slow emergence of botany as its own discipline, the continuities with tradition held their power. These continuities are both authorial and thematic. Throughout all the states of sixteenth- and seventeenth-century studies of plants, medicine and botany kept on invoking the authority of Dioscorides, Galen, and Aristotle. By highlighting the patterns of thought connecting Renaissance and early modern botany, by noting the circumstances where traditions and innovations intersect, the contributions to this volume reveal the continuing salience of medical botany even when a more modern science of plants becomes visible in the seventeenth century.

Sarah R. Kyle
A More Modern Order: Virtual Collaboration in the *Roccabonella Herbal*

The illustrated *herbarius* (now Venice, Biblioteca Nazionale Marciana, *Lat.* VI, 59 [coll. 2548]) of Venetian physician Nicolò Roccabonella (1386–1459) remains mysterious in part because of what he calls its "more modern order", the significance of which has been unexplored in modern scholarship.[1] Created by Roccabonella and the Venetian painter Andrea Amadio (*fl.* 15th cent.) in the mid-fifteenth century,[2] penned in ink predominantly

---

[1] See the classic literature on Roccabonella's herbal: Morelli 1802: 398–405; Cicogna 1824–1853: 2.151–153; Valentinelli 1868–1873: 5.61–67; Teza 1897–1898; De Toni 1919–1925; Pächt 1950; Minio 1952–1953; Ineichen 1959; Kristeller 1963–1997: 1.96; and Baumann 1974: 126–128; and the more recent literature by Pesenti 1984; Mariani Canova 1986: 514 and 1988; Cappelletti 1989; Marcon 1988, 2002, 2003; Paganelli and Cappelletti 1996; Pitacco 2002; and Pelusi 2011. Note that in the nineteenth- and early twentieth-century literature the codex's creator is referred to as Benedetto Rinio (1485–1565) – a later owner of the manuscript who inscribed his name as an *ex libris* on the opening folio (*praefatio*, f. 1 recto). On the Rinio family and the fate of the *Roccabonella Herbal* after the death of Roccabonella's son and heir, see Pitacco 2002 and related archival materials now Venice, Archivio delle Istituzioni di Ricovero e di Educazione, DER E 189.1–6 (Rini). The pioneering archival work of Giuseppe Praga (1893–1958), whose unpublished notes are now Venice, Biblioteca Nazionale Marciana, *Carteggio e note manoscritte*, Marc. It. VI, 505 (coll. 12299), fasc. 20: "Roccabonella e Amadei, erbario del sec. XV", reminded modern scholars of the author's identity – Nicolò Roccabonella of Conegliano, as named in the *praefatio* (f. 1 verso) and in the index's *excipit* (f. 15 recto). Michelangelo Minio (1952–1953) subsequently published Praga's findings in the mid-twentieth century. Early Modern scholars, however, seem to have attributed the work correctly. For instance, when Ulisse Aldrovandi (1522–1605) consulted the manuscript ca. 1571–1572, he made notes on its contents and identified both Roccabonella and Andrea Amadio as its creators (Bologna, Biblioteca universitaria, Aldrovandiano 136, tomo V, c. 208) (Minio 1952–1953: 50 and Marcon 2002: 116). De Toni points to Aldrovandi's notes and identifications, but still attributes the codex's creation to Rinio (1924: 180n1).

[2] In recent scholarship, the dating of this manuscript usually follows that proposed by Mariani Canova who locates the manuscript's creation to 1445–1448 (1988: 21–28). Mariani Canova notes that in his preface Roccabonella mentions his son, Jacopo (1420–1479), then a medical student at the University of Padua (1988: 26). Jacopo completed his medical studies in 1447, providing a potential *terminus ante quem* (for Jacopo's records at the University of Padua, see Zonta and Brotto 1922: 362, 374–375, 379, 423). While I concur with Mariani Canova that Jacopo's graduation provides a *terminus ante quem*, particularly for the preface (f. 1), I suspect that the creation of the entirety of the codex – given its

---

Note: The research for this project was supported by the Renaissance Society of America – Kress Foundation Centro Vittore Branca, the American Philosophical Society, and the Friends of the University of Wisconsin-Madison Libraries. I am grateful to these institutions for their generous support. I also wish to thank Ilenia Maschietto (Biblioteca Fondazione Giorgio Cini), Laura De Rossi (I.R.E. and Fondazione Venezia), and Elisabetta Lugato (Biblioteca Nazionale Marciana) for their invaluable guidance to the Venetian *fonds* and collections.

on paper,[3] and illustrated in watercolour, the *Roccabonella Herbal* contains 458 chapters on individual plants used for medicinal purposes. While the general organisation of each chapter (image on one side of the folio and text in two columns on the other) and its subject matter (medicinal plants) remain consistent throughout the manuscript, the textual and visual contents of the codex, to my knowledge, do not consistently conform to any recognisable or traditional system of classification or organisation. Both the textual and visual content of Roccabonella's book confront the reader with the uncanny feeling of engaging with different systems or parts of systems – from multiple times and places – simultaneously. So, what is its "more modern order"?

In the preface (f. 1),[4] Roccabonella identifies his motive for the creation of his ambitious and expansive codex. He informs the reader that after travelling to "various regions" (varias p[er]lustra[n]do regiones) in his youth and completing his medical studies at the University of Padua,[5] he settled in Venice to practice medicine in 1415.[6] Roccabonella explains that during his many years as a practicing physician he witnessed critical – and even fatal – errors made by apothecaries in their drug recipes. The cause of these errors, Roccabonella alleges, was their unfamiliarity with the

---

length and breadth of its contents, which represent the culmination of Roccabonella's medical career – spanned a much longer period, likely from the early 1430s through to Roccabonella's death in 1459.

3 Marcon identifies four possible watermarks within the codex that date to the first half of the fifteenth century, providing a possible *terminus post quem* for the creation of the codex (1988: 154). The majority of the watermarks were in use primarily during the 1430s and include: a balance (scales) similar to Briquet 2413 (Dijon 1417/Utrecht 1434); a lion's head similar to Briquet 15846 (Florence 1431), Briquet 15847 (Venice 1437), and Piccard *Raubtiere* 1409, 1411 (Innsbruck 1431, Ulm 1453); three mountains within a circle similar to Briquet 11889, –92, –95, –97 (1385–1407 and widely used across France as well as in Brussels, Genoa, and Siena); three mountains (not encircled) similar to Briquet 11663 (Genoa and Florence 1434). Marcon considers the encircled-mountains watermark on the paper comprising the index as close to Briquet 11895 (Florence 1434). See Marcon's complete codicological description (1988: 154–155).

4 Ineichen (1959: 464–466) provides a transcription of the *praefatio*. My citations of the *praefatio* primarily follow this transcription. Morelli (1802: 398–405) and Valentinelli (1868–1873: 5.61–67) provide partial transcriptions as well as descriptions of the codex. All translations of the *praefatio* are original and unpublished; with my deepest thanks to Tyler J. Fyotek for sharing his expertise in medieval Latin.

5 Roccabonella graduated from the University of Padua in January 1410 (Zonta and Brotto 1922: 13).

6 Another manuscript penned in Roccabonella's hand, now Montpellier, Bibliothèque universitaire, 277, contains a compilation of medical treatises as well as a personal record of his children's names, birthdays, locations of their births, and the identities of their godparents (f. 161 recto). This record reveals that during the first 16 years of his professional practice (the final entry of the record dates to 1431) Roccabonella and his family kept their residence in several different quarters of the city. Initially, on first relocating to Venice in 1415, Roccabonella lived in the modest quarter around the church of San Salvatore (1415–1419); from there the family moved to increasingly posh neighbourhoods – from the quarters associated with the church of San Lio (1421), to Santa Maria Formosa (1425), and to San Giovanni Crisostomo (1426–1429). By 1431 the family lived in the quarter of San Giovanni Nuovo, suggesting Roccabonella's prosperity (Sigerist 1935: 161).

"vocabulary and foreign idioms of the ancient authors of medicine" (p[er]egrina vocab[u]la exteraq[ue] ydiomata). The apothecaries, Roccabonella charges, did not apply "their studies as the more specific foundation of this sort of medicine" (h[uiu]scemodi p[ar]tic[u]laria medicinę fu[n]dame[n]ta medicis studia adhibe[n]tib[us]). They chose, unlike the exemplary (moral) physicians of the ancient and recent past, whose lives and legends Roccabonella recounts in his preface (and among whom the physician locates himself and his family), to disregard as trivial or useless the study of logic, history, language, and of medicine's canonical authorities.

Such rhetoric may evidence the jockeying of physicians and apothecaries for authority as they increasingly vied for pieces of the medical marketplace's pie,[7] and we should take Roccabonella's criticisms of apothecaries' study habits and business practices with a grain of salt. Nevertheless, this real or supposed error prompted Roccabonella to complete what he calls his life's "great labour". Roccabonella explains,

> Ea p(ro)p(ter) me(n)te cu(m) op(er)e decrevi, quę ia(m) vetustate fere deleta s(un)t aute(n)tice dicta de si(m)pliciu(m) notitia, i(n) luce(m) q(uan)tu(m) valuero serie quada(m) moderniori revocare, ut sexagenarij mei Nicolai Rochabonella Coniclane(n)sis phisici ętas frustra no(n) elaba(tur) novave no(n) pauca michi memorię [. . .].[8]

> On this account I became resolved in my mind after great labour to call back into the light in some more modern order, to the extent I was able, authoritative facts from my own knowledge of [plant] simples which already had almost been lost because of their antiquity, in order that my age [–] that of Nicolai Rochabonella of Conigliano, natural philosopher in his sixties [–] not slip away in vain or that the numerous new facts not be locked away in my memory [. . .].

Roccabonella seeks both to illuminate and preserve traditional knowledge from "antiquity" and his own "new" knowledge for future generations of medical professionals and to present and organise this information in an innovative "more modern" way.[9]

In addition to combating apothecaries' erroneous conduct, Roccabonella longs to leave a useful and remarkable gift – a key to his own memory – not just for the wayward apothecaries, but also for his eldest son Jacopo (1420–1479), who, following in his father's footsteps, studied medicine at Padua and pursued a career as a practicing physician,[10] and for his father Lodovicus (d. before 1410), to honour and memorialise the physician who, Roccabonella explains, first "supplied the greatest benefits in this

---

7 On the dynamics of the medical marketplace in Early Modern Italy and in Europe more broadly, see Palmer 1984, 1985a and 1985b; Eamon 2003; Gentilcore 2003a and 2003b; Cavallo and Gentilcore 2007; Egmond 2008; Welch 2008; Bamji 2016; and Pugliano 2017.
8 *Praefatio*, f. 1 verso (transcription per Ineichen 1959: 466).
9 Ann Blair argues that safeguarding hard-won knowledge from what she calls the "traumatic loss of ancient learning of which [scholars] were keenly aware" motivated later sixteenth-century compilers and their increasingly encyclopedic projects (2013: 382).
10 For records of Jacopo's studies at the University of Padua, see Zonta and Brotto 1922: 362, 374–375, 379, 423. See also Mariani Canova 1988.

science to me his own son" (m[ichi] nato suo Lodovicus medicus doctissim[us] maxi-[m]a i[n] hac sci[enti]a p[re]stitit b[e]n[e]ficia[11]). Roccabonella closes his preface with praise for his collaborator, Andrea Amadio, whom he identifies by name and calls the "magnificent Venetian painter" (Venetu[m] picture[m] s[u]blime[m]) who completed this great work "with not a little favour" (no[n] parvo m[ichi] collato favore) to Roccabonella, suggesting a close, essential collaboration between the physician and the artist in the creation of this monumental codex.[12] Taken together, the acknowledgments in Roccabonella's *praefatio* emphasise the centrality of the act of collaboration – literal and metaphorical, and past and present – to the creation of the codex and its "more modern order".

The herbal's chapters consist of single paper folia comprising full-page illustrations of the plant in watercolour on one side (usually the recto) accompanied on the other side (usually the verso) by texts on the plant in question executed in an untidy, abbreviated gothic cursive hand and organised in two columns in the manner of university textbooks – *libri di banco*. Amadio's plant representations vary in style, ranging from novel and extraordinarily realism to simple schema to imaginative stylistic hybrids. His imagery emulates and incorporates examples found in earlier illustrated herbals, particularly the images from the late fourteenth-century illustrated vernacular translation of the *Liber Serapionis aggregatus in medicinis simplicibus* (Serapion's Book of Aggregated Simple Medicines) now known as the *Carrara Herbal* (London, British Library, Egerton 2020), deliberately connecting his work back to older illustrative and allegorical traditions of plant representation.[13] Amadio offers new perspectives on plant representation that potentially defy past illustrative traditions, as well – for instance, he represented plants previously unknown in earlier illustrated herbaria.[14]

In Roccabonella's accompanying texts, readers encounter, in the left column, an evolving list of synonyms – names by which the plant and its parts are known in other languages, chiefly in transliterated Greek and Arabic and in Latin, but also in several vernacular languages. In the right column, Roccabonella includes a brief description of the

---

11 *Praefatio*, f. 1 verso (transcription per Ineichen 1959: 466).
12 *Praefatio*, f. 1 verso (transcription per Ineichen 1959: 466).
13 On the *Carrara Herbal* and its illustrative cycle, see Kyle 2017.
14 Amadio and Roccabonella added to and amended older sources. For instance, Amadio's representation of what would later be called *Atropa belladonna* L., nightshade or belladonna ("De faba inversa", f. 16 recto), and the *Anemone hepatica* L., common hepatica ("De taura herba", f. 104 recto), are the first appearances of these plants in illustrated *herbaria* (De Toni 1919: 203–204 and 250, and Marcon 2002: 116); *Alchemilla xanthochlora* Rothm., lady's mantle ("De herba stella", f. 3 verso), *Filago germanica* L., common cottonrose ("De Cartafilagine", f. 19 recto), and *Tragopogon pratensis* L., goat's beard ("De herba licia", f. 307 recto) are also new plant drugs (Paganelli and Cappelletti 1996:112–113). De Toni (1919: 249 and 271) and Olariu (2018: 158n8) further identify the representation of sesame ("De sesame", *Sesamum orientale* L., f. 103 recto) and balsam apple ("De carança", *Momordica balsamina* L., f. 140 recto) as previously unknown in illustrative traditions.

medicinally useful parts of the plant and when to harvest them; but the majority of the text in this column is a bibliography of sorts in which Roccabonella lists by name the ancient and more recent medical authorities who comment on the plant, the titles of their works, and the chapters in which further information about the therapeutic applications of the plant in question could be found.

As the reader may expect given Roccabonella's prefatory remarks, the codex does not follow the order of plants included in the Greek, Arabic, or Latin canonical botanico-medical works, nor does it adhere to the order of earlier medieval lexica (glossaries or *synonyma*) of materia medica. While the influence of Dioscorides' (*fl.* 1st cent. A.D.) first-century pharmacopeia Περὶ ὕλης ἰατρικῆς (*De materia medica*) looms large in Roccabonella's codex,[15] the chapters themselves do not follow the original order of medicinal substances found in Dioscorides' Greek work.[16] Neither are the entries arranged alphabetically, nor are the images of plants grouped stylistically or arranged morphologically; although some groupings of visually similar or "related" species of plants are placed in sequences, the majority of plants are not grouped in what we now call "families".[17] Consequently, the material seems jumbled and puzzling rather than didactic and useful, per Roccabonella's intent. In its apparent disorder the codex

---

15 The literature on Dioscorides' work and on its translation and transmission is vast. As a starting point and for a brief and accessible account of the translations, redactions, and transmissions of *De materia medica* prior to the first printing of its Latin translation (Colle di Val d'Elsa: Johannes Allemanus de Medemblick, 1487), see Funk (2016) and Scarborough's introduction to Beck's recent English translation of *De materia medica* (Beck 2005: xiii–xxi).

16 *De materia medica*'s five books include: 1) aromatic oils, salves, trees and shrubs and their products; 2) animals, parts of animals, animal products, cereals, pot herbs and sharp herbs; 3) roots, juices, herbs, and seeds; 4) roots and herbs not previously mentioned; and 5) wines and minerals (Riddle 1970 [repr. 2008]: 120, and Scarborough and Nutton 1982: 191). Dioscorides tells his reader that he deliberately eschewed the alphabetical order, preferring to arrange his materials "according to the natural properties of each one of them" (*De materia medica*, Preface, §5 [tr. Beck 2005: 3]). In this "new arrangement", he seems to have grouped plant medicines first into large categories, dividing them according to the "three kingdoms of nature" (vegetable, animal, mineral), and then subdividing them according to their broad appearance and then again by their physiological effects on the body. On Dioscorides' organisation of the five books of his *De materia medica* and its many interpretations, see Riddle 1970 [repr. 2008]: 120 and 1985, especially chapter 3; Scarborough and Nutton 1982: 191; Nutton 2004: 129; and Hardy and Totelin 2016: 80–82. For an alternative interpretation of Dioscorides' organisational system as following the *scala naturae* – the scale of natural properties in which positive and negative values are attributed to certain characteristics of the substances – see Touwaide 2007: 43–44.

17 For example, several species of the plant family of fruiting vines now known as *Cucurbitaceae* appear together in sequence at ff. 41–45 ("De angularia" [*Citrullus vulgaris* Schrad., watermelon], "De citrulo" [*Cucumis sativus* L., garden cucumber], "De cucumere" [*Cucumis flexuosus* L., serpent melon], "De popone" [*Cucumis melo* L., melon], "De cucurbita" [*Cucurbita Lagenaria* L., bottle gourd]). Likewise, four members of the *Poaceae* family (at ff. 94–97) are grouped together. But "De spelta" (f. 150), perhaps *Hordeum distichum* L., two-rowed barley (De Toni 1919: 277), and "De frumento" (f. 316), perhaps *Hordeum hexastichum* L., common barley, following the entry in the *Carrara Herbal* (see Ineichen 1959: 451), or *Triticum sativum* Lam., wheat, following Roccabonella's Latin synonym "Triticum"

seems unfinished and assumes characteristics of miscellanies, commonplace books, or notebooks popular among humanists and educated professionals and suggests Roccabonella's individual compilatory and compositional methods. But could this apparent disorder be Roccabonella's desired "more modern order"?

Artful *bricoleurs*, the co-creators repurposed and redeployed earlier textual and visual knowledge systems to engage the readers in novel ways and, in doing so, to stimulate the creation of new knowledge systems. Instead of a systematic engagement with plants and their medicinal properties that would fit into organisational and theoretical models well-established in the late medieval period (such as those found in the textual and artistic source materials for the codex), when readers encounter Roccabonella's herbal they engage in the creative *process* of knowledge-*building* and system-*seeking* that characterised Roccabonella's career and the changing roles of medicine and natural philosophy during his lifetime. Through the apparent disorder and experimentation in their text, imagery, and organisational method, Roccabonella and Amadio establish a dialogue between not only the canonical medical authorities and their works but also between the readers and the codex's diverse textual and visual information. They implicitly ask their readers to *choose* where to go, which linguistic path to follow or source to consult, while – at the same time – pointing in all these directions at once. By bringing together, and so bringing out of chronological time and regional place, a wide array of naming and illustrative conventions and of authorities and their works, Roccabonella and Amadio create a virtual collaboration in a virtual library.[18]

Reading Roccabonella's Herbal

To read Roccabonella's *herbarius* requires physical space and ample time. A large, heavy codex, measuring 28.5 x 20.5 cm, its 458 chapters on individual plants are written on paper and prefaced by a parchment frontispiece on which Roccabonella inscribed his *praefatio* and identified both himself and his artist collaborator Amadio (f. 1).[19] An expansive alphabetical index of plant names in the various languages represented in the codex's entries and their associated chapter-numbers follows over the subsequent fourteen folia (ff. 2–15), providing a finding aid for the reader in search of a specific plant name. The modern folio numeration begins anew after the index,

---

(De Toni 1924: 361), are also members of that plant family and are visually similar to their fellow flowering grasses, yet they are kept sequentially distant in the codex.

18 For the term "virtual collaboration", I am indebted to the work of developmental and comparative psychologist Michael Tomasello (Tomasello 1999) and philosopher of cognitive science Alexander James Gillett (Gillett 2018).

19 For a complete codicological description, see Marcon 1988. For details of the manuscript's watermarks, see n3 above.

hence the first chapter, "De Meliloto" (*Lotus corniculatus* L., bird's foot trefoil), is understood as the first folio of the codex's contents proper.

Figure 1: *De Meliloto* (*Lotus corniculatus* L., bird's foot trefoil), by Andrea Amadio, in *Liber de simplicibus* (Roccabonella Herbal). Venice, Biblioteca Nazionale Marciana, *Lat.* VI, 59 (coll. 2548): f. 1 recto. 28.5 x 20.5 cm, watercolour on paper, Venice, ca. 1430–1459. Su concessione del Ministero della Cultura – Biblioteca Nazionale Marciana. Divieto di riproduzione.

As is the case in this and in all other chapters, Amadio's watercolour illustration of the plant is isolated from Roccabonella's text on the opposite side of the folio. Amadio's illustration of "meliloto" shows attention to naturalistic detail as well as a compositional playfulness characteristic of many of his plant representations, portraying the plant as though it were pressed into this very book – its stems bent back upon themselves in graceful arabesques. However, in a mode more didactic than botanically accurate, Amadio represents the plant as it would appear in two different seasons at the same time: the plant bears its bright yellow-orange, sweet pea-like flowers (which bloom throughout the summer), as well as its long seedpods, which grow in a cluster reminiscent – as the plant's common name suggests – of a bird's foot (and which mature in the fall). He also carefully draws attention to the plant's alternating compound leaves – their trio ("trefoil") of clover-like leaflets that grow from short stems with additional leaflets at the base of each stem. Yet Amadio has only vaguely suggested the plant's roots and their structure. Instead of bird's foot trefoil's taproot, Amadio fictionalises indistinct, schematic rootlets flowing from the base of the principal stem.

From this opening image of what Roccabonella calls in his preface the "primary form" (formę p[rim]e[20]) of the plants, the artist echoes the physician's textual compilations and practice of pointing in several historical and authorial directions at once. In the ensuing imagery, he continues to deny viewer expectations for consistency in style and technique. In addition to its blending of schematic and more naturalistic styles of representation, the image of "meliloto" is, in fact, a copy – a visual translation – of an illustration found in the late fourteenth-century luxury *herbarius* known as the *Carrara Herbal*, which was commissioned by and belonged to the last prince of Padua, Francesco II "il Novello" da Carrara (1359–1406; ruled 1390–1405).[21] "De meliloto" is the first example of Amadio's visual translation into his much more extensive illustrative cycle of all the plant imagery from the *Carrara Herbal*. Its presence in the Roccabonella *herbarius* connects the reader back to earlier illustrated herbals and their histories and it signals to the readers the intergenerational nature of the codex's visual and textual contents.[22]

---

20 *Praefatio*, f. 1 verso (transcription per Ineichen 1959: 466).
21 Art historian Otto Pächt (1902–1988) first identified the relationship between the imagery in the *Carrara Herbal* and the *Roccabonella Herbal* (see Pächt 1950: 30).
22 In their first chapter Roccabonella and Amadio misinterpret the plant (De Toni 1919: 190), perhaps following the visual precedent established in their illustrative source; an image of *Lotus corniculatus* L. accompanies the *melilotus* chapter in the *Carrara Herbal*, as well ("Del meliloto", f. 15 recto). Bird's foot trefoil (*Lotus corniculatus* L.) resembles species of sweet clover (*Melilotus*), which may account for these errors. Both bird's foot trefoil and clovers (trefoils) are now considered members of the same *Fabaceae* family of flowering plants. Prigioniero et al. recently identified Dioscorides' μελίλωτος, *Melilotus* (III.40 [tr. Beck 2005: 197]), as *Melilotus neapolitanus* Ten. (Prigioniero et al. 2020: 63). Later in the codex, at chapter 152, "De trifolio interciso sive medica" (f. 151), Roccabonella also identifies sweet clover (*Melilotus officinalis* [L.] Pall.) as "lotos" (De Toni 1924: 277).

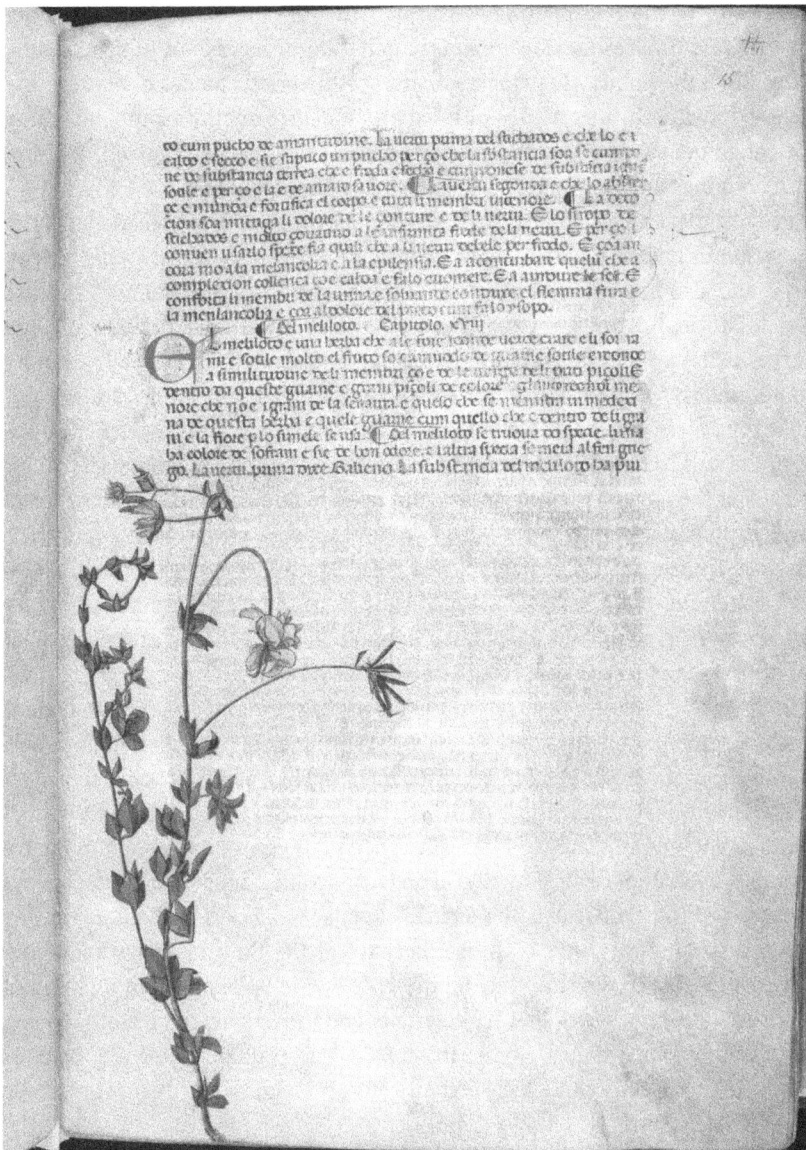

Figure 2: *Del Meliloto* (*Lotus corniculatus* L., bird's foot trefoil), in *Carrara Herbal*. London, British Library, Egerton 2020: f. 15 recto. 35 x 24 cm, gouache on vellum, Padua, ca. 1390–1405. © The British Library Board Egerton 2020

Amadio's translation of the *Carrara Herbal*'s imagery can be atmospheric rather than prescriptive, but his debt to the earlier herbal is evident and intentional. Unlike his anonymous artistic predecessor, who in the *Carrara Herbal* combined styles to effect a type of salubrious reading experience and to promote an image of the codex's owner as a healer of the body politic of Padua,[23] Amadio plays with a diversity of styles to puzzle and amaze his viewers and to generate new ways of understanding and thinking about the plants. He presents readers with various modes of plant representation, and so with various modes of *seeing and identifying* the plants, including both historical and entirely new ones, mirroring in his method Roccabonella's desire to showcase both the "antique" and "new" knowledge about plants.

Amadio's technical and formal oscillation in plant imagery draws readers deeper into the question of both the plants' many names and identities across times, places, and cultures, and also into their long history of therapeutic and medical uses in other traditions – folkloric, religious, mythic, and legendary – traditions often referenced by the sources listed in Roccabonella's accompanying texts, and occasionally even by Roccabonella himself.[24] For instance, Roccabonella alludes in his entry "De populo" (*Populus nigra* L., black poplar, f. 156 verso) to the Ovidian myth of the Heliades, the daughters of the sun god and Clymene and sisters to the ill-fated Phaethon. After their brother scorched the earth as he unsuccessfully tried to pilot his father's sun-chariot across the sky, the sisters mourned at his grave on the bank of the river Po. In their grief, they were transformed into poplar trees (see Figure 3), their tears turning into amber (Ovid, *Metamorphoses*, 2.344–366). Of course, in reality amber does not run like tears from poplar trees, and Roccabonella notes this inaccuracy and posits that perhaps another type of tree unknown to him must be its source.[25]

In a similar way, in his representation of chicory ("De cicorea", *Cichorium intybus* L., f. 207 recto, Figure 4), Amadio again blends styles, bringing his attention to naturalistic details – like the saucer-shaped flowers with their ray-like petals or the rough, dentate basal leaves which resemble those of dandelions – together with details associated with illustrative and mythic traditions less concerned with verisimilar imitation. Above the plant, at the folio's upper edge, Amadio added a sun with a smiling face, whose rays reach down toward the plant, while the formally similar petal-rays of the blue flowers extend upward seeking the sun's embrace. Because of the flowers' remarkable and memorable relationship to the sun (they faithfully follow it, opening when it rises and closing as it reaches midheaven), literary references and myths about chicory abound. In the Latin tradition, for instance, Horace (65–8 B.C.), Virgil (70 B.C. – A.D. 19), Ovid (43 B.C. – ca. A.D. 18), and Pliny (A.D. 23/24–79) all reference chicory in their works.[26] By

---

[23] See Kyle 2017.
[24] On how late medieval illustrated herbals connected medicinal uses of plants to their many associated myths, legends, or folk practices and reminded their readers of the alternative cultural contexts in which plants and their healing properties feature, see Kyle 2022.
[25] For a transcription of Roccabonella's comment, which accompanies his list of synonyms for the resinous sap produced by black poplars, see De Toni 1924: 277.
[26] Horace described chicory as a staple in his own diet (along with olives and mallows) (*Odes* 1.31. ll. 15–16); Virgil, in his first *Georgic*, mentions the danger of its bitter root to other crops (1. ll. 118–121); Ovid, in Book 8 of his *Metaphorposes* features chicory in the tale of Baucis and Philemon (a devout, if impoverished, couple) who prepare a modest meal of chicory and radishes for Jupiter and Hermes (themselves disguised as

Figure 3: *De popolo* (*Populus nigra* L., black poplar), by Andrea Amadio, in *Liber de simplicibus* (Roccabonella Herbal). Venice, Biblioteca Nazionale Marciana, *Lat.* VI, 59 (coll. 2548): f. 156 recto. 28.5 x 20.5 cm, watercolour on paper, Venice, ca. 1430–1459. Su concessione del Ministero della Cultura – Biblioteca Nazionale Marciana. Divieto di riproduzione.

---

peasants) (8. ll. 665–667), and Pliny, in his encyclopedic *Naturalis Historia*, notes the medicinal properties of chicory, including its powers of attraction (as the flowers are themselves attracted to the sun) (20.30).

Figure 4: *De Cicorea* (*Cichorium intybus* L., chicory), by Andrea Amadio, in *Liber de simplicibus* (Roccabonella Herbal). Venice, Biblioteca Nazionale Marciana, *Lat.* VI, 59 (coll. 2548): f. 207 recto. 28.5 x 20.5 cm, watercolour on paper, Venice, ca. 1430–1459. Su concessione del Ministero della Cultura – Biblioteca Nazionale Marciana. Divieto di riproduzione.

including a visible reference to the plant's nyctinastic traits in his image, Amadio invites these legendary associations as another avenue of knowing chicory's identity, histories, and uses.

Amadio's inclusion of both naturalistic images and unrealistic, deliberately archaising, or schematic ones illustrates the taxonomical and linguistic problems articulated by Roccabonella in his preface and visible in the physician's exhaustive lists of synonyms and sources. Indeed, physicians, humanists, apothecaries, and other players involved in the quest to rebuild ancient knowledge from original source texts and from personal experiences may have looked to Amadio's images to help tackle the very problem that Roccabonella's list of synonyms identifies: there was no consistent identification of the plants by name.[27] This disorienting viewing experience, like its complementary reading experience, makes visible the methodological – and moral – crisis that characterised this period of transition in medicine and the natural sciences. Through imitation of older models and the use of a diversity of styles, the illustrations echo the very process of accrual, experimentation, and collaboration seen in the textual contents and illuminate the stakes in the changing landscape of medical and natural-historical knowledge. Like the lists of synonyms, the diverse illustrations generate new ways of understanding, identifying, and thinking about the medicinal plants.

Turning the page, the reader encounters Roccabonella's textual entry on the plant in question (see Figure 5).[28] The text consists primarily of a glossary of plant names – in the left column – and a bibliographic cross-reference to the works of medicine's canonical authorities – in the right column. The creation of *synonyma* or works that compile equivalents (synonyms) of medicinal substances in the same and different languages, has a lengthy history among physicians and translators.[29] This practice was central in advancing what historian Charles Burnett terms the "globalization of knowledge", particularly in the field of pharmacotherapeutics.[30] Roccabonella's work advances this practice in new directions. His synonym list focuses on the plant names in Greek, Arabic, and Latin – the traditional languages of the medieval medical authorities – a focus that follows the format used by Simon of Genoa (*fl.* late 13th cent.) in his influential *Synonyma medicinae seu Clavis sanationis* (Medical Synonyms or the Key of Healing), which Simon likely composed while serving as papal curia chaplain and physician to Nicolas IV (Girolamo Masci; pontificate 1288–1292) and his court.[31] Simon's *Clavis* draws on the *Breviarium* of Stephan of Antioch (*fl.* first half of the 12th cent.),[32] another source consulted by Roccabonella. Notably Simon, together with

---

27 On the strategies of Early Modern physicians, apothecaries, natural historians, botanists, and members of the *Respublica literaria* for coping with these inconsistencies, see, as a starting point, Ogilvie 2003; Margócsy 2010; and Egmond 2012. For an overview of pre-Linnean botanical naming practices and classification systems, see Reveal 1996.
28 For a transcription of the entry for chicory, see Teza 1897–1898: 26–27.
29 Burnett 2016: 132.
30 Burnett 2016: 131.
31 Touwaide 2013b: 48, and Burnett 2016: 131.
32 Burnett 2016: 132. The *Breviarium* consists of a list of 583 medicinal materials organised into three columns of text, with the Greek term on the left, the Arabic equivalent on the right, both transliterated into the Roman alphabet, and the Latin in the center (Burnett 2016: 135). The terms, however, follow

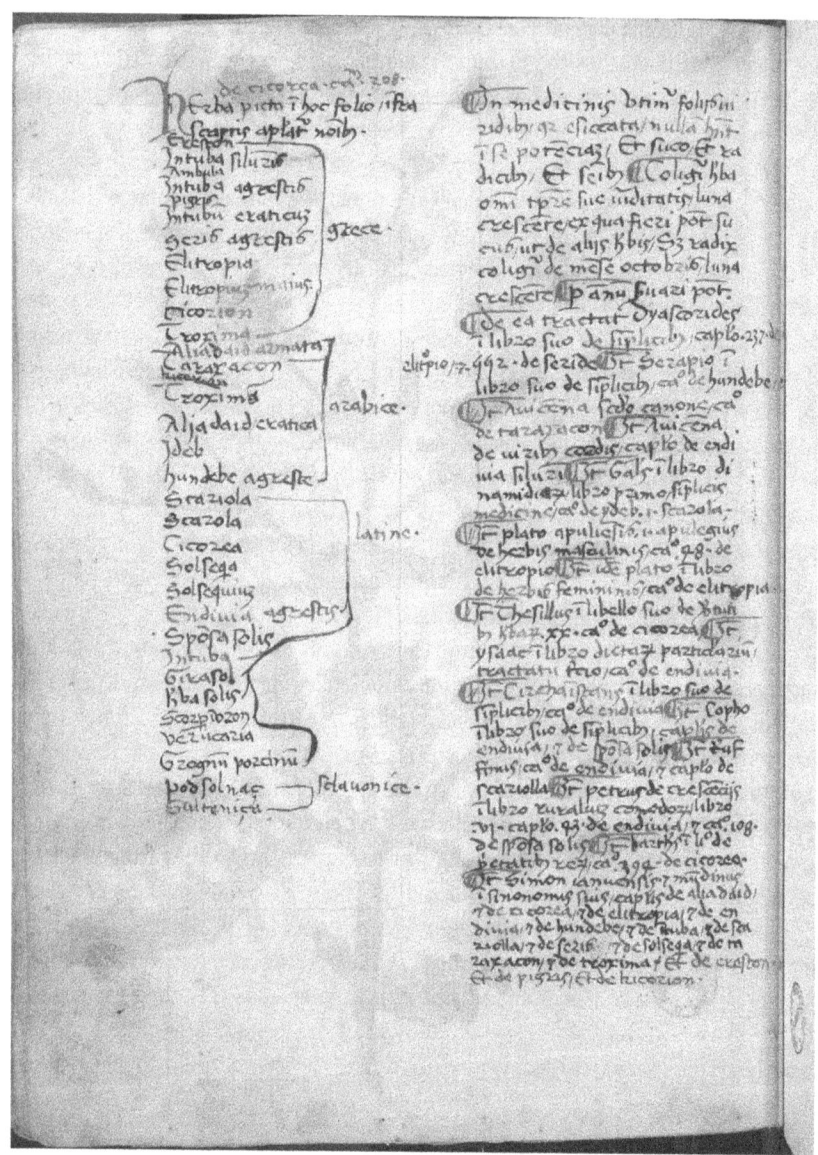

Figure 5: *De Cicorea* (*Cichorium intybus* L., chicory), by Nicolò Roccabonella, in *Liber de simplicibus* (Roccabonella Herbal). Venice, Biblioteca Nazionale Marciana, *Lat.* VI, 59 (coll. 2548): f. 207 verso. 28.5 × 20.5 cm, ink on paper, Venice, ca. 1430–1459. Su concessione del Ministero della Cultura – Biblioteca Nazionale Marciana. Divieto di riproduzione.

---

the Greek alphabetical order, perhaps according to an index found in one family of Greek Dioscorides manuscripts (Burnett 2016: 135). Simon's *Clavis* is alphabetical, following the Latin order.

Abrāhām ben Shēm-Tōb of Tortosa (*fl.* late 13th cent.), translated the work of Serapion the Younger (*fl.* mid-13th cent.) from Arabic into Latin as *Liber Serapionis aggregatus in medicinis simplicibus* (Serapion's Book of Aggregated Simple Medicines);[33] Serapion's work is another critical source for Roccabonella and, in its illustrated vernacular translation in the *Carrara Herbal*, for Amadio as well. Simon's *Clavis* is also a principal resource for Roccabonella who closes most of his bibliography entries by noting the chapters in Simon's work where the plant is identified and discussed. However, while Simon organises his lexicon's entries alphabetically, Roccabonella does not follow this suit in his *herbarius*.

Simon's method of ascertaining information may also have influenced Roccabonella. As Petros Bouras-Vallianatos notes, Simon advocated for an understanding of plants based on knowledge of both ancient texts and personal experience. In his preface, Simon makes clear his approach:

> Triplici existente omnium medicinarum simplicium genere plantarum videlicet animalium et mineralium non satis ad eorum cognitionem solis scripturis innitendum deputaverim quando frequenti visitatione ac diligenti investigatione ad id studium impendatur, quam maxime circa plantarum distantias [. . .].[34]

> Regarding all simple medicines that are classified into three categories, namely plants, animals, and minerals, it is not enough to rely only on our knowledge of the writings, when we could apportion frequent visits and careful inquiry to support this study, as it is very much related to the diversity among plants.

That Simon cites his meeting with a woman from Crete, with whom he was able to confirm the identities of some herbs, testifies to his method.[35]

It is likely from similar collaborative experiences and encounters with physicians, apothecaries, merchants, and traders in Zara (now Zadar, Croatia) that Roccabonella grew his lexicon in ways that exceeded the earlier efforts of Simon and Stephan of Antioch, who had confined their lexica to Greek, Arabic, and Latin. Other vernacular languages than these classical three appear in Roccabonella's glossary. In 1449, Roccabonella took his *herbarius* with him to Zara where he served two, two-year contracts as a practicing physician in the city while assisting his youngest son, Lodovico (1422–1453), in establishing a trading business.[36] Here, Roccabonella forged a community and network of professionals from diverse linguistic and cultural backgrounds that may have helped the physician to craft and refine his lexicon. After Greek, Arabic, and Latin, the other languages most often represented in Roccabonella's *synonyma* are those

---

[33] There is no extant Arabic version of Serapion's text preceding its translation into Latin. The work's transmission rests on this single translation (Dilg 1999: 226). Since Simon cites Serapion in his *Clavis*, he may have translated Serapion's work around the same time or earlier (Dilg 1999: 226).
[34] *Clavis*, Preface §5 (tr. in Bouras-Vallianatos 2013: 36 and 36n27).
[35] Bouras-Vallianatos 2013: 36.
[36] Minio 1952–1953: 53.

he labels *sclavonice* or *schavonice* (likely Serbian-Croatian) and *teotonice* (German).[37] Roccabonella's son Lodovico was a merchant mariner who traded supplies from across the Venetian territories and whose father-in-law, Giovanni Carreto (*fl.* mid-15th cent.), a treasurer of the Venetian militias, was connected to an apothecary from northern Europe, likely from Mainz, Germany. This apothecary, Giovanni di Ermanno Reinhardt (Iohannes filius Hermanni Renardi, *fl.* mid-15th cent.), partnering with another German-speaker, established a shop in Zara. Apothecary shops – early pharmacies – sold all kinds of goods, from foodstuffs to medicines to the pigments for making paint and served as lively meeting spots or sites of information exchange.[38] In this shop, Roccabonella likely encountered not only Reinhardt and his compatriot, who would have been familiar with the German and "Slavic" terms for the plants in the herbal,[39] but also other citizens of or visitors to Zadar, which was a stronghold of commercial exchange in Dalmatia and the wider Venetian territories.

The other languages in which plant names appear in Roccabonella's glossary, albeit with less frequency, include, dialects of Italian (*vulgare italico*)[40] and French (*galice*),[41] and perhaps in a North African dialect (*punice* or *sec punicos*, according to the Punics [Carthaginians]) and a transliterated Coptic (?) (*sec egipcios*, according to the Egyptians).[42] Roccabonella's linguistic categories often repeat; in several instances, Roccabonella creates another section in his left-hand column for a language already

---

[37] Some scholars allege that the entries in Croatian and German were added later and by different hands (Teza 1897–1898: 28; Marcon 1988: 155; Pelusi 2011: 129 and 129n65). While some of these entries do appear inconsistent with the principal hand of the manuscript, in the clear majority of all vernacular entries the hand remains consistent, albeit sometimes penned in a different ink, suggesting an interval of time between Roccabonella's synonym entries, perhaps in accord with the physician's travels between Venice and Zadar. For examples of "sclavonice" entries in alternate hands, see Roccabonella's chapter on "De Bardana" (burdock, *Arctium lappa* L., f. 25 verso), where "Ripagl" is listed as an equivalent in a different hand, and, in Roccabonella's chapter on "De branca ursina" (acanthus or bear's breeches, *Acanthus mollis* L., f. 54 verso), "Medvidia ztopa" is given as an equivalent. Notably, in her more recent scholarship, Marcon changes her position and acknowledges that many of the "sclavonice" terms are in Roccabonella's hand (2002: 113).
[38] See Matthew 2002; De Vivo 2007; DeLancey 2011; and Kostylo 2015.
[39] Minio 1952–1953: 55.
[40] For example, "De camaleonta nigra" (bull thistle, *Cirsium lanceolatum* L., f. 186 verso) is identified as "Carlina" in "vulgare italico".
[41] For instance, "Humulus" is noted as the equivalent in "galice" for "De lupulo" (common hops, *Humulus lupulus* L., f. 79 verso).
[42] For example, in the entry for "De rosmarino" (rosemary, *Salvia rosmarinus* Spenn., f. 135 verso), Roccabonella lists "Çibbir" as the equalivant in "punice"; in the entry for "De nasturcio" (nasturtium, *Lepidium sativum* L., f. 317 verso), Roccabonella lists alternative names "according to the Punics" ("sec punicos") as "Cosinin" and "according to the Egyptians" ("sec egipcios") as "Cinocardamum" (De Toni 1924: 362); for "De fragaria" (strawberries, *Fragaria viridis* Weston, f. 147 verso), Roccabonella pens the name in "teotonice" at the lower edge of the page beneath the illustration and, beneath the illustration for "De popolo" (black poplar, *Populus nigra* L., f. 156 recto, Figure 3), Roccabonella or Amadio noted "populus idest talpon" (populus is talpon) – "talpon" is Venetian dialect for poplar ("il pioppo"

represented, suggesting a process of accrual and evolution in the glossary in accord with Roccabonella's medical peregrinations.[43] Also in the glossary column, Roccabonella notes plant names as used by certain authorities, principally Pliny, Stephan of Antioch, and 'Ali ibn al-'Abbas al-Majusi, known in the Latin West as Haly Abbas (d. ca. 994),[44] and sometimes seeks to correct earlier authors' use of plant nomenclature as well.[45]

A Rollcall of Medical Tradition

Roccabonella's "more modern order" is, in part, a new way of relating to ancient tradition. His herbal provides a comprehensive up-to-date cross-reference of medical authorities and invites an interactive *conversation* with them. The sources listed in Roccabonella's second column of text include many of the authorities whose works students would encounter in medical school and that would be instrumental in their future medical practices. In addition to the works of Aristotle (384–322 B.C.) and those attributed to Hippocrates (ca. 460 – between 375 and 351 B.C.), medical students were required to read the works of Greek physician Galen (A.D. 129 – after [?] 216), and Persian physicians and philosophers Rhazes (al-Rāzī, ca. 854–925 or 935) and Avicenna (Ibn Sīnā, ca. 980–1037).[46] They also consulted compilations of case studies (*consilia*) and of medical recipes, sources that – during their future professional lives – would

---

in modern Italian) (Nardo et al. 2009: 536); on Amadio as author of this phrase, see De Toni 1924: 277 and Marcon 1988: 155.
43 On the *peregrinatio medica* (foreign medical travel) and its importance to Early Modern physicians, see Cunningham 2010; and Egmond 2021.
44 For example, Roccabonella cites Pliny ("sec Plinium", according to Pliny) among his glossary entries for "De salvia domestica" (sage, *Salvia officinalis* L., f. 324 recto) and "De caulibus" (wild cabbage, *Brassica oleracea* L., f. 194 verso); he cites Haly Abbas' *Liber regalis* among the lexicon entries for "De abscinthio romano" (Roman wormwood, *Artemisia pontica* L., f. 118 verso) and Stephan of Antioch's *Breviarium* for "De piro" (pear, *Pyrus communis* L., f. 62 verso). Haly Abbas' *Kitāb al-malikī* (The Royal Book) was translated by Stephan into Latin as the *Liber regalis* and to which Stephan appended his glossary of materia medica known as the *Breviarium*. On Haly Abbas and Stephan of Antioch, see Burnett 2016: 134.
45 For instance, in the entry "De edera" (ivy, *Hedera helix* L., f. 144 verso), Roccabonella argues that Dioscorides, Serapion, and Avicenna all confused the names *cistus* and *cissus*, Latinisation of the Greek for ivy (edera). He also seeks to correct Avicenna at "De memite" (yellow-horned poppy, *Glaucium flavum* Crantz, f. 8 verso). Modern historians and philologists have charted Roccabonella's and Amadio's errors of identification and attribution. See De Toni 1919–1925, and Ineichen 1959, especially.
46 Students read the Hippocratic *Aphorismi* (Grendler 2002: 177), *Prognostica*, and *De fracturis* (Grendler 2002: 314–318). On the medical curriculum at Italian universities, see Grendler 2002: especially chapter 9.

often be exchanged and shared.[47] In terms of medical botany, the primary resources – in addition to Galen – were Dioscorides and Pliny.[48] Roccabonella builds on these canonical, scholarly foundations and includes other sources of significant professional use in medical practice.

The majority of Roccabonella's bibliography-columns begin with references to the works of four canonical authorities (see Figure 5): Dioscorides' *De materia medica*, Serapion's *Liber aggregatus*, Avicenna's *Liber canonis* (*al-Qānūn fī 'l-ṭibb*; The Canon of Medicine) and *De viribus cordis* (*al-Adwīa al-qalbīya*; On the Faculties of the Heart), and Galen's *De simplicium medicamentorum temperamentis ac facultatibus* (On the Mixtures and Properties of Simple Medicines) and *De dinamidia* (The virtues of medical substances), which is now considered Pseudo-Galenic. While Dioscorides and Serapion provided essential pharmacopeias, during Roccabonella's lifetime, Avicenna's *Liber canonis* was the basic medical theory taught at Italian universities;[49] in this work, the Persian physician advanced a view of medicine that complemented the Aristotelian natural philosophy taught at universities and integrated it with tenets of Galenic medicine.[50] Gerard of Cremona (ca. 1114–1187) likely translated Avicenna's work into Latin in the late twelfth century; it served as the principal medical theory taught at Italian universities until well into the Early Modern period.[51] By the time Roccabonella graduated from the University of Padua in 1410, Avicenna's work had been an established part of the curriculum there for nearly a century.[52] Arnaud de Villanova (d. 1311), professor at Montpellier, translated *De viribus cordis* (On the Faculties of the Heart), the other Avicennian work to which Roccabonella refers, into Latin in the late-thirteenth or early fourteenth century either from the Arabic original or from an earlier Hebrew translation.[53] This text, often bound together with the *Liber canonis* in later printed editions of Avicenna's works, addresses issues of emotional and mental health.[54]

---

[47] For details of the university medical curriculum as recorded in the earliest extant statutes from the University of Bologna (1405), see Grendler 2002: 314–24. For a list of subjects specifically taught at Padua, and with which Roccabonella and his son would have been familiar, see Grendler 2002: 24. On *consilia*, in particular, see Siraisi 1981, 1987b (repr. 2001b), 1994 (tr. 2001, especially in relation to the growing interest in Avicenna's medicine) and 2007: 65–79; Agrimi and Crisciani 1994; and Crisciani 2005.
[48] Grendler 2002: 343.
[49] Siraisi 1987a, 1987b: 111 (repr. 2001b: 162–163) and 1994 (tr. 2001); and Crisciani 2005: 308.
[50] Grendler 2002: 320.
[51] While scholars have long attributed the translation of Avicenna's *Canon* to Gerard of Cremona, Touwaide suggests that Gerard of Sabloneta (*fl.* mid-13th cent.) may be the translator (2013a: x).
[52] Giovanni Mondina da Cividale (ca. 1275–1340), physician and professor at the University of Padua, completed his commentary on the first book of Avicenna's *Canon* in 1316 (Pesenti 1999: 63).
[53] On the dissemination of Avicenna's text into Western Europe, see Siraisi 1987a.
[54] These three works are bound together in several early printed editions of their Latin translations, including editions printed in Basel (per Joannes Hervagios, 1556) and Venice (apud Iuntas, 1582).

Most of the remaining sources referenced in Roccabonella's entries are medieval or more modern works that incorporate, translate, or comment on the canonical medical authorities and represent a veritable rollcall of medical tradition from the fourth- through fourteenth centuries. Following generally the order in which they are listed (although their order of appearance can vary and not all authorities appear in all entries), these sources principally include: Salernitan physician Matthaeus Platearius (d. 1191), compiler of the widely circulated, unillustrated pharmacopeia known by the opening words of its prologue "Circa instans"; Apuleius Platonicus, an unknown fourth or fifth-century author whose *Herbarium* served as the primary treatise on materia medica in the Latin West from late Antiquity through the thirteenth century and to whom Roccabonella attributes Pseudo-Dioscorides' *Ex herbis femininis* (From Female Herbs);[55] Macer Floridus, likely the pseudonym for Odo de Meung (fl. late 11th cent.), a French monk, physician, and author of *De virtutibus herbarum et aromatum* (On the Power of Herbs and Spices), a poem on materia medica set in Latin hexameter verse; "Isaac" refers to the celebrated Egyptian-Jewish physician Isaac Israeli ben Solomon (or, in the full form of his name, Abu Ya'qub Ishaq ibn Suleiman al-Isra'ili, d. ca. 932), better known under the Latinized form of his name used in the Middle Ages, Isaac Judaeus (ca. 832 – ca. 932), whose treatise on "universal" and "particular" dietetic therapeutics, *De dietis universalibus et particularibus* (On Universal and Particular Dietetics), was translated from Arabic into Latin by Gerard of Cremona or Constantine "the African" (d. ca. 1098) and entered into the medical curriculum of European universities in the thirteenth century;[56] Rufinus (fl. second-half of the 13th cent.), Italian physician whose *herbarius*, *De virtutibus herbarum* (On the Power of Herbs), dates to after 1287 and is known only in a single extant fourteenth-century copy (now Florence, Biblioteca Medicea Laurenziana, Ashburnham 116 [189–121]);[57] Petrus de Crescentiis (ca. 1233–1321), Piero de' Crescenzi, Bolognese juror and author of the popular early fourteenth-century treatise on agriculture *Liber ruralium commodorum* (On Rural Life);[58] Copho, a twelfth-century physician and anatomist (and perhaps student of Matthaeus Platearius[59]) working in Salerno, whose "libro suo de Simplicibus", as noted by Roccabonella, is likely *Ars medendi* (The Medical Arts), which contains information on materia medica (both simple and compound medicines) and instructions for their therapeutic uses;[60] and Frater Bartholomeus

---

55 Riddle notes that in several manuscript copies Apuleius' *Herbarium* runs directly into the Pseudo-Dioscorides' text without any change of attribution (1981: 46 and 46n7). In his bibliographical entries, Roccabonella often calls Apuleius' work *De herbis masculinis*, suggesting that the physician followed the tradition in which Apuleius' *Herbarium* was associated with "masculine herbs" (Riddle 1981: 47).
56 Jacquart and Micheau 1990: 114. On Constantine "the African", in particular, see Jacquart 1996.
57 Thorndike 1932: 64, and 1946; McVaugh 1975/2008: 601.
58 In addition to presenting information on curative properties of plants, de Crescenzi's twelve-volume work discusses the practicalities of animal husbandry and plant cultivation. On the treatise's contents and its fortunes from the later Renaissance, see Ambrosoli 1992 (tr. 1997): especially 41–95.
59 Kristeller 1945: 156.
60 Ventura 2009: 12.

(Bartholomeus Anglicus, ca. 1203–1272), Scholastic and Franciscan monk, and author of the encyclopedic compendium of things divine and mundane *De proprietatibus rerum* (On the Properties of Things, ca. 1240), which includes sections on materia medica and their applications for health and healing.[61] As mentioned above, Roccabonella's bibliographic columns of medical authorities usually close with reference to "Simon Januensis", Simon of Genoa, compiler of the *Clavis sanationis*, a multilingual lexicon of materia medica produced in Rome in 1290 that includes the names by which medicinal materials were known in Greek, Arabic, and Latin. Simon and his work may have served as both an important resource and an inspirational model for Roccabonella and the physician's own project.

Humanist Habits

Roccabonella's preoccupation with exhaustive compiling, cross-referencing, and correcting authoritative medical works and synonyms of plant names may also illuminate his choice to arrange his chapters in a seemingly random way. Both humanists and physicians practiced several types of note-taking and list-making, which intersected with patterns of excerption and (dis)organisation found in other compilatory genres of writing, like miscellanea and florilegia.[62] Particularly in the era before widespread access to printed books, physicians' notes often consisted of several types of note-taking used in combination with one another to manage and sort information crucial to their professional lives.[63] In his recent study of the unpublished personal notes and notebooks of several – primarily German – Early Modern physicians, Michael Stolberg identifies three principal types of medical note-taking, all of which include elements of empirical observation. He argues that medical students and physicians circulated in and between diverse social and cultural spheres, resulting in exceptionally variable forms of note-taking.[64] Stolberg further posits that while these different forms of

---

[61] Notably, several of these authors and their treatises are found in the two other extant manuscripts that belonged to Roccabonella (now Montpellier, Bibliothèque universitaire, 277, and Florence, Biblioteca Medicea Laurenziana, Ashburnham 1448). These manuscripts are compilations of medical, alchemical, astrological, and natural historical texts copied – and in some cases translated – by Roccabonella. On Ashburnham 1448 and Roccabonella's translation of *De dinamidia*, see Marcon 2002: 114. On Montpellier 277, see Sigerist 1935, and, as Roccabonella's manuscript, Olariu 2018: 161n8.

[62] In the last several years there has been a surge of interest in note-taking and archiving practices of scholars and humanists (though, much more rarely, of physicians) during the Early Modern period. See, as a starting point, Blair 2004, 2010a and 2010b; Yeo 2014; and Vine 2019. On the intersection of list-making, antiquarianism, and collecting efforts of humanists, physicians, natural historians, and botanists, see Keller 2014. Keller argues that what she calls a "rapid essayistic style" of writing, note-taking, and list-making best suited "a natural order which was itself in flux" (2014: 426).

[63] Stolberg 2016: 254.

[64] Stolberg 2016: 245 and 248.

note-taking mirrored humanists' practices, they served different functions and purposes for the professional medical community – they were "working tools" invaluable to professional practice.[65] For Roccabonella, whose *herbarius* reflects these note-taking practices, they were not only "working tools", but the building blocks of a new tool with which to think and to collaborate.

The three types of note-taking that Stolberg identifies include plain notebooks, also known as *adversaria*, commonplace notebooks (often organised in what Stolberg terms the "sequential" or "textbook" approach), and practice journals or casebooks. Plain notebooks served as sites for unsorted information ordered only temporally, according to when the note-taker encountered and entered the notes. Physicians would sometimes sort their plain notebooks by their notes' subject category – that is, the notebooks' contents, while still ordered only temporally, were unified by their subject matter in individual notebooks; this form may be visible in the structure of Roccabonella's *herbarius*, since, although its entries are ordered in a seemingly random fashion, all of them share the subject matter of medicinal plants. This form of note-taking, whether collected haphazardly in a single notebook or divided rudimentarily by subject matter into individual notebooks, needed to be further processed either through indexing or through copying the miscellaneous notes into a more structured notebook, a process Stolberg terms "secondary notetaking".[66] Roccabonella's exhaustive index evidences this process.

Commonplace notebooks, conversely, contain diverse notes on topics assigned to a shared heading (i.e.: a "common place" or *locus communis*). Physicians easily adapted the practice of commonplacing to medical knowledge, and this malleable form of note-taking is central to treatises on medical therapeutics and to collections of empirical observations. It served not only as a primary form of note-taking, but also as a secondary form for adapting plain notebooks into more accessible and useful resources.[67] Roccabonella's gathering of bibliographical references and names for each plant (as the entry's *locus communis*) could be considered a kind of commonplacing.

Commonplacing was central to long-standing educational pedagogies, and both physicians and humanists would have encountered this method of note-taking in their "preuniversity" and university studies.[68] Humanists routinely used this form to collect excerpts from and references to ancient works, and while such "rhetorical

---

[65] Stolberg 2016: 264.
[66] Stolberg 2016: 250 and 261.
[67] Stolberg 2016: 252.
[68] I borrow the term "preuniversity" studies from Paul Grendler, who analyses the transition of humanist teachers from what he calls "preuniversity" studies to the university (2002: especially 199–229). Stolberg notes that teachers trained students to collect notes in common-places in the Latin schools and that this practice would have been universally familiar to Early Modern scholars (2016: 252). For overviews on commonplacing, especially in early printed sources, see Blair 2010, and Moss 1996.

kinds" of commonplace notebooks were also found in physicians' collections of notebooks, they were not representative of the principal subject or use of physicians' commonplace notebooks.[69] Rather, physicians adapted what Stolberg aptly terms this "quintessential humanist tool for knowledge management" to suit their purposes.[70]

Physicians' commonplace notebooks manifested in what Stolberg terms the "sequential approach", in which physicians entered information in temporal order, but assigned a heading to each note, a heading that was repeated in future notes with the same topic and that was later used in the construction of an index; the "textbook approach", in which physicians' organised notes thematically, reserving pages of individual albums for notes on specific subjects identified in headings penned along the top of each page in the manner of a textbook's chapter headings;[71] and the "alphabetical approach", the least used among physicians, in which physicians ascribed their notes onto pages pre-assigned with alphabetised headings of different topics or, more simply and flexibly, onto pages pre-assigned with initial letters, descending in alphabetical order.[72] Physicians often included in their commonplace notebooks examples of these three types of note-taking, inconsistently grouped together.[73]

Stolberg argues that each type of commonplacing suited a different aspect of physicians' professional and scholarly identities. For instance, the "textbook approach" seems well-suited to students in medical school and physicians in the early phases of their careers. Perhaps following their teachers' instructions, beginners could find in this approach an easy way to maintain a structure that enabled them to draw together the many fruits of their studies, particularly on topics well-established in the curricula, and to better learn and memorise this information.[74] As a note-taking method, it also helped students to find their own point of view and encouraged questioning.[75] The "sequential approach", Stolberg posits, may have been better suited to advanced students and note-takers, those whose notes would address diverse and irregular topics, rather than the more familiar or standard topics one would encounter in medical school.[76] For instance, when physicians learned new information – about a novel plant or medical therapy – they could make note of it and assign it to a new and suitable heading; but this new information would be challenging to retrieve at a later date since in this form of note-taking, much like in plain notebooks, entries on any particular topic, although anchored by a heading, were located throughout the notebook. To respond to this challenge, physicians created indices as finding aids, a

---

69 Stolberg 2016: 252.
70 Stolberg 2016: 251; Stolberg 2013 and 2014.
71 Stolberg 2016: 253.
72 Stolberg 2016: 254.
73 Stolberg 2016: 265.
74 Stolberg 2016: 255.
75 Stolberg 2016: 261.
76 Stolberg 2016: 256 and 261.

practice informed by the well-established use of indices in medieval manuscript culture as guides for the reader.[77] Such a tack is visible in Roccabonella's *herbarius*, providing an invaluable tool to access the codex's contents and reflecting Roccabonella's professional engagement with practical, contemporary modes of systematising information for himself and for future readers.

The third type of note-taking practice Stolberg identifies among early modern physicians encompasses what he terms "practice journals" and "casebooks"; in these notebooks, physicians recorded details of their daily professional practices and kept entries about their individual patients and associated diagnoses and therapeutic applications.[78] This practice echoes, in part, the late-medieval compilation and exchange of *consilia* – case studies circulated among fellow physicians and among patients from at least the early fourteenth century.[79] Casebooks and practical journals follow a temporal order of the physicians' daily work and include observations and developments in patients' health, concerns, and patterns of care, and some of them included indices of diseases and of individual patient-names.[80] In this way, casebooks and practice journals could serve as what sociologist Alberto Cevolini calls "forgetting machines";[81] akin to merchants' account ledgers, which were necessary tools for maintaining good business hygiene, these notebooks contained information on the daily business of physicians' practices – details about individual patients, their treatments and fees, medicines, and dosages.[82] As such, these notebooks stand in contrast to the commonplace and plain notebooks, which were tools – "working tools" – with which to learn or to memorise.[83]

History, Biography, and the Virtuous Lives of Famous (Medical) Men

In the body of his codex, Roccabonella's entries on plants reflect not only his training and familiarity with humanist note-taking and memorisation strategies, but also his dedication to the humanists' and physicians' shared project to reclaim and preserve ancient Greek and Latin sources and their shared focus on ancient sources and issues

---

[77] Stolberg 2016: 256–257. Stolberg grants that the lines between commonplace notebooks following the "sequential approach" and temporally ordered plain notebooks can be blurry, especially when physicians created indices for notebooks of both types (2016: 257n43).
[78] Stolberg 2016: 258–259.
[79] Siraisi 2007: 13 and 65–79, and Crisciani 2005. See also note 47 above.
[80] Stolberg 2016: 259. Physicians also used these books to track any outstanding and received fees or dues associated with their patients – that is, alongside the details of professional practice, these notebooks served as an ancillary form of bookkeeping (Stolberg 2016: 258).
[81] Cevolini 2016.
[82] Stolberg 2016: 263.
[83] Stolberg 2016: 264.

of translation. Growing access to the original Greek texts of the medical authorities, especially to those of Galen and Dioscorides, revealed to humanists and physicians alike the many translation and transmission errors in canonical medical texts and commentaries, which for physicians and apothecaries, in particular, resulted in increased anxiety and questions about the "true" identities of the plants in important medical recipes.[84] This quest prompted collaborative explorations – and sometimes fierce debates – in search of the plants' identities.[85] As Roccabonella rightly notes in his *praefatio*, "when a human life in any way is in the power of a doctor, if [the doctor] does not have experience of many writings, he could be judged a killer of men rather than a healer of illnesses" (Cu[m] e[n]i[m] humana vita q[uo]da[m] modo i[n] potestate sit medici, ni multa[rum] habuerit p[er]itia[m] script[ur]a[rum], occisor potius ho[m]i[n]u[m] q[uam] eg[ri]tudinu[m] curator pot[uer]it iudicari).[86]

In his *praefatio* Roccabonella further signals and illustrates his engagement with humanistic pursuits – not just with the important issues of translation and identification but with subject matter traditionally associated with humanism, particularly the lives and histories of illustrious men. Using language characteristic of humanists' treatises on the lives of (primarily ancient Roman) famous men (*viri illustres*), a genre popularised anew by the poet Petrarch and other humanists in the fourteenth century but tied to a much older literary tradition,[87] Roccabonella recounts the lives, histories, and virtues of medicine's *viri illustres* – from Apollo, the Greek god of medicine, and Aesculapius, demi-god and legendary founder of medicine, to Hippocrates, Galen, Avicenna, and a host of other faithful followers in the Greek, Latin, and Arabic traditions. He upholds these legendary and historical physicians and philosophers from across vast expanses of time as exemplars worthy of imitation and admiration, establishing a "moral blueprint" for physicians to follow.[88] Roccabonella asks his reader:

---

[84] The scholarship on translation efforts of Renaissance physicians and humanists is vast. See, as a starting point, Reeds 1976; Siraisi 1990 and 2007; Nutton 1985a, 1985b, 1988, 1995 and 1997; and Ogilvie 2006: 11–12 and 121. On the anxiety that these efforts elicited, particularly among physicians and apothecaries, and on the practical responses in the medical marketplace, see Pugliano 2017 and 2018.

[85] Perhaps the most famous example of such a debate unfolded in the generation after Roccabonella's death when Niccolò Leoniceno (1428–1524), professor of medicine and moral philosophy at the University of Ferrara, sought to emend and correct Pliny's *Naturalis Historia* in his *De Plinii et plurium aliorum medicorum in medicina erroribus* (On the Errors in Medicine of Pliny and Many Other Medical Practitioners, Ferrara 1492) and Pandolfo Collenuccio (1444–1504), historian and jurist, came to Pliny's defense in his *Pliniana defensio* (Plinian Defense, Ferrara 1493). See the classic literature on this debate, Thorndike 1923–1958: 4.593–610, and Castiglioni 1953, and the more recent studies in Ogilvie 1997: 89–112; Touwaide 2000 and 2008a; Findlen 2006: especially 438–442; Fortuna 2007; and Tomlinson 2012. For a discussion of intellectual context for these debates, see Nutton 1997.

[86] Roccabonella, *Praefatio*, f. 1 verso (transcription per Ineichen 1959: 465).

[87] See Stacey 2007; and Witt 2009.

[88] On humanists' biographies of medical *viri illustres*, see Siraisi 1987b.

No(n)ne i(n) medico decet e(ss)e modu(m) maturitatis mo(rum), verbo(rum) facetia(m), corporis castitate(m), ad i(n)firmos m(u)lta(m) salutis p(ro)missio (n)e(m), freq(ue)nte(m) visitatione(m) & lit(er)atura(m) gra(m)matice, p(ro)fessio(n)e(m) dialetice philosoficeque discipline, geometrie, musice & astronomie, ut ad vera(m) p(er)ve(n)ire valeat medicine notitia(m) p(er) aute(n)tico (rum) libro(rum) studia, maxime Ypocratis, Galieni & Avice(n)ne[89][?]

Is it not fitting in a doctor that there be a measure of maturity of character, a refinement with words, a purity in body, a great assurance of health for the sick, frequent visitation, as well as an education in philology, a competence in the dialectical and philosophical discipline, the geometrical, the musical and astronomical, so that he is able to reach the true personal knowledge of medicine through the study of authoritative books, especially those of Hippocrates, Galen, and Avicenna?

In other words, the ideal physician – upon whom Roccabonella models himself and encourages his readers similarly to emulate – knows the lives and histories of medicine's greatest practitioners and possesses personal virtue, decorum, rhetorical skill, and what we might today call a professional "bedside manner"; moreover, he is educated not only in medicine, but more broadly in the liberal arts – the disciplines of philology, logic, philosophy, geometry, music and astronomy. The ideal physician is a *humanist*-physician.

Like their humanist counterparts, physicians trained in the liberal arts first in their "preuniversity" studies and Latin schools and then in university, as students of the faculty of arts to which the medical schools belonged.[90] Emphasising his connection to these traditionally humanist pursuits advanced Roccabonella's sense of himself as a moral, educated physician and highlighted his erudition, illustrating what Sarah Gwenyth Ross calls the "cultural legitimacy" of the physician and his family among the middle-rank, upwardly mobile citizens of Venice.[91] Ross argues that Venetian physicians, in particular, had an acute form of "bibliophilia"; their efforts to advance socially and economically and to grow visibility of their literary collections and connections to humanist enterprises were "mutually sustaining".[92] Physicians, as both Ross and Stolberg point out, leveraged their liberal arts training, fashioning themselves as humanist scholars by cultivating their non-medical interests, particularly in history and poetry of both classical and more modern origins, and realising these interests in their collections of books, manuscripts, and other artifacts.[93] Roccabonella's

---

89 Roccabonella, *Praefatio*, f. 1 verso (transcription per Ineichen 1959: 465).
90 Grendler 2002: 199–229. On humanism and its role in university-studies, in particular, see Lines 2006.
91 Ross 2016: 5. On the relationships between identity, "taste", social mobility, and the collection and display of objects associated with the idea of humanism in the Early Modern Rome, especially, see also Ago 2006. On the idea of "literary cachet" and education as catalysts for socio-economic mobility in Early Modern Florence, see also Maxson 2014: especially 60–4, 79–81, 151, 180.
92 Ross 2016: 22.
93 Stolberg 2014 and 2016: 245. Analysing bequests, testaments, and estate inventories of middle-rank Venetians, Ross finds that physicians were the most "consistent and omnivorous [book] collectors",

praefatio, with its commentary on education and its accounts of medical *vitae*, suggests his participation in these efforts. While no inventory of Roccabonella's collection or estate remains extant, Roccabonella did indeed ascend the social ladder and gain prosperity and prestige during his lifetime. His changing addresses reveals that, as his family grew and his career progressed, Roccabonella relocated to increasingly upmarket *sestieri* (neighbourhoods) in Venice.[94]

A Model Miscellany

Roccabonella's and Amadio's organisational and artistic strategies may also reflect the growing popularity among humanists in the first half of the fifteenth century of Aulus Gellius' second-century "miscellany" *Noctes Atticae* (Attic Nights). This work provided humanists – and humanistically trained physicians – with what Anthony Grafton terms a "methodological model" for the organisation of knowledge and the process of learning, a model that grew organically out of their shared commonplacing and note-taking practices.[95] While writers from Macrobius (fl. ca. A.D. 400) to Augustine (A.D. 354–430) and Petrarch (1304–1374) consulted, mined, and praised Gellius' work, humanists transformed it in the fifteenth century, concurrent with Roccabonella's and Amadio's creation of their codex.[96]

Gellius (ca. A.D. 125 – after 180), a Roman author and grammarian, allegedly began his sole extant work, *Noctes Atticae*, during a spell of cold winter nights in Athens (an anecdote that provides the work with its title) and continued it upon his return to Rome. The *Noctes Atticae* consists of 20 Books (Book 8 is not extant), in which Gellius collects excerpts from works by various Greek and Latin authors on topics ranging from grammar, history, and philosophy to geometry, antiquities, and

---

dedicating a large part of their budgets to the purchase of books the majority of which were not directly associated with their profession, privileging instead ancient and modern literature (2016: 30, 39, and 41).

94 See n6 above.
95 Grafton 2004: 321 and 324; Blair 2010b: 72.
96 On the fortuna of the *Noctes Atticae*, its printed editions, translations, and commentaries, see Holford-Strevens 2014. Prior to the fifteenth century, *Noctes Atticae* was known in two large fragments – Books 1–7 and 9–20, respectively (Grafton 2004: 320). In 1429, Nicholas of Cusa (1401–1464) began to circulate across his wide scholarly network what he considered to be the complete version of Gellius' text; in this version, Nicholas relocated the preface from the end of Book 20 to its rightful place at the start of Book 1. Shortly thereafter, the famed pedagogue Guarino Guarini (1374–1460) and other humanist scholars emended and systematically corrected the text – cross-referencing Gellius' Greek excerpts with those found in their corresponding extant source texts – and brought the work's two halves together. *Noctes Atticae* emerged in more complete and corrected versions throughout the 1420s and 1430s, offering a model for organising knowledge that both complemented and advanced the humanist agenda of encountering the past and bringing it forth anew in their present.

philology, and then knits these excerpts into a narrative framework through use of personal reflections and accounts of his experiences and encounters, as well as fictive dialogues among his protagonists and critiques of their discussions and works.[97]

The text's miscellaneous organisation, Gellius informs his reader in his preface, stems from his practice of jotting down anything he deemed worth remembering as he encountered it in his reading and then writing it in his notebook for later ease in jogging his memory. Gellius wrote,

> Vsi autem sumus ordine rerum fortuito, quem antea in excerpendo feceramus. Nam proinde ut librum quemque in manus ceperam seu Graecum seu Latinum uel quid memoratu dignum audieram, ita quae libitum erat, cuius generis cumque erant, indistincte atque promisce annotabam eaque mihi ad subsidium memoriae quasi quoddam litterarum penus recondebam, ut quando usus uenisset aut rei aut uerbi, cuius me repens forte obliuio tenuisset, et libri ex quibus ea sumpseram non adessent, facile inde nobis inuentu atque depromptu foret.[98]

> For whenever I had taken in hand any Greek or Latin book, or had heard anything worth remembering, I used to jot down whatever took my fancy, of any and every kind, without any definite plan or order; and such notes I would lay away as an aid to my memory, like a kind of literary storehouse, so that when the need arose of a word or a subject which I chanced for the moment to have forgotten, and the books from which I had taken it were not at hand, I could readily find and produce it.

In other words, much like plain- and commonplace notebooks, Gellius preserved the order in which he encountered the material – haphazard as it might appear to his readers – and built a "literary storehouse" for future use. Its order, then, maps Gellius' path of learning and the mind and values of the individual author; yet the popularity of Gellius' work – both in the ancient world and in the Early Modern period – belies the esoteric particularity Gellius' method implies. Rather, it appealed to readers as a font of knowledge curated and narrated in memorable ways.[99]

Using another strategy that his fifteenth-century imitators would employ, Gellius transformed his sources into a cast of characters and put them into staged conversations. These fictive dialogues, situated in vivid settings, often in a garden or beneath a tree, provided memorable context for the excerpted material;[100] the setting served as a "stage" for the speakers' erudition, framing the imaginary conversations with memorable "sets" upon which the discussions of learned facts unfold. For humanists like Angelo Decembrio

---

[97] For a useful introduction to Gellius, his work, and the context in which he lived and wrote, see Howley 2018: 1–63. Howley's analysis of the authorial intent behind Gellius' apparent miscellaneous ordering is especially insightful.
[98] Gellius, *Noctes Atticae*, Pr. 2 (Page, Kapps, and Rouse [eds] and Rolfe [tr.] 1927: 1.xxvi–xxxvii; also cited by Grafton 2004: 324 and 326).
[99] On reading and memory practices, from antiquity through to the Renaissance, see, as a starting point, Yates' classic study (1966). For the relationship of imagery to reading and memory practices in the medieval period, in particular, see, as a starting point, Carruthers 1990.
[100] Grafton 2004: 329.

(1415–after 1467), whose literary dialogues in *De politiae litterariae* (On literary polish) were inspired in many ways by *Noctes Atticae*,[101] Gellius' work and his mission to preserve, emend, and correct his sources represented the seeds of an "ideal library" – the kind that fifteenth-century humanists longed to perfect.[102] Roccabonella's *herbarius*, with its parallel focus on collecting plant names and identities, connecting them with the relevant passages from the works of medicine's traditional and more modern authorities, and associating them with remarkable imagery, constitutes such an "ideal library" in miniature. It is a vast repository of visual and textual information meant to be accessed, engaged, and discussed. Roccabonella adopted Gellius' method to curate the contents of his over-thirty years of medical study and experience in ways that invite collaboration and generate new knowledge while reinforcing the readers' memory of critical information about medicinal plants and the authoritative sources who analyse and use them.

In keeping with Amadio's atmospheric, translational approach to the plant imagery in the codex, and unlike the humanists following Gellius' model more directly, Roccabonella creates dialogues that remain *implicit*, neither fictively recreated nor recollected in a narrative progression. His collected synonyms and bibliographical lists suggest dialogues with fellow physicians and apothecaries, past and present. These dialogues unfold as Roccabonella both seeks to identify the names by which plants are known in other languages and to enumerate the relevant sections from the works of medical authorities in which to find additional information on the plants' therapeutic applications. The naming and cross-referencing imply a conversation among traditional and new authorities, presenting them as collaborators and inviting the readers – the next generation of medical experts – to participate. Moreover, Amadio's stylistically diverse imagery, with which the readers engage *before* encountering the collaborative "dialogues" found in the lists of synonyms and cross-references, locates these dialogues in a memorable "setting". In the readers' minds, the authorities "speak" in a vivid setting that enables this knowledge to be more easily recalled.

A Virtual Library, a Virtual Symposium

By the time Roccabonella and Amadio created the *herbarius* in the mid-fifteenth century, the role of "talkative" interlocutors who invited readers and viewers into

---

[101] Celenza 2004: 57, and Grafton 2004: 319.
[102] For an example of Gellius' corrections and textual investigations, see *Noctes Atticae* 2.3.5 where Gellius examines corrections entered into a copy of Virgil's *Aeneid* (Page, Kapps, and Rouse [eds] and Rolfe [tr.] 1927: 1.128–131). See Grafton 2004: 334–335 for this and other examples of Gellius' efforts, as Grafton notes, to "defend the classical writers against such inept . . . emendation", efforts continued by fifteenth-century humanists and physicians concerned with the corruption of classical texts by copyists (2004: 334). For Decembrio's "ideal library", see Grafton 2004: 329, and Celenza 2016.

conversation was a firmly established practice in literature and visual art[103] – and popular among humanists and physicians alike. The practice of including portraits of medicine's canonical authorities in illustrated herbals and other medical texts illuminates Roccabonella and Amadio's intentions and emphasises the importance of conversation and collaboration among ancient and modern scholarly luminaries and their readers. In earlier illustrative traditions of treatises on materia medica, we find several examples of frontispieces or series of prefatory illustrations that contain portraits of physicians, often from vastly different chronological and geographical realities, who are shown together at work or grouped in conversation – that is, as active collaborators.[104] A series of such prefatory illustrations opens the influential *Tractatus de herbis et plantis* of Manfredus de Monte Imperiale (*fl.* mid-14th cent. [?]), an otherwise unknown physician working in southern Italy. Manfredus' *Tractatus de herbis* (now Paris, Bibliothèque nationale de France, *Lat.* 6823, ca. 1330–1340) may have inspired, in part, the commission and creation of the *Carrara Herbal*, one of the sources for both the textual and visual contents of Roccabonella's *herbarius*.[105]

In Manfredus' herbal, the series of prefatory illustrations includes a scene of Manfredus speaking to his students, who hold plant specimens up for identification and discussion (f. 1 recto), followed by two pages with scenes of canonical medical authorities shown conversing with one another. These pages contain eight portraits of medical *auctores* and their commentators who were central to the curriculum of the Neapolitan school and medical community at Salerno to which Manfredus likely belonged. The eight figures are portrayed in pairs: each figure sits on an individual bench and gestures to the figure directly across from him. Excerpts from their most recognisable works stream from their mouths across the empty page toward their conversational partners.

Of the eight figures represented, four are named. The cast of characters includes Hippocrates or Constantine "the African" who recites the incipit from the *Prognostica*

---

103 Webb 2011: 428.
104 Important examples of such portraiture exist across the earlier traditions of illustrated Greek and Arabic works on medicinal materials, including Dioscorides' *De materia medica*. For instance, portraits of legendary and historical physicians adorn two of the prefatory frontispieces of the volume of collected texts of Greek medicine, which includes the alphabetised Greek version of Dioscorides' *De materia medica*, now Vienna, Österreichische Nationalbibliothek, *medicus graecus* 1, ff. 2 verso and 3 verso, Constantinople, traditionally dated to ca. A.D. 512/13; portraits of physicians at work appear in the Arabic copy of Dioscorides' *De materia medica*, now Istanbul, Süleymaniye Kütüphanesi, Ayasofia 3703, f. 2b, 33 x 24 cm, Baghdad (?), 1224, and also in the copy of the *Kitab na't al-hayawān* (Book on the Characteristics of Animals) by Nestorian Jibra'il ibn Bakhtīshū', now London, British Library, Or. 2784, f. 101 recto, North Jazira (?), ca. 1220.
105 On the relationship between Manfredus' *Tractatus de herbis* (Paris, Bibliothèque nationale de France, *Lat.* 6823, ca. 1330–1340) and the *Carrara Herbal* (London, British Library, Egerton 2020, ca. 1390–1405), see Kyle 2017: 47–53.

Figure 6: *Portraits of Hippocrates and Johannitius (above) and Hippocrates and Galen (below)*, by Lippo Vanni or Roberto d'Oderisio, in *Tractatus de herbis et plantis* (Herbal of Manfredus de Monte Imperiale). Paris, Bibliothèque nationale de France, Lat. 6823: f. 1 verso. 34.5 x 24.7 cm, Naples, ca. 1330–1340.
Source: Bibliothéque nationale de France

(Prognostics) to Ḥunayn b. Isḥāḳ (known in the Latin West as Johannitius, 809–873), the Arab translator; Johannitius, in turn, recites lines from his *Isagoge in Artem parvam Galeni* (Introduction to Galen's Art of Medicine). Beneath this pair, Hippocrates is

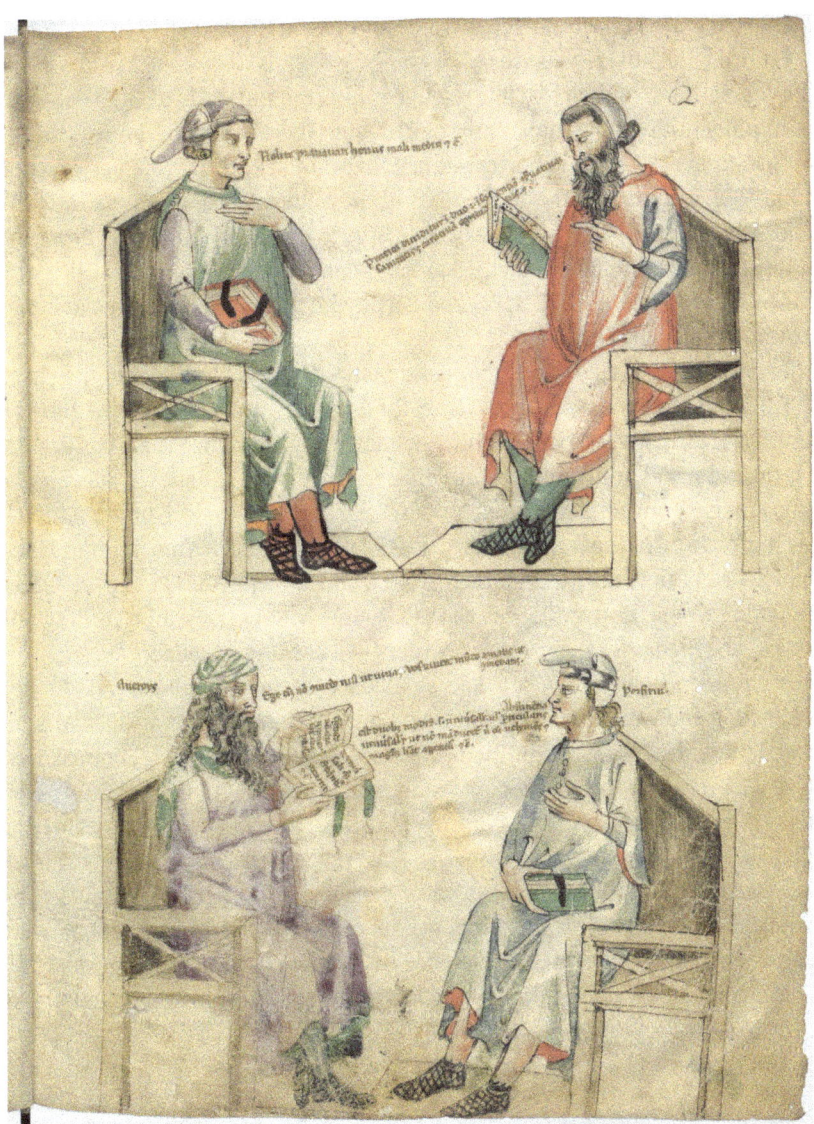

Figure 7: *Portraits of Mesue and Bartolomeo of Salerno (above) and Averroes and Porphyry (below)*, by Lippo Vanni or Roberto d'Oderisio, in *Tractatus de herbis et plantis* (Herbal of Manfredus de Monte Imperiale): f. 2 recto. 34.5 x 24.7 cm, Naples, ca. 1330–1340.
Source: Bibliothéque nationale de France

represented again, now reciting an excerpt from his *Aphorismi* (Aphorisms) to Galen who returns the Hippocratic maxim "Intendo enim manducare ut vivam, alii intendunt vivere ut manducent" ("Truly, I intend to eat so as to live; others live but to eat"), perhaps demonstrating his debt to Hippocrates and to the Hippocratic school's focus on diet as central to health (see Figure 6).[106]

On the facing folio, and so visually grouped with the portraits of the first four speakers, an unidentified physician, perhaps Yuḥannā Ibn Māsawayh (known in the Latin West as Mesue, 777–857), is portrayed in conversation with Bartolomeo of Salerno (fl. 1150–1180), a commentator on the texts of the authorities depicted on the preceding folio. Bartolomeo, in turn, recites the incipit from his commentary, *Practica*. Beneath Mesue and Bartholomeo, the Aristotelian commentators Ibn Rushd (known in the Latin West as Averroes, 1126–1198) and Porphyry (234–ca. 305) face one another in conversation (see Figure 7).[107]

Like the authorities listed in Roccabonella's bibliography-column and in the characters who feature in Gellius' *Noctes Atticae*, these figures are separated by history and geography. Their conversations unfold outside of physical time. Their position in the codex, on the pages following the opening frontispiece that shows Manfredus teaching his students, suggests that these timeless conversations occur in the mind of Manfredus himself and in the minds of his careful readers. The portraits refer to Manfredus' knowledge of the *auctores* and their works, which he puts into dialogue in his own work and teaching, illustrating his contribution to and participation in the lineage of medical knowledge and inviting his readers to learn and to collaborate.[108]

The textual and visual contents of Roccabonella's *herbarius* similarly reflect this labour of learning and valuation of collaboration and teaching. Roccabonella alerts us to the didactic intent of his work in the *praefatio* – to guide wayward apothecaries and to serve as a resource for future physicians, encouraging all readers to study and to emulate the lives of medicine's *viri illustres*. Further, the codex contains evidence of Roccabonella's literal collaboration with physicians, apothecaries, and merchants in Zara, which resulted in building his lexicon in directions that surpassed its literary precedents; it also contains evidence of his metaphorical (virtual) collaborations with the authorities named and cited in his reference list, which attests to the many learned voices that contributed to the physician's knowledge and informed his medical practice

---

106 On the physicians' identities, see Collins 2000: 292n113, and Touwaide 2013b: 46–7. Touwaide (2013b: 46) argues that the physician facing Johannitius may be Constantine "the African", another scholar-physician who translated Arabic treatises into Latin.
107 Collins 2000: 292n113. On the identification of Mesue, see Touwaide 2013b: 46.
108 Kyle 2017: 49–52, and 2022: 178–180. In a similar way, the portraits of respected authors adorn the fifteenth-century *studioli* of the educated elite, including, perhaps most famously, those of Federico da Montefeltro (1422–1482), the Duke of Urbino, and Isabella d'Este (1474–1539), Marchesa of Mantua (see Campbell 2004). In Federico's *studiolo*, for instance, portraits of authors with their associated works or other identifying attributes appear to converse among themselves from within their different architectural frames, inviting viewers into a conversation and also into a comparison between words and images. Art historian Jennifer D. Webb argues that these conversations inspire the viewer to imagine and then to imitate these dialogues (2011: 427).

(and the creation of this very codex). Amadio's emulation and incorporation of the plant imagery from earlier sources and his stylistic variance likewise evidences an artistic virtual-collaboration, engaging in conversation with past illustrative traditions while advancing new models and exemplars for future study and providing "sites", conceptual anchors, for conversation about the plants' identities and medicinal uses by the authorities referenced in each of the codex's entries.

Conclusion

The codex's "more modern order" documents acts of collaboration, both literal and metaphorical or virtual. Together with Amadio's plant imagery, Roccabonella's names of the plants-represented and of the authorities who specialise in the plants' therapeutic uses allude to and initiate conversations, both real and imagined, alerting readers to patterns of learning and to possible collaborators. The unusual reading experience, which highlights Roccabonella's note-taking practices and attachment to humanistic pursuits, encourages the readers to remember the plants' (many linguistic and cultural) identities and to map those identities onto the relevant authoritative texts – establishing their own "modern order". Engaged in such complex mental mapping, the readers increase their understanding of medicinal plants' identities and the associated medical literature to effect successful therapeutic or medical interventions. Through this act of mapping – of thinking, remembering, associating – they lay the foundation for building new knowledge about the plants and their medicinal usefulness, a foundation critical to the changes and developments in the fields of medicine, botany, and natural history in the sixteenth century.

Perhaps adhering to Gellius' tack of building his work "haphazardly", Roccabonella compiled the herbal's contents according to his individual experience of the material and his study of plant medicines throughout his lengthy career. Like the portrait-series in Manfredus's *herbarius*, the authorities Roccabonella consulted engage in conversation in his mind – a conversation he records in his bibliography - entries and into which he invites his reader. Amadio's imagery creates a "setting" for Roccabonella's entries, facilitating the conversations between medical authorities – implied in the list of their names and works – and between the unnamed physicians, apothecaries, and other readers involved in the identification, harvest, preparation, and trade of plant medicines evident in the list of synonyms. As the readers encounter these conversations made memorable by Amadio's diverse imagery, they too become active participants in a dialogue. They become collaborators, growing knowledge about the plants and their applications and knitting this information into their memories for future access and use.

Roccabonella's *herbarius* represents an innovative engagement with its traditional subject matter: the plants, their imagery and histories, and their medicinal and moral virtues. Its inconsistencies are generative; in an era before the wider access to ancient and more modern medical works in print, they spark new modes of understanding and thinking about materia medica – in collaboration. Roccabonella's and Amadio's textual and visual entries emphasise that this herbal, while superficially complete, remains unfinished. It is a work in progress, a conversation that invites additions, subtractions, and corrections, and reveals the frontiers of medical and artistic theories and practices in transition during the mid-Quattrocento. As its lengthy history of display, citation, and use after Roccabonella's death – and long into the era of print – would continue to illustrate, through its apparent disorder, unpredictability, and inconclusiveness the codex served as a catalyst, helping to shape the new landscape of pharmacological and natural historical knowledge at the dawn of the Early Modern era.

Barbara Di Gennaro Splendore
Mediterranean Botany. Making Cross-Cultural Knowledge about Materia Medica in the Sixteenth Century

Introduction

In 1591, physician Prospero Alpini (1553–1617) published *The Dialogue on Balsam* (*De balsamo dialogus*), a treatise on a medicinal plant called balsam. At the time, many believed this exceptionally expensive commodity was extinct. The dialogue was uniquely set in Egypt between three characters: Abdella and Abdachim, respectively characterized as "Egyptian physician" and "Jew", and Prosper Alpinus, the "Italian".[1] In the text, the three physicians describe the plant, explain how to recognize its precious resin, and argue that balsam originally came from Syria and Arabia, where it was still harvested, and was not native to Egypt. A graduate of the School of Medicine in Padua, Alpini served as physician to the newly appointed Venetian consul in Cairo between 1580 and 1583. Back in Venice, in 1591 and 1592, Alpini also published *The Medicine of the Egyptians* and *The Plants of Egypt*.[2] These works were the first scientific works about Egypt published in Europe and remained influential for centuries.

A shrub species found in nature only in selected areas of Eastern Africa and the Arabic peninsula, today the balsam plant (*Commiphora gileadensis*) is little used and of limited commercial importance.[3] From Antiquity onwards, instead, balsam was an expensive commodity in the Mediterranean.[4] A freshly harvested balsam could cost twice the price of silver, or even as much as its weight in gold, as it did in Majorca in the

---

[1] Alpini, *De balsamo*. Alpini, *De plantis Aegypti*: 62r. "Abdella Medicus Aegyptius, Abdachim Hebreus, Alpinus Italus".
[2] Alpini, *De medicina Aegyptiorum*. Alpini, *De plantis Aegypti*. The second edition of *De balsamo dialogus*, or rather the reprint, is printed at the end of *De plantis Aegypti*, and the numeration of the pages continues throughout with the title page on page 58r of the volume published by Francesco de Franciscus from Siena in 1592.
[3] Ben-Yehoshua et al. 2011; Mandaville 2011: 167.
[4] See Milwright 2001 and 2003.

Note: My deepest gratitude to the Institute of Research (ICUB) at the University of Bucharest, Romania, where I was honoured to be a Fellow in 2022/2023. A preparatory version of this essay was published in Italian (Di Gennaro Splendore 2018). I would like to thank all the participants in Francesca Trivellato and Alan Mikhail's seminar at Yale University where this research started as a research paper. I also would like to thank all friends and colleagues at the Early Modern Interdisciplinary Graduate Lunch at Yale who offered comments and support. Unless otherwise indicated, all translations are mine.

fourteenth century.⁵ Numerous therapeutic qualities were attributed to the plant's bark, fruit, resin, and seeds. Among its many properties, balsam cured sores and wounds, was an antidote against plague, venoms, fevers, and it was also a remedy against wrinkles and kidney stones. The most prized part of balsam was undoubtedly *opobalsamum*, an aromatic resinous extract (often erroneously defined as a juice or sap) obtained by making an incision in the bark. Prompted by balsam's illustrious history, recent medical and scientific studies confirmed that the plant has several medicinal properties.⁶ At the time of Alpini, interest in balsam had grown since its resin, wood, and seeds figured among the ingredients of theriac. Perhaps the most famous pharmaceutical formulation in Western medicine, in the second half of the sixteenth century theriac underwent a renovated success. Naturalist such as Ulysses Aldrovandi (1522–1605) in Bologna and apothecary Ferrante Imperato (ca. 1525–ca. 1615) in Naples embraced enthusiastically the quest to identify the "true" ingredients of theriac.⁷ Several physicians and learned apothecaries hoped to find "true balsam" to make the best possible theriac.

Through the analysis of his 1591–1592 works on Egypt and especially of the treatise on balsam, this essay contributes to the historiography exploring the production of new knowledge on materia medica at the end of the sixteenth century. I explore the role that Alpini assigned to classical texts, first-hand experience, and indigenous informants in the creation of new natural knowledge about materia medica. My claim is threefold. First, I argue that when making new natural knowledge about balsam, Alpini was strongly influenced by its very object. Balsam commercial value as well as its cultural and religious meanings played a critical role in Alpini's choice of rhetorical strategies. Second and immediately consequent, I argue that in the balsam dialogue Alpini presented natural knowledge on balsam as a cross-cultural scientific production. He gave prominence and authority to his Egyptian informants, if only rhetorically, on the opposite not only religion did not constitute a barrier to create natural knowledge, on the opposite new knowledge was trustworthy because it was shared by people following different religions. Third, I suggest that Alpini's strategies varied depending on the discipline: making botanical knowledge required different strategies than making medical knowledge. As Brian Ogilvie has argued, by the 1590s medicine and botany were fully autonomous.⁸ Indeed, the two disciplines called upon different authorities for validating new knowledge. When discussing Egyptian medical practices, as Nancy Siraisi has argued, Alpini had to connect his findings to both the classics and medical theory.⁹ When writing about plants, instead, the classics could be more or less reliable, while knowledge of plants implied physical proximity, observation, and experience. The ancient classics offered an inescapable point of reference, but their authority came second to first-hand observation.

---

5 Pegolotti, *La pratica della mercatura*: 124.
6 On the medical properties of balsam, see Iluz et al. 2010; Al-Sieni 2014: 23–27.
7 Findlen 1994: 241–284 (Chapter 6).
8 Ogilvie 2006.
9 Siraisi 2007: 233–246.

The quest for botanical species for medical or commercial use was an important force behind the development of early modern global events. In their studies of how interest in the natural world and commercial interests were intertwined, historians have shown the myriad ways in which the movement of people and things contributed to the creation of new knowledge about the natural world.[10] The idea that modern science was an exclusive product of European culture, which spread from Europe throughout the rest of the world (George Basalla's diffusionist model), prevailed among historians in the Western-dominated world for most of the twentieth century. More recently, Jorge Cañizares-Esguerra, David Arnold, Kapil Raj and others have widely criticized and disproven this model for South America and South Asia by using empirical evidence.[11] Rather than a one-directional flow from Europe to other continents, the history of science must be understood as a continuous and multi-directional exchange. Connections, appropriations, co-productions and negotiations among different cultures were inevitable when constructing knowledge and when building empires. Thus, my question is not "whether" other cultures contributed to science, but rather what was the nature of such exchanges in the sixteenth-century Mediterranean.

Historians of botany and medicine have focused their attention primarily on exchanges of natural knowledge in colonial settings. Exploring the connection between the developments in botany, Europe's commercial expansion, and colonial politics, Londa Schiebinger and Claudia Swan have concluded that in each European empire the science of botany developed in relation to state politics, between either individual or territorial initiative and colonized peoples. Each encounter between Europeans and non-Europeans led to different botanical practices. Since botany and colonial politics were intertwined, botany was also interwoven with ideas of supremacy, and the most common model of European botany erased "cultural complexities" and "objectified" specimens. European scholars appropriated local knowledge in order to create global goods and universal knowledge. In order to include indigenous plants in the European classificatory system, scholars stripped them of their native uses and cultural narratives. Harold Cook has asserted that the contribution of indigenous informants must be retrieved from the "erasures" of botanical texts.[12] Alex Cooper has argued that these appropriations occurred at home as well, at the expenses of wise women and popular herbalists.[13] Such an objectifying model of botany, however, is not comprehensive. Historians have also described less one-directional processes, where indigenous knowledge "accompanied" plants on their voyage. For instance, instruments used to prepare a beverage from cocoa beans journeyed together with the

---

10 Among others see, MacLeod 2000; Smith and Findlen 2002; Cook 2007.
11 Cañizares-Esguerra 2001; Arnold 2001. For a thorough discussion of these themes, see Raj 2007.
12 See especially Schiebinger 2005 and Cook 2005.
13 Cooper 2007.

plant revealing the "failures" and "internal contradictions" of the Spanish colonizers in Central and South America.[14]

Alpini's *Dialogue on Balsam*—and to a lesser extent *The Plants of Egypt*—is a sample of a distinct botany, a Mediterranean cross-cultural botany. Several aspects set Mediterranean botany apart from colonial botany. First, in the sixteenth century, the Ottomans had the upper hand in the balance of power within the Eastern Mediterranean. At the end of the sixteenth century, Egypt was a territory of the Ottoman Empire; it was not a colony of Venice, nor had it ever been one. Second, there were long steady relationships. For centuries, as the Venetians travelled, traded, and sojourned in the Eastern Mediterranean, they won and lost battles against Arabs and Turks. From as early as the fourteenth century, physicians, apothecaries, and surgeons accompanied Venetian consuls on their posts. The permanent presence in Constantinople of a consular medical practitioner dates back to the sixteenth century, while a similar presence in Syria and Egypt pre-dates it.[15] Despite this long relationship, Alpini's texts were the first scientific studies published in Venice on Egyptian flora and medical practices. Third, the people of the Mediterranean shared cultural roots: both religious and medical. However belligerent, they all shared Galenic medicine as well as a number of religious books, prophets, and holy places. Finally, together with Leonhard Rauwolf (ca. 1535–1596), Pierre Belon (1517–1564) and Andrea Alpago (ca. 1450–1521/1522), Alpini belonged to the last generation of travellers who were not sent abroad explicitly to explore other places and cultures. Their accounts of the lands they visited were the outcome of personal initiatives, not official requests.[16] For all these characteristics such a botany could only have been realized in the Mediterranean.

This essay starts with the commercial, medical, and religious history of balsam: balsam was unlike any other plant, and it mattered. Section two shows that Alpini gave authority to his Egyptian informants and section three contextualizes and analyzes Alpini's *Dialogue on Balsam*, considering the intellectual context in which Alpini wrote and what were the possible models for his unusual interreligious dialogue on natural knowledge. Section four draws a parallel among the works of Alpini dedicated to Egyptian botany and the one focused on medicine. Finally, I discuss the impact of Alpini's works on his career and on the commerce of balsam.

---

14 Norton 2006.
15 At least two of Alpini's fellow inhabitants went to Egypt: Antonio Matteazzi and Cornelio de' Bianchi. Bianchi practiced in Marostica—Alpini's hometown—until Alpini was 23 years old and it is probable that the two knew each other personally. For Marco Antonio Matteazzi, see Bartolommea and Cavagna 1834; for Cornelio de' Bianchi, see Lucchetta and Lucchetta 1986. For the presence of Venetian physicians in the Levant, see also Siraisi 2007: 225–260; Lucchetta 1997.
16 Olmi 2009.

A Medical and Religious Good

In addition to its valuable medicinal properties, balsam was also an expensive good for religious reasons. An account from the seventh century AD stated that opobalsam was used to anoint Hebrew kings.[17] Possibly, opobalsam is the Balm of Gilead mentioned in the Hebrew Bible, but this is still a subject of debate due to the absence of definitive archaeological findings.[18] From the sixth century onward, balsam was an ingredient in chrism, the ointment used in the Christian rite of confirmation.[19] Already used to embalm and preserve corpses, in the Middle Ages balsam acquired new symbolic meanings.[20] At least until the French Revolution, the kings of France were also anointed with balsam.[21] For centuries, Christian pilgrims and travelers on their way to Mount Sinai visited the garden of Mataria, where Egyptians traditionally cultivated and harvested the most prized quality of balsam. According to a legend, possibly dating back to the sixth century, the Holy Family rested during their flight to Egypt in this garden in the village of Ayn Shams, built on the ancient city of Heliopolis near Cairo.[22] The garden of Mataria was well known to the early modern world and, in the fifteenth and sixteenth centuries, it began to appear on both hand-drawn and printed Italian maps of Egypt, occasionally with the caption "balsam grows here".[23]

Mataria and balsam also held a place in Muslim religion and culture. Mataria was a Muslim place of prayer. During the Mamluk sultanate of the thirteenth and fourteenth centuries, opobalsam was extracted for the sultan at an annual public ceremony held in the garden of Mataria.[24] At least two Muslim historians, the Syrian al-'Umari (1301–1349) and the Egyptian al-Maqrizi (1364–1442), reported that the Mamluk Sultan donated a part of Mataria balsam to hospitals in Egypt and Syria.[25] Appreciated on all shores of the Mediterranean, balsam often figured as a diplomatic gift because of its high symbolic and commercial value.[26]

---

17 Milwright 2001: 7.
18 For balsam in the Hebrew Bible, see: Genesis 37:25, 43:11, Jeremiah 8:22, 46:11, 51:8, Ezekiel 27:17, Kings 10:10. For a discussion concerning the identification of balsam, see Ben-Yehoshua et al. 2011: 47–50, who, unlike Groom 1981, has maintained that the balsam given to Solomon by the Queen of Sheba was *Commiphora opobalsamum*.
19 Moroni 1840: 4.64–65.
20 Truitt 2009.
21 Milwright 2001: 15.
22 Heyd and Raynaud 1923: 575. On the garden of Mataria and its associated legends, see Jullien 1904; Halikowski Smith 2008.
23 See for example the particular of Cairo in the *Tabulae Novae* of Ptolemy's *Geography* by Pietro del Massaio (Florence, 1470).
24 Milwright 2001: 8.
25 Milwright 2003: 201–202; Halikowski Smith 2008: 109.
26 Milwright 2001: 10.

In the early modern period, part of the scientific interest and price of this plant depended upon the supposed scarcity of balsam supply. At the turn of the fifteenth century, several European travelers to Egypt, notably Pero Tafur (ca. 1410–ca. 1484) in 1436–1439 and Peter Martyr d'Anghiera (1457–1526) in 1503, reported that balsam plants at Mataria had dried up and become extinct.[27] In the mid-sixteenth century, influential scholars such as Pietro Andrea Mattioli (1501–1578) and Bartolomeo Maranta (1500–1571) wrote that the opobalsam brought to Italy and sold by apothecaries did not come from the balsam plant and that there was no hope of ever finding any genuine balsam again. Maranta, in his text *About Theriaca and Mithridate*, published in 1572, listed a number of possible substitutes for balsam in the production of theriac, while Mattioli published a recipe for artificial balsam.[28] Substitution was an integral part of Galenic pharmacy, a complex practice according to which missing materia medica could be substituted with analogous matter.[29] Suitable substitutes were sought after and could be the object of speculation.

Balsam's medical and religious significance coupled with its perceived scarcity, for example, caused an American plant to acquire popularity. In 1528, the Spaniard Antonio de Villasante (1477–1536) from Santo Domingo seized the monopoly of sales of a local medicinal plant which he called "balsam." Villasante's choice of name was dictated by mercantile interests as well as by the analogy of the curative effects of the new plant with those attributed to Old World balsam. Yet another iteration of the fact that Europeans assimilated New World discoveries through the lens of their classical models.[30] The Spanish crown was inclined to accept Villasante's claim that New World balsam was the same as Old World Balsam, while Spanish physicians held different opinions on the matter. Nevertheless, a number of experiments carried out in Spain demonstrated the plant's efficacy in healing certain types of wounds.[31]

Sponsored by influential physicians, New World balsam quickly entered into use as a substitute of Old World balsam. In his *Medical study of the products imported from our West Indian possessions* (*Historia medicinal de las cosas que se traen de nuestras Indias Occidentales*, 1565), the Spanish physician Nicolás Monardes (ca. 1512–1588) claimed that the discovery of New World balsam alone justified the pains of Cristoforo Colombo (1451–1506). He hyperbolized, "Because balsam, which used to be found in Egypt, died out long ago: the vine ceased to yield and can be found nowhere else in the world." New World balsam was not the same as Old World balsam but could serve as its substitute.[32] The market greeted the news with enthusiasm and when the first shipment of balsam arrived in Rome from the New World it could fetch up to 100 ducats an

---

27 Tafur 1874: 85, 576; Barrera-Osorio 2002: 166; Milwright 2001: 206; Halikowski Smith 2008: 112.
28 Mattioli, *Discorsi*: 46–48; Maranta, *Della theriaca*: 10.
29 Touwaide 2012: 44.
30 This argument was first put forward by Elliott 1970.
31 Barrera-Osorio 2002: 163–165.
32 Monardes, *Dos libros* (trad. it. Ziletti, *Delle cose*): 20.

ounce. However, the enthusiasm was short-lived, and by 1565 the price for balsam from the Americas had fallen drastically to 3–4 ducats an *arroba*.[33] In 1571, Pope Pio V granted the Church in South America permission to use the local balsam, relieving them of the necessity to import balsam from the Mediterranean.[34] In Tuscany in 1573, Cosimo I de' Medici (1519–1574) informed Francesco I de' Medici (1541–1587) that the apothecary guild had already given permission to a single apothecary to produce theriac with balsam "from the Indies" and added, "In my opinion it should be determined that in the future, whoever makes theriac, mithridatium or any other electuary which needs opobalsam should be allowed to use the balsam from the Indies, since it appears that these physicians have given it their approval".[35]

When Alpini returned to Venice from Cairo, the uncertainty surrounding Old World balsam extinction together with the arrival of New World balsam undermined the trade of balsam in Venice. As is the case for most materia medica, it is not possible to determine the market volume of opobalsam in the sixteenth century. Partly because of its overall limited volume, materia medica were rarely mentioned in detail in ships' documents accompanying their cargo. While traces of balsam can be found in churches and private royal collections, hardly any exist among archeological findings.[36] Alpini brought back to Venice seeds of several plants, a live plant of balsam that "several people saw at my house", and an undisclosed quantity of balsam to sell.[37] On December 1585, a year after his return from Egypt, Alpini urged his colleague Ottaviano Rovereti—his successor as physician to the Venetian Consul in Cairo—not to invest in balsam since all of his own stock remained unsold.[38] The widespread opinion that the plant was extinct—endorsed by respected scholars such as Mattioli and Maranta—jeopardized sales. Thus, while the scientific interest in balsam remained high, commercial interest waned because of the uncertainty regarding its identification and the competition of New World balsam.

Alpini's Dialogue on Balsam

Alpini's choice to publish a dialogue on balsam was the outcome of commercial as well as medical interests. On the one hand, Alpini hoped to reignite the commerce of balsam. On the other hand, he aimed at an academic position in Padua. He was open

---

33 Monardes, *Dos libros*: 39. The *arroba* is a unit of weight of varying value (11.5–14.7 kgs) in use on the Iberian Peninsula.
34 Heyd and Raynaud 1923: 580; Hanbury and Ince 1876: 293.
35 Florence, Archivio di Stato, Mediceo del Principato vol. 241, f. 84: Cosimo I de' Medici to Francesco I de' Medici, January 7, 1573 (Medici Archive Project). Also, in Bellorini 2016: 144–145.
36 Milwright 2001: 10.
37 Alpini, *De medicina Aegyptiorum*: 138r.
38 Ongaro 2011b: 294.

about his motivations. As historian Giuseppe Ongaro has suggested, Alpini had publication plans that changed suddenly as he did not get the position of *medico condotto* in his hometown, as he had hoped.[39] He anticipated the publication of the dialogue and dedicated it to the Reformers of the Studio of Padua, Venetian officials responsible for overseeing the University. With such a dedication Alpini implicitly nominated himself for an academic position.[40] As for the commercial motivations, a note at the very beginning of the *Dialogue* informed that some "learned apothecaries" sought "true knowledge" on balsam and urged Alpini to hasten the treatise's publication knowing that it was already penned. While this kind of statements were often conventional, Alpini added that Venetian apothecaries sought the approval of the Medical College of Venice to allow them the use of balsam in their upcoming productions of theriac.[41] Indeed, during the 1590s, the Medical College ruled at least twice on the subject of balsam (and specifically of xilobalsam, or balsam wood).[42] Whether the apothecaries really hastened Alpini or not, the topic of balsam was commercially and medically timely.

The main argument of the dialogue on balsam was highly contentious, as it intervened in a critical debate, positing an original point of view about the geographical origin of the plant. According to the *Dialogue on Balsam* the balsam plant came from Syria and Arabia, it was neither native to Egypt nor to Mataria, where it was transplanted. Changing the place of origin of balsam meant affirming that balsam was not lost, as many affirmed. Even if Mataria balsam had dried up it was still possible to get balsam from other regions. In the emporiums of the Eastern Mediterranean, merchants could buy balsam from Syria and Arabia, with which Venetian apothecaries could make "true" theriac in Venice. Writing a treatise with such an argument was a bold move on Alpini's part, who at the time was a physician without a position. In the *The Plants of Egypt*, Alpini showed that he was aware of the contentious nature of his treatise: "Your claim—says Guilandino, Alpini's interlocutor in the dialogue—is so astounding that the physicians will have a hard time accepting it".[43]

Let us now turn the pages of the *Dialogue on Balsam* to analyze Alpini's rhetorical choices. Where did authority lie? The full title of the treatise leaves no doubt as to where authority was lacking: *Dialogue on Balsam. In which now shines forth the true knowledge of the balsam plant, opobalsam, carpobalsam and xilobalsam, unknown to the majority of ancient and present-day doctors.*[44] Ergo, according to Alpini neither his

---

39 Ongaro 2009: 7.
40 Ongaro 2009: 8.
41 Alpini, *De plantis Aegypti*: 61v.
42 Marc. It. VII, 2361 (coll. 9717): f. 138.
43 Alpini, *De plantis Aegypti*: 20r.
44 Alpini, *De balsamo*, Title page: "In quo verissima balsami plantae, opobalsami, carpobalsami, & xylobalsami cognitio, plerisque antiquorum atque juniorum medicorum occulta, nunc elucescit". Opobalsam is the resin of balsam, carpobalsam the fruit and xilobalsam the wood.

peers—understood to be his colleagues in both Italy and Europe—nor ancient physicians could claim to have "true knowledge" of balsam. In the *Dialogue*, Alpini uses the Latin word *nugis* ("trifles" or "of no significance") to define the statements by contemporary physicians who insisted that balsam was extinct.[45] According to Alpini, the doctors in Italy and Europe simply did not know.

Alpini credited two Egyptian scholars as his main sources of knowledge on balsam, while at the same time creating archetypes that represented local knowledge. In the dialogue, Abdella and Abdachim, respectively Muslim and Jewish physicians, provided all the answers about balsam. Alpini's role in the dialogue was to facilitate their conversation, thereby allowing such knowledge to emerge. He questioned the two Egyptians, while maintaining a mild skepticism—probably anticipating the likely objections his Italian and European readers would raise—which encouraged his interlocutors to elaborate on their responses. Alpini also presented Egyptians and Arabs in Cairo, Muslim pilgrims to Mecca, and even the "living memory of the Arabs" (Arabes veterum memorijs) as sources of superior knowledge because of their continuous proximity to the plant, and their first-hand experience of it.[46]

> In the knowledge, use and experience of those medicaments with what audacity can anybody favor us Italians over the Egyptians, the Syrians, and the Arabs?[47]

Not only Egyptian specialists of any faith, but also ordinary people who habitually used the plant were to be favored over Italian doctors with no direct experience.

While direct experience is where authority laid, the classics still played an important role. Abdella and Abdachim's deep knowledge of the classics helped validate the knowledge that the Egyptians brought to the conversation. The people of the Mediterranean shared the same natural environment as well as ancient cultural roots. Several passages in the *Dialogue* are dedicated to a close and detailed analysis of the classics. The text abounds with descriptions and citations taken from the Greco-Roman tradition as well as from Arab and Jewish authors, such as Pausanias (*fl.* 5th cent. CBC), Strabo (ca. 64/63 B.C.-ca. 24 A.D.), Titus Flavius Josephus (ca. 37-ca. 100), Avicenna (ca. 980–1037), and Symeon Seth (*fl.* 11th cent.).[48] Following the careful examination of classical texts, the three physicians reached the conclusion that none of the ancient authors writing about the balsam plant had ever actually seen one. Abdella stated: "I always deemed a serious mistake to obtain knowledge about this plant and its products from Dioscorides (*fl.* 1st cent. A.D.), Theofrastus, and other classical authors, when their contrasting descriptions

---

45 Alpini, *De plantis Aegypti*: 64r–64v.
46 Alpini, *De plantis Aegypti*: 64v–65r, 66r–67r, 69r.
47 Alpini, *De plantis Aegypti*: 60v: "An vero quis Aegyptijs, Syris, & Arabibus in corundem medicamentorum cognitione, usu, & experientia praeferendos contendet Italos nostros?".
48 Alpini also cited: Serapion, Theophrastus, Alexander Polyhistor, Diodorus Siculus, Dioscorides, Cornelius Tacitus, Plinius the younger, Galen, Justin, and Gaius Iunius Solinus.

are a certain argument for their ignorance".[49] The classical knowledge Alpini attributes to Abdella and Abdachim made them authoritative and credible to the reader, as they shared a common scientific tradition. In *The Dialogue* classical texts were still an inescapable point of comparison when making new natural knowledge. But the authority previously assigned to classical texts shifted towards new ways of making knowledge, namely local empirical knowledge and direct observation.

Not only did Alpini credit his Egyptian informants, he also kept and enriched the cultural meanings attached to balsam. Contrary to other European authors of botanical works, Alpini grounded his claim about the geographical origins of balsam in both cultural and religious references connected to the plant, stressing the cross-cultural and religious aspects of balsam.[50] For example, Alpini explained that pilgrims on the road to Mecca knew of a sandy and infertile hill close to the city of Bedrunia where many balsam plants grew only thanks to a miracle performed by Muhammad (*miraculo Mahemetis*).[51] Thus, in Alpini's tale balsam is explicitly a Muslim religious plant. Similarly, Alpini wrote of a Muslim commander, Messinor, who every Friday went to pray in the garden of Mataria, the place where balsam grew and where both Christians and Muslims believed there was a stream with miraculous properties.[52] Alpini noted that in 1575 a eunuch by the name of Messir ordered the transplantation of forty new balsam plants to Mataria to replace those which had dried up.[53] By drawing these stories together, Alpini stressed the commonalities of balsam for the two religions. By telling a Muslim legend and other events related to Muslims in support of his thesis, Alpini created a cross-cultural production: not only he did not strip the plant of its cultural meaning, he also diffused Muslim knowledge by attaching positive meaning to it.

An interreligious dialogue

Dialogues abound in Renaissance literature, but it is rare to find a dialogue between people of different religions in a treatise on natural history, or any scientific topic for that matter. Among the different types of dialogue identified by Peter Burke (catechism, drama, dispute, and conversation), Alpini's text can be characterized as a "conversation".[54] Conversations usually take place in a garden, in this case the garden of Mataria, and the speakers are not endowed with complete personalities: there is no

---

49 Alpini, *De plantis Aegypti*: 69r.
50 For scholars who "erased" their indigenous informants, see Cook 2005.
51 This road is described as the road from Damascus to Egypt in Alpini, *De medicina*: 21v, and from Mecca to Bedrunia in Alpini, *De plantis Aegypti*: 63v, 67r.
52 Alpini, *De plantis Aegypti*: 20v, 66v, 69v.
53 During the conversation Abdella claims to have been present, together with the French consul Paolo Mariano, when the balsam was transplanted, see Alpini, *De plantis Aegypti*: 66v.
54 Burke 1989: 3.

real contrast between the speakers' positions. Instead, each personage contributes to the formation of the concluding idea in order to make the case for the author's objective. In the *Dialogue on Balsam* all three interlocutors agree on the origin of the plant, leaving the reader without a diversity of opinions.

In the *Dialogue*, Alpini presents exponents of the three religions of the Book agreeing on a specific topic and thus putting aside all of their other differences. A scientific dialogue between a Christian, a Jew, and a Muslim was quite unusual for the time. Unfortunately, Alpini did not explain his decision to structure the dialogue in this fashion, but the exceptionality of the *Dialogue* invites a few speculations on the possible origin of such a choice. In this section, I suggest that the *Dialogue* was a testimony of the openness of the Venetian intellectual atmosphere in the late 1580s and early 1590s.

In the Middle Ages, Jews and Christians held several *disputations*, while in sixteenth-century religious literature, interreligious dialogues between Protestants and Catholics discussing theological matters were not unusual.[55] Alpini might have been familiar with some of these works and, as a result, was influenced by them. Most likely Alpini did not know Garcia da Orta's (1501–1568) *Colóquios dos simples e drogas da India* [*Colloquies on the Simple and Drugs of India*] (1563) a dialogue about materia medica with characters from different religions. Alpini never cited da Orta and the Latin translation of the *Colloquies* was not in the form of a dialogue.[56] A near contemporary work to *The Dialogue on Balsam* is the *Colloquium Heptaplomeres* by Jean Bodin (ca. 1530–1596), which takes place between a Catholic, a Lutheran, a Calvinist, a Jew, and two non-confessional figures, a Deist and a skeptical naturalist. Bodin likely wrote his text between 1593 and 1596, after the publication of *The Dialogue*, ruling out the possibility that Bodin's work influenced Alpini (while the opposite might have happened).[57]

An intriguing possibility is that the *Book of the Kuzari*, a controversial text containing a dialogue about religion between a Muslim, a Jew, and a natural philosopher served as model for Alpini.[58] The *Book of the Kuzari* is a medieval text connected with *Qol Mebasser*, also known as *The Khazar Correspondence*, another medieval text published by a Jewish scholar, Yitzhak Abraham Aqrish (ca. 1489-ca. 1578), in Istanbul in 1577. Aqrish bought the book from a rabbi "who was physician to the Turkish Governor of Egypt under the Sultan Sulayam".[59] Thus, the *Book of the Kuzari* or the *The Khazar Correspondence* might have circulated in Egypt during Alpini's sojourn there. When in Cairo, Alpini regularly associated with local scholars of different faiths, which likely favored a tolerant approach towards their cultures as well as sharing knowledge and books. Later on,

---

55 For a recent study of dialogues between religions, see Weidenbaum 2009.
56 On da Orta's *Colóquios*, see Fontes da Costa 2015.
57 On *Colloquium* see, Turchetti 2015; Suggi 2005. Kuntz 1975: xxxvii, affirmed that Bodin wrote the *Colloquium* in 1588. For a thorough discussion of Kuntz's English translation, see Yates 1976; Rose 1977.
58 Garosci 1934: 110n4; Dunlop 1967: 117, 128n33. The manuscript of *Qol Mebasser* is preserved in the Bodleian Library. Golb 1997/2013: 142n17. See also Kuntz 1975: xxxviiin58.
59 Dunlop 1967: 128.

Alpini recalled his Egyptian experience with great enthusiasm.[60] The intrinsic cross-culturality of the meanings attached to balsam also invited Alpini to transfer his positive Egyptian experience onto the written page, validating local knowledge because of (and not regardless of) religious differences.

Alpini most likely based his two interlocutors Abdella and Abdachim on people he met during his appointment in Cairo. In *The Medicine of the Egyptians*, Alpini recounts that, a few days before his return to Venice, he paid a visit to a Muslim savant:

> Mohammed, by far the most learned of all the Egyptians in the *res herbaria*, whom I knew very well, and to whom every day new medicaments were sent by friends from Arabia, Ethiopia, India and other places [. . .].[61]

Alpini used almost the same words in *De balsamo* to describe the fictional character of Abdella: "I do not believe there to be an Egyptian or an Arab with more expertise in the field of botany than our Abdella here".[62] In the *Dialogue*, Abdella is placed at the center of a network of Egyptian, Turkish, Arab, and European contacts. What is more, some of the descriptions of this network are so detailed that they make it possible to imagine Abdella was indeed a real person. For example, Abdella claims to have been together with Paolo Mariano, the French consul, when in 1575 some balsam plants were reintroduced in Mataria. Abdella also affirmed to have treated both Haly Bey, a leader of soldiers and pilgrims, and Horrem Bey, another leader of a pilgrim caravan. Finally, Abdella also stated that he gave balsam to Francesco Priuli, the Venetian consul in Cairo.[63] If the character Abdella interacted with real people, it seems possible that he too was a real person.

For the other character in *The Dialogue*, Abdachim, similar detailed references are missing. However, in *The Medicine of Egyptians*, Alpini writes of an unnamed Jewish physician conversant both in Arabic and Italian who helped Alpini consult a book written in Arabic.[64] This unnamed Jewish physician could have been modeled upon—an otherwise unknown—Diodatus, a scholar with whom Alpini undertook the translation of Averroes's (1126–1198) *De Animalibus* from Hebrew.[65] It is easy to imagine that Diodatus and the unnamed Jewish physician who helped Alpini translate the recipe from Arabic were either the same person or served as the models for Abdachim.

---

60 Siraisi 2007: 234; Ongaro 2011b: 287.
61 Alpini, *De medicina Aegyptiorum*: 133v: "Nullo meo consilio id accidit, sed casu atque ab alio Mehemete mihi valde familiari, rei herbariae longe omnium Aegypttiorum doctissimo, cui cotidie ex Arabiae, atque Aethiopiae, Indeaeque; locis aliqua nova medicamenta ab amicis mittebatur [. . .]".
62 Alpini, *De plantis Aegypti*: 63r: "Neminem Aegyptiorum, vel Arabum esse nunc arbitror nostro Abdella in re plantaria magis veratum".
63 Alpini, *De plantis Aegypti*: 66v.
64 Alpini, *De medicina Aegyptiorum*: 133v. See also Siraisi 2007: 239. For the identification of the book, see Alpini, *De medicina Aegyptiorum* (Fr. transl. Fenoyl, *La médecine des Égyptiens*): 1.378n260.
65 See Pagallo 2011: 161–230.

Along with Alpini's scholarly acquaintances and positive experiences in Cairo, the open-mindedness of Venetian intellectual circles toward Egypt and new ideas also encouraged Alpini to conceive of such a dialogue. Recent studies have shown how ancient Egypt during the fifteenth century was very present in the imagination of learned Europeans, Italians in particular. According to Brian Curran, Egypt in the Italian Renaissance (especially in the fifteenth century) held a significant place in the imaginary of the learned. Egyptian legacies served as claims for authority.[66] Once back in Venice, Alpini continued to enjoy a stimulating intellectual life as he was part of an especially open-minded intellectual circle. Alpini's patrons were Antonio and Giovanni Morosini, members of an influential Venetian aristocratic family. Another authoritative member of the Morosini family, Andrea (1558–1618), held Alpini in high regard and considered him a close friend.[67] Andrea was the host of the so-called "Ridotto Morosini", where Galileo Galilei (1564–1642), Paolo Sarpi (1552–1623) and Giordano Bruno (1548–1600), among others, discussed nature and philosophy. At the time Alpini was publishing his works (1591–1592), Andrea Morosini testified in favor of Giordano Bruno in front of the Inquisition. If Alpini began the draft of his dialogue on balsam in Cairo – as he claimed to have done – and in it wished to emphasize the influence that his Egyptian friendships and associations had had on him, the Venetian company of scholars and patrons of the arts would surely have encouraged him, or at least not discouraged him, in expressing his appreciation of Egypt and Egyptian natural knowledge.[68]

Different Rhetorical Strategies for Medicine and Botany

In the same year *The Dialogue on Balsam* appeared, Alpini also published *The Medicine of the Egyptians*, and one year later *The Plants of Egypt*. *The Plants of Egypt* is a botanical work of considerable importance, which for the first time described coffee and introduced sex identification of date palms to European readers. Both works are in the form of a dialogue, in which Alpini conversed with his former professor, Melchior Guilandinus (1520–1589) prefect of the Botanical Garden in Padua. A comparison of the *Dialogue* with these coeval works on Egypt shows that Alpini relied on different rhetorical strategies depending on the subject matter. For medicine Alpini relied more on the classics and medical theory, while for botany he used first-hand observation as the main validating practice.

---

66 Curran 2007.
67 Ongaro 1961–1963: 117.
68 Alpini, *De plantis Aegypti*: 20r.

In *The Medicine of the Egyptians*, the medical experience of the Egyptians was insufficient to provide an authoritative basis for his claims. In this text, Egypt appeared in decline following the Ottoman invasion of 1517. According to Alpini, Egyptian physicians' medical practices were not based on theory, but on empirical facts.[69] This inevitably qualified Egyptian physicians as empirics, a step below Italian physicians who differentiated themselves from empirics because their practice was based on rationality and theory. Nonetheless, Alpini was impressed by Egyptian phlebotomy and imported some of these techniques into his practice. To mediate between these two positions, as Nancy Siraisi has argued, when discussing medical practice Alpini felt compelled to establish a direct link between Egyptian practices and ancient Egypt, or with authoritative Greek and Arab medical texts to justify the introduction of Egyptian innovations into his own practice.[70] Although he strongly believed that a number of Egyptian practices were beneficial to patients his observation and Egyptian tradition in the case of medicine were not sufficient to justify his position.

Unlike medicine, the subject of plants allowed Alpini to present himself as the main authority on the matter he discussed: he did not need to root botanical knowledge in theory and put his Egyptian informants into the background. In *The Plants*, Alpini relied almost completely on his first-hand observations of both plants and Egyptian medical practices. Often, he used the phrase "I saw." When agreeing with the ways in which Egyptians used plants, Alpini wrote "they use" or "the women consider it to be." If he was uncertain, he wrote "I have heard it said." The use of these strategies to indicate levels of certainty recalls those used by Aristotle in his works on animals (*De partibus animalium* and *De generatione animalium*). We know that Alpini knew well these works as he translated Averroes's *De animalibus*—the most authoritative compendium of the two Aristotelian treatises—together with the Jewish philosopher Diodatus.[71] Whereas in *The Dialogue* Abdella and Abdachin were the main authorities, in *The Plants*—with the exception of the pages dedicated to balsam—Alpini did not include any direct citation of Egyptian informants. Nor did he supply any specific references other than his own first-hand experience. Yet, in order to accumulate such a quantity of information, it would stand to reason that he had recourse to more than one local informant. What is more, the size and precision of the chapter on gynaecological remedies and illnesses shows that Alpini must have had one or more women among his informants. But the only local informants cited in *De Plants* are a Greek farmer and an apothecary from Zakynthos.[72] At the same time, however, a long passage in *The Plants'* entry about balsam is a detailed account of the political relationships between the Ottoman Empire and the governor of Mecca at the time of the yearly pilgrimage. While this passage was

---

[69] Alpini, *De medicina Aegyptiorum*, for Alpini's opinions on Egyptian medicine and physicians see especially book 1.
[70] Siraisi 2007: 236.
[71] See Lehoux 2017; Pagallo 2011.
[72] Alpini, *De plantis Aegypti*: 128r.

not necessary to Alpini's argument about balsam, the author expected that his readers would appreciate it.[73] Thus, in *The Plants* Alpini presented himself as the main authority, concealed indigenous informants, but deemed non-botanical information about Egypt of interest to his readers.

Another difference between *The Medicine* and *The Plants* is the role of ancient and medieval classics. In the latter, the relationship with the classics is associated with the character of Guilandinus. Through Guilandinus's remarks, in *The Plants* Alpini established a link with the literature and positioned his work in relation to classical authorities. For example, Guilandinus compared Alpini's assertions with the ancient classics: "Dioscorides—who you will not deny observed true opobalsam—said that its color is akin to milk".[74] Guilandino also posed questions regarding single authors or inconsistencies and summarized Alpini's arguments:

> But tell me, are you asking us to take your word on balsam over that of Dioscorides, Theophrastus, Pliny, Serapion and other authoritative authors that in teaching the discipline of plants are preferred to others and are considered more trustworthy?[75]

The classics did not offer a starting platform for the discussion of *The Plants*, but—as it was in *The Dialogue*—they were indeed an inescapable point of comparison.

Certain Authority, Uncertain Knowledge on Balsam

The impact of Alpini's works, which this section analyzes, was multifaceted. His works on Egypt secured for Alpini a new position, prestigious relations, and lasting notoriety all over Europe. The choice to give authority to Egyptian physicians was well received, but knowledge on balsam remained uncertain.

Alpini's works on Egypt received immediate recognition. In 1592, Ulysses Aldrovandi sent a man from Bologna to Venice to procure balsam from Alpini, who in turn seized the opportunity to start up a correspondence with the famous naturalist.[76] In 1594, two years after the publications on Egyptian plants and medicine, Alpini became *lettore dei semplici*—an *ante-litteram* pharmacology chair—at the University of Padua. Then, in 1604 Alpini was appointed prefect of the Botanical Garden and "exhibitor of the simples" in Padua.

Alpini enjoyed a remarkable reputation well after his death in 1616, as it is testified by several translations, posthumous editions, and re-editions of his works. In 1619,

---

[73] Alpini, *De plantis Aegypti*: 21rv.
[74] Alpini, *De plantis Aegypti*: 22v.
[75] Alpini, *De plantis Aegypti*: 20r, 22v.
[76] Bologna, Biblioteca universitaria, Aldrovandiano 136, tomo XIX, undated letter from the volume containing letters from 22nd November 1592 to 23rd May 1593, c. 248r.

Antoine Colin (*fl.* 17th cent.), an apothecary from Lyon, translated *De balsamo dialogus* in French, publishing it together with Monardes' *Historia Medicinal*.[77] In a comment on chapter seven of Monardes' *Historia*, Colin clearly articulated his opinion that Alpini's word on balsam was final.

> With all due respect for his scholarship, I am greatly astonished at Monardes' claim in this passage concerning genuine balsam, frequently used in ancient times, in which he declares it to have dried up and become extinct. The contrary has been shown to be true in a treatise [The Dialogue on Balsam . . .], which presents an authoritative and sound argument that today, as in the past, balsam is to be found in Arabia, and in large quantities; there are daily journeys there by way of traders' caravans coming from Mecca.[78]

Towards the end of the 1630s, Johann Vesling (1598–1649), professor of medicine at the University of Padua, had Alpini's two volumes on botany reprinted, thereby confirming the conclusions concerning balsam.[79] In the eighteenth century, the reputation of Alpini and his works grew throughout Europe, largely due to the high regard in which he was held by the Dutch physician Herman Boerhaave (1668–1738).[80] A new edition of his dialogue on balsam appeared in Leiden, in 1719, and all of Alpini's works were republished between 1710 and 1735.[81]

Numerous citations of *The Dialogue* in works published across Europe confirm that Alpini's dialogue was an inescapable point or reference on the subject across the continent.[82] A few examples show that the choice to give authority to Egyptian physicians was well received by the readers of the dialogue. In a short 1640 publication the Apothecary College of Naples mentioned by name the Muslim and Jewish Egyptian physicians in Alpini's dialogue as if they were real people: "[Alpini] together with the Arabian physician Abdella and the Jewish physician Abdachim examines all the difficulties relative to the transaction [of balsam] and dissolves them with clarity".[83] In 1663, physician and author Giuseppe Donzelli (1596–1670) described the relationship between the three physicians an "affectionate correspondence".[84]

Despite the success of Alpini's works and methods, trade of opobalsam did not take off and knowledge about its identification remained highly disputed. In 1611, German physician Heinrich Smet (ca. 1535–1614) lamented that it was not possible to get

---

77 Alpini, *De Balsamo* (Fr. transl.: Colin, *Histoire du baulme*).
78 Alpini: *De Balsamo* (Fr. transl.: Colin, *Histoire du baulme*: 31).
79 Vesling, *De plantis aegyptiis*. The title page of *De balsamo* is dated 1639, and *De plantis* 1638.
80 Rinaldi 2011: 242–243.
81 Weissmann, *Balsamum*; Heinsius, *Dissertatio medica De opobalsamo*; Ely, *Dissertatio*. On the circulation of *De balsamo* between the sixteenth and seventeenth centuries, see Rinaldi 2016. On the reputation of Alpini in the eighteenth century, see Rinaldi 2011.
82 See Rinaldi 2011.
83 Collegio degli Speziali di Napoli, *Parere*.
84 Donzelli, *Petitorio*: 121.

any part of the plant.[85] On the occasion of a production of theriac held in 1639 in Rome, a heated diatribe arose on the subject of opobalsam.[86] At the center of the dispute was the question of whether the Roman apothecaries actually used genuine opobalsam among the ingredients of theriac. Renowned physicians and apothecaries from Rome, Florence, Naples, Padua and other cities opposed criticism voiced by a peripheral group of Roman physicians who denied that the opobalsam used in the production was indeed opobalsam. For a decade, numerous texts both in Latin and in the vernacular were published across Italy. While Alpini's texts retained a central position in the long and heated dispute, the quantity of opobalsam in circulation was so little that most authors could recall only occasionally seeing opobalsam. Opobalsam was still expensive and its supply extremely scarce. The scarcity of opobalsam made it an object elusive to scientific knowledge.

Conclusion

Alpini resorted to different sources of authority in his three 1591–1592 publications depending on his claims and objectives. In *The Plants* and *The Medicine* Alpini tended towards a Venetian appropriation of Egyptian medicinal and botanical knowledge, where Egypt dissolved into the background, and the real scientific conversation no longer took place on location but in Padua's Botanical Garden between Alpini and Guilandinus. In *The Medicine* Alpini's goal was to bring new medical knowledge to Europe, and to do so he anchored authority to familiar ancient writers; specific informants and events were alluded to, but Alpini did not consider the experience of contemporary Egyptians a reliable source of authority for his readers. In *The Plants*, Alpini did not mention any specific Egyptian source and relied more on his own observations, but he added political and economic information about Egypt and Mecca he thought his readers might appreciate.

In *The Dialogue*, instead, Alpini gave authority to his Egyptians informants, if only rhetorically. Classical authors were held in high regard, but their word was not final. *The Dialogue* was built on the history of balsam itself, a shared Mediterranean history to which all three religions contributed. In this dialogue, different religious faiths did not represent a barrier to the transfer of information. Quite the opposite, the accordance of opinion among the three physicians served to further validate the natural knowledge presented, as if it were the fruit of a cross-religious endeavor. Because the three religions shared common sacred places as well as a common medical culture, Alpini presented the dialogue as a cross-cultural production which neither objectified nor stripped balsam of its cultural meanings. Rather, the treatise *re*-placed balsam at

---

85 Rinaldi 2011: 324.
86 Rinaldi 2011: 301–302.

the center of a network of relationships in the Eastern Mediterranean. Alpini implicitly defended a number of religious, cultural, and commercial practices that, for centuries, had been the basis of strong bonds between Christians, Muslims, and Jews. *The Dialogue* offered an agreeable conversation—a Mediterranean dialogue—in a place of tranquillity, a garden where scholars of different religious backgrounds could meet and share knowledge about the natural world, ancient texts, and medical practices. In the late sixteenth century, Alpini's coeval use of different modes of knowledge production shows that when making new natural knowledge about non-European plants the appropriation model was not absolute, a different model was still possible. The power of the mythical history of balsam evoked the Mediterranean as a place of peaceful and learned dialogue.

Tassanee Alleau

A Bridge to the Underworld?
An Explanation of the Act of Digging up Plant Roots in Early Modern Medical Fictions

The Underground Myth: Defining the "Medical Fiction"

This chapter explores why the underground or the subterranean is a recurring theme in several 'medical fictions' of the early-modern period. First, though, a definition of 'medical fiction' is needed. If, as Margaret Healy stresses in her book *Fictions of Disease in Early Modern England*, disease is a "recurring nightmare of great fiction"[1], a persistent theme of the texts she discusses is the Underworld. These medical fictions were, according to her, discourses "which could simultaneously embrace and (by prescribing prevention and cure) intervene in multiple areas of life", such as political and religious matters[2]. Medical fictions might be verse, epic poetry (derived from oral tradition), plays or even myths containing legendary and ancient medical notions. These texts give us clues that inform a cultural understanding of diseases and their remedies. Some of these fictions offer substantial information on how medicine was theorized and practiced in the early modern period. This study will focus on the example of the mandrake, as it exemplifies the extent to which a subterranean plant, or the parts of a plant found underground, that is the roots, served as key elements in the definition and description of plants and their use as cures in the early modern period. By concentrating on the rhetorical aspects of texts about the mandrake, this case study will uncover the broader metaphors used to convey information on vegetal medicine.

This case study follows Lévi-Strauss on fictions: "Each mythological system is inspired by an opposition, one might describe as dietary"[3]. In this sense, the difference between subterranean plants and those that grow on the surface, is thought of as being a reflection of their opposing properties, which in turn influence their use as remedies or food. To fully understand the construction and meaning of early modern botanical and medical theories concerning roots and plants, historians must grasp the symbolic

---

[1] Healy 2001: 115–116, 157–156.
[2] Healy 2001: 47.
[3] Lévi-Strauss 2009: 32, "Chaque système mythologique s'inspire d'une opposition, pourrait-on dire, diététique" [translation is mine].

Note: I would like to thank my thesis director Pascal Brioist for the first proofreading, and my co-director Concetta Pennuto for her precious advice.

and folkloric narratives that surrounded them. This cultural dimension served as a tool in the writing of natural histories of plants such as the mandrake or the deadly nightshade (*Atropa belladonna* L.). Descriptions of their harvesting constituted a narrative *topos* often used in herbals or *herbaria*[4], naturalist treatises which classified plants by category of use or by name. These descriptions were meant to foster devotion, faith, and morality in Renaissance pharmacopeias, as remedies for specific illnesses, particularly those related to sexuality, fertility, melancholy, and madness. Thus, the underground plant is linked to the concept of the underworld, using imagery of the *axis mundi*—the root being the bridge to the *mundus subterraneus* (i.e., the plant organ with access to the inverted world). The act of digging up the roots, rhizomes, or tubers was structured through magical rituals and gestures. These gestures were dedicated to warding off threats of death or evil spirits.

The mandrake is a root belonging to the Solanaceae or nightshade family which includes other very poisonous plants (belladonna, datura, henbane, to name a few). It was often associated with fertility. This ancient belief can be found in the Book of Genesis and was later described by Dioscorides (*fl.* 1st cent. A.D.). The curse of the mandrake was the deadly shriek it was thought to possess: It was believed that only a dog was capable of hearing this. This is why we find many representations of mandrake roots with a dog tied to its stem, whilst a man waits for the animal to die so that he can uproot the plant. The herbal of Pseudo-Apuleius contains a page with an illustration of a mandragora (the mandrake plant) and a dog (see Figure 6), giving instructions to readers on how to dig up the root and how to use it following the Doctrine of signatures[5]. Another common image of the mandrake root was as a pair of male and female roots, depicted with human traits – both as a parallel with Adam and Eve and as a symbol of fertility.

There are two types of medical fiction which will serve as examples for the present analysis. The first can be categorized as texts which were not used by physicians in the practice of medicine, but which provide insight into the cultural foundations of those practices. Works such as Fracastoro's poem on syphilis or Machiavelli's play about the mandrake were written as pure fictions intended to criticize the occult[6], and irrational or false medicine. The second type is represented by examples such as the myth of Demeter and Persephone or Laurent Catelan's (ca. 1568–1647) portrait of the mandrake in *Rare et curieux discours de la plante appellée Mandragora, de ses especes, vertus et usage* (1638); these provide evidence for the use of certain plants and serve as both medical authorities *and* socio-cultural references for early modern botanical theories.

---

4 An herbarium was a book that contained collected plant specimens which were dried and preserved with names and other relevant information.
5 See Cule 1997: 99–100; Repici 2009: 77–94; and Baldassarri 2022b.
6 On occultism, see Brancher 2015: 71.

This chapter, following Margaret Healy, uses the mandrake myth as a starting point to understand how natural history was constructed and how plants were conceptualized in the early modern era. The investigation relies on the historical and cultural analysis of printed primary sources across a broad timeframe, from the early sixteenth to the end of the seventeenth century. The focus will be on plants rather than diseases, on cures rather than illnesses. First, the idea of the plant and its correspondence with the notion of an underground microcosm will be analyzed. Then it will explore several symbols and allegories of the underworld that were used in medical or botanical treatises as well as books of secrets, demonstrating that medical theories and practices were rooted in occult images and magical beliefs[7]. Three important concepts gave inspiration to these fictions: The doctrine of signatures, from which Giovanni Battista Della Porta (1535–1615) drew his inspiration[8], the Paracelsian principle of sympathy and antipathy[9] and the Hippocratic-Galenic humoral theory. However, the humoral theory came to be criticized by early modern physicians. The latter part of the chapter will show how the mandrake root was anthropomorphized and how the notion of its 'body' related to that of the subterranean[10]. Finally, we will look at the relationship between the act of digging up or harvesting the plant and magical rituals, such as the use of a dog (with its symbolic connection to the underworld). This will demonstrate that the mandrake root was used as a panacea in the pharmacopeias of the early modern period.

Underworld and Underground Symbolism

The Vegetal and the Underground Microcosm

The primary question of this investigation is this: How were ancient symbols read by physicians and naturalists in early modern times? The mandrake has been known and used since Antiquity. To decipher the mandrake's myth, and those of other plant-based remedies of the *mundus subterraneus*, we need to consider their related early modern symbols with the help of cultural history and philosophical ideas. The Renaissance was a period where ancient, medieval and contemporary ideas were mixed to form new theories. As Margaret Healy explains, "the juxtaposition of these epics foregrounds the socio-culturally constructed nature of explanations of disease,

---

[7] Books of secrets were compilations of various texts, in particular medicinal recipes that contained magic formulae or guides for rituals. Ruscelli, *Les secrets*, and Della Porta's *Magiae naturalis* are good sixteenth century examples of the genre.
[8] Panese 2003: 6–24; Brancher 2015: 60; Baldassarri and Matei 2018.
[9] Pennuto 2008.
[10] Brancher 2015: 74–101.

and literature's important participation in that process"[11]. Symbols of the underground or the subterranean appear frequently across a variety of literary sources. From the fifteenth to the seventeenth centuries, it is possible to find such symbols in collections of allegories such as the *Emblemata* or *Hieroglyphica*, or in books of secrets, inspired by Ancient and Medieval Greco-Arab symbology. Hélène Duccini writes that "knowledge of allegories certainly did not pretend to the creation of new forms, but rather to a codification which could make allegorical language universal"[12]. This language of symbols was strongly influenced by syntheses, the rereadings of magical texts from a flourishing period of experimental sciences during the Middle Ages. It was driven by the development of printed books, astrological images, talismans, iconographic amulets, or ritual images, which strike the viewer with rich visual references to the secrets of the occult sciences and to biblical texts. The magician-philosopher Marsilio Ficino reused the 'conceptual tools' of the ancient Hermetic scholars. The Hermetic texts reflect the conception of nature as a form of esoteric understanding. The early modern texts were then strongly influenced by a "quasi-religious" revival and were "replete with magic"[13]. William Eamon has said that the "literary tradition behind recipe books dates back to Hellenistic times. Its roots are in the alchemical tradition"[14]. We can understand 'hermeticism' as a group of eclectic and diverse doctrines including alchemy, magic, astrology, medicine, philosophy, and theology. Unlike witchcraft, the occult and hermeticism[15] were not suspected as forms of heresy by the Church[16].

Renaissance alchemist-physicians and astrologer-physicians also seized upon these powerful metaphoric, occult and emblematic images in their theoretical work, such as that of Giordano Bruno (1548–1600). In 1579, publisher and engraver Gérard de Jode (1509–1591) produced a series of prints to accompany a book by the poet Laurent Van Haecht Goidtsenhoven (1527–1603), or, in the Latinized form of his name, Laurentius Haechtanus, entitled *Parvus Mundus* (Μικρόκοσμος). Among these, one can find the image of the *Homo arbor inversa*[17], which depicts man as an inverted tree, captioned thus: "*omnis enim arbor que non facit fructum bonum excidetur & in ignem mistetur*" ("Any tree that does not bear good fruit will be cut down and destroyed by fire")[18].

---

11 Healy 2001: 3.
12 Duccini 2008: 13.
13 Eamon 1994: 24.
14 Eamon 1994: 30.
15 Brancher 2015: 49.
16 Kahn 2007: 8–9.
17 Van Haecht Goidtsenhoven, *Parvus Mundus*.
18 According to a phrase from *The Holy Bible*: Matt. 7:19.

Figure 1: *Man as an inverted tree*, in Laurent Van Haecht Goidtsenhoven, Μικροκόσμοσ *Parvus mundus*. Antwerp: Plantin, 1579. French translation: *Le Microcosme contenant divers tableaux de la vie humaine*. Amsterdam: chez Théodore Pierre, 1613: p. 35.

These prints are full of hermetic symbols which were believed to reflect astral influences. The hermeticism of the sixteenth and seventeenth centuries indeed represented a vision of the world structured around the 'universal sympathy' of things and the theory of signatures. Hermeticism, as a philosophical theory, was intended to discover the secrets of nature; as Della Porta put it "the very root of the greatest part of [the] secret and strange operations of natural magic"[19]. The human body was considered the reflection of the macrocosm. The microcosm[20] in turn was perceived as a mirror of the celestial world. Gérard de Jode depicts the human being as a tree whose head is its roots and whose feet are its branches, an image that was borrowed from Plato (428/427 or 424/423–348/347 B.C.)[21]. The image of the man concealed in the hollow of a tree is a complex visual metaphor: Everything is hidden, and one can only understand the world by deciphering the secrets of nature.

Contrary to this notion of sympathetic influences between things, there were antipathetic influences (a Paracelsian concept *par excellence*). The universe was made up of male/female dichotomies, of good/bad dualisms. Accordingly, what was "on the surface" was the opposite of what lived "underground", just as the moon was the opposite image of the sun. Inversion is therefore another hermetic mode of thinking since, if everything is a mirror, everything must thus exist in its inverted image. This esoteric concept was taken from the mystery cults, a notion of things as being eternally hidden; figuratively—sometimes even literally—concealed underground. According to Bernard Tannier, this use of secrets and the occultation of the world's meanings was of a rhetorical nature: "It is that of Saint Augustine speaking of the Scriptures: 'This obscurity was willed by God in order to combat the scorn of our judgement whereby that which is easily understood often seems devoid of any value'"[22].

We can find references to the underground and Underworld in alchemistic texts, as in a quote found below the frontispiece to the *Dyas chimica tripartita* (1625), signed by H. C. D. or Heramnnus Condeesyanus, probably the pseudonym of Johann Grasshoff (ca. 1560–1623), with the Latinized name Johannes Grassaeus (or Crassaeus) a Pomeranian jurist and alchemist[23].

The quote reads:

Quae sunt in superis, haec inferioribus insunt:
Quod monstrat coelum, id terra frequenter habet.
Ignis, Agua et fluitans duo sunt contraria: felix,
Talia si iungis: sit tibi scire satis[24].

---

19 Eamon 1994: 214.
20 Brancher 2015: 23.
21 Napolitano Valditara 2007; Repici 2012.
22 Tannier 2018: 85-97 [translation is mine].
23 According to the notice on the Bibliothèque Nationale de France website, https://data.bnf.fr/fr/12067473/johann_grasshof/.
24 [Grasshoff], *Dyas chymica*: f. 3 [translation is mine].

The things that are in the upper world, are present in the lower parts of these things: often the sky shows what the earth has. Fire and floating water are two opposites: happy, such that if you connect them you know enough.

Figure 2: *The metaphor of the underground, cave or grotto*, by Théodore de Bry (engraver), in [Grasshoff (ed.)], *Dyas Chymica Tripartita*. Frankfurt: Luca Jennis, 1625: frontispiece.

The engraving is thought to have been made by Theodore de Bry, an editor and engraver to whom we owe a number of alchemical illustrations, including those of the *Atalanta fugiens* by the alchemist-physician Michael Maeier (1568–1622)[25]. It exemplifies perfectly

---
25 Maeier, *Atalanta fugiens*.

the alchemical reasoning and the early modern hermetic intellectual currents of the *philosophia occulta*, which were no longer satisfied with Galenic explanations. The forms of rationality offered by these discourses should not be considered as in opposition to the development of scientific thought during the sixteenth and seventeenth centuries[26].

For alchemists, the *chymicall homunculus* that appears in Paracelsus (ca. 1493–1541) *De natura rerum* (1537) resembled a small woman or a small man. This alchemical figure looks like the illustrations of male and female mandrakes in the *Gart der Gesundheit* (1485), attributed to Johannes de Cuba (1430–1503). The roots, shaped like two small humans, offer an emblematic image of the Promethean myth which was linked to the Underworld. As Plato explains in his *Protagoras*:

> PROTAGORAS. – [. . .] There was once a time when there were gods, but no mortal creatures. And when to these also came their destined time to be created, the gods moulded their forms within the earth, of a mixture made of earth and fire and all substances that are compounded with fire and earth. When they were about to bring these creatures to light, they charged Prometheus and Epimetheus to deal to each the equipment of his proper faculty. Epimetheus besought Prometheus that he might do the dealing himself; "And when I have dealt", he said, "you shall examine." Having thus persuaded him he dealt; and in dealing he attached strength without speed; to some, while the weaker he equipped with speed; and some he armed, while devising for others, along with an unarmed condition, some different faculty for preservation. To those which he invested with smallness he dealt a winged escape or an underground habitation; those which he increased in largeness he preserved by this very means; and he dealt all the other properties on this plan of compensation. In contriving all this he was taking precaution that no kind should be extinguished; [. . .] Then he proceeded to furnish each of them with its proper food, some with pasture of the earth, others with fruits of trees, and others again with roots; and to a certain number for food he gave other creatures to devour: to some he attached a paucity in breeding, and to others, which were being consumed by these, a plenteous brood, and so procured survival of their kind.[27]

The oral traditions and philosophy of Antiquity were not the only inspiration for early modern alchemists and naturalists. The Bible was another source, and we come across many "underground", "Underworld" and "root" motifs in canonical scripture. In Biblical cosmology, the underground is presented in a very Manichean manner: It is depicted as the fertilizer of the earth; the source of the water that runs through the Garden of Eden, or, opposingly, as an evil force represented through its dangerous "underground fires" and threatening mountain roots[28], Subterranean streams, lakes of fire, or the pits of Gehenna (where dead bodies are disposed of), were all associated with Chaos or with the subterranean "gates of death".

Renaissance intellectuals deployed Underworld symbolism across various kinds of writings. It was through these types of discourse, which melded biblical references,

---

[26] Joly 2007: 167–184.
[27] Plato, *Protagoras*: 320c-321d (Engl. transl. Lamb 1967: 129–131).
[28] As in the French translation of *The Holy Bible*: Jonah 2:6: "Je suis descendu jusqu'aux racines des montagnes" [To the roots of the mountains I sank down].

mythological narratives, and alchemical or occult discourses, that the underground took on a different meaning from that which it had in Antiquity. Alchemical and hermetic imageries of the *axis mundi* were associated with plants and trees, such as the tree of knowledge depicted in Athanasius Kircher's (1602–1680) *Ars magna sciendi* (1669), as well as his philosophical tree[29]. The authors of botanical discourses and medical fictions acquired and used these symbols as theoretical tools to explain newfound knowledge and to overcome the theoretical constraints brought about by new discoveries.

Plants and the Underworld in Botanical Discourses and Medical Fictions

Now let us consider the various examples of fictions that describe plants or vegetal elements. Intellectual discourses about the Underworld were inspired by occult, alchemistic and hermetic theories. Paracelsian ideas nourished the Hippocratic and Galenic humoral and elementary theories, particularly in the field of medical practices. These were not in perpetual opposition, but the humoral doctrine came to be undermined by new concepts, and was challenged and questioned by physicians such as Jean Fernel (1497–1558) in *De abditis rerum causis* (1548), and by Girolamo Fracastoro (1478–1553)[30]. Authors deployed metaphors and founding myths from the Bible or ancient mythology, or would elaborate texts with the status of an etiological myth[31]. Medical fictions were the discourses that often involved cultural and symbolic images and magical beliefs that some physicians, such as Fracastoro and Leonhart Fuchs (1501–1566), labelled superstitions[32]. The fictions were imbued with medical symbolism. As Thomas S. Sozinskey has written, "a symbol is an illustration of a thing which, to use a poetic phrase, is 'not what it seems'"[33].

These symbols originated in the *scala naturae*[34]. The position that was dedicated to herbs and roots, from a biblical perspective, comprised the ingredients of therapeutic remedies as well as recipes for food. At that time, roots, herbs, and bulbs were considered "crawling beings"[35] or "êtres rampants" in French. They were considered

---

29 Kircher, *Mundus subterraneus*. Kircher, *Ars magna sciendi*.
30 Bayle and Gauvin 2019: 87.
31 Bayle and Gauvin 2019: 93–94.
32 Bayle and Gauvin 2019: 111–115.
33 Sozinskey 1891: 1.
34 The *scala naturae* or 'Great Chain of Being' is an organizational theory that classifies the living beings and objects of both the Earth and celestial spheres according to a biblical hierarchy; originating in an underground sphere and eventually ascending to that of Heaven. See Grieco 1993; Brancher 2015: 29–30.
35 As in Genesis 1:24.

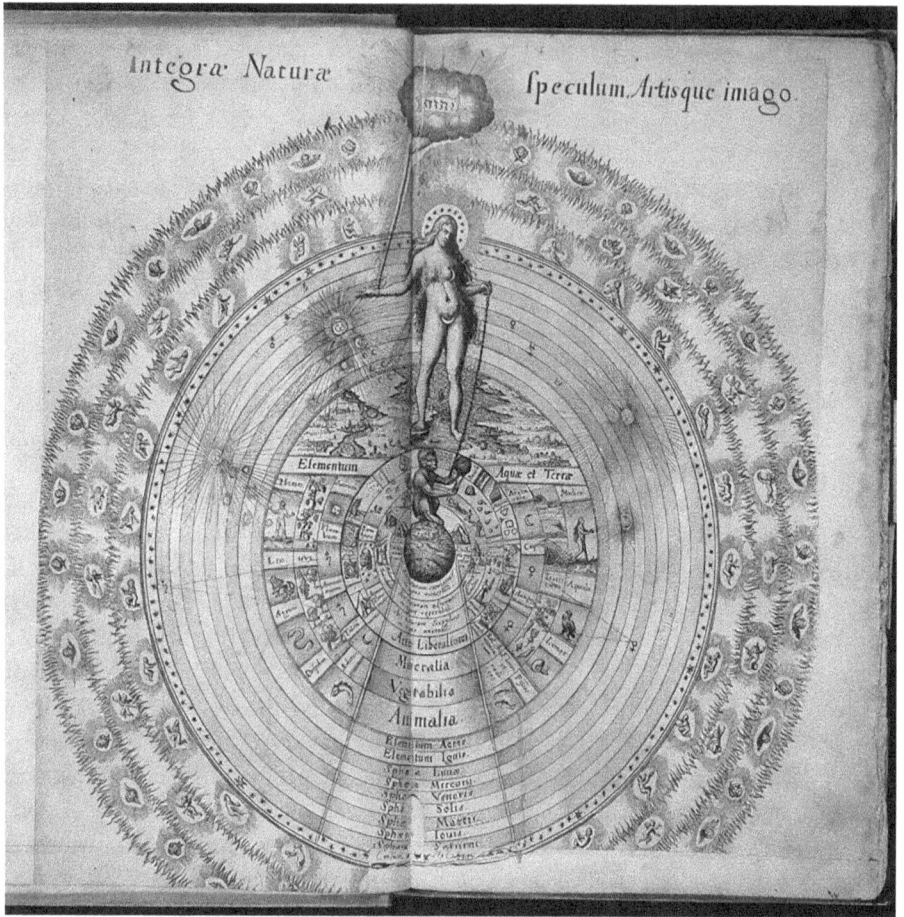

Figure 3: *Nature's correspondences in the Microcosm and macrocosm: "Integrae Naturae speculum Artisque imago"*, in Robert Fludd, *Utriusque cosmi maioris*. Oppenheim: Hieronymi Galleri, 1617, planche hors texte.

aspects of the earth, as connected to the underground and to the inanimate. The use of roots was largely spread through the revival of ancient natural philosophies such as those of Dioscorides and Theophrastus (ca. 371-ca. 287 B.C.). This was nevertheless perfectly coherent with the description given by Robert Fludd (1574–1637) of the *Integrae naturae speculum* in his book *Utriusque cosmi maioris scilicet et minoris Metaphysica, physica atque technica Historia* (1617).

In this visual representation of the relationship between microcosm and macrocosm, Robert Fludd depicted every being in a hierarchical order, illustrated by circles, through their terrestrial and celestial correspondences, which meant that each being corresponded to one of the four elements, or with stars and planetary orbs[36]. The human body, too, is depicted as corresponding to different elements. In Fludd's diagram the body's head and the reasoning mind are connected to the heavens beneath God (represented by a cloud), while its feet are rooted to the ground, next to less-valued elements such as roots, minerals, and flowers. This bond of the body with the Underworld was viewed as both dangerous and attractive for the physician or healer. On the one hand, it was considered dangerous because plants had a direct relation with the world of death. On the other hand, plants were appealing for their ambivalence, and were considered as containing two opposite effects: The power to cure or to poison. This ambivalence also found its origins, in part, in a controversy of the Renaissance period in which naturalists debated where to place roots and herbs on the *scala naturae*[37]. In fact, Ancient medical practitioners, and early modern ones, such as Paracelsus, in writing about the theory of poison, remarked that in each ingredient and in plants, there resides both a poison and its antidote; the differentiation between both aspects lay in the dosage. This characterized the ambivalence of the *pharmakon* (remedy).

Renaissance scholars not only took their inspiration from occultism, alchemy, and hermeticism, but created a whole new mythology surrounding the plant-object. These inspirations were very noticeable in the ancient herbals and plant treatises. Dioscorides, Theophrastus, Pliny, and other renowned naturalists were compilers of knowledge of both past and present. They gathered as much information as they could on a specific plant, deploying the same narrative tools as early modern historians. Indeed, herbals were sometimes even called *historiae*, with reference to the interdisciplinary nature of the naturalists. Writing a *historia* required that the naturalist take up the tools of a historian, inquiring both in the fields of natural phenomena *and* human deeds, "including the nature of man"[38]. The plants were described not only physically but historically. A writer would use etymology, philology, medical theories, topographical and geographical data. He would also gather stories as a form of evidence, including the testimonies of ancient practitioners. These testimonies were mostly structured upon the ancient use of plants, their etymology, and organized around ancient myths. This was the case for Leonhart Fuchs' *De historia stirpium* (1542).

The poppy (or *Papaver somniferum* L.) is a good example of a plant that Renaissance scholars continued to envision as having been designed by the Creator in order to send a human to sleep. The 1549 French translation of Fuchs's herbal reports that: "*In French, cultivated poppy, or simply poppy. The Greeks called it Mecon, from the verb meconein, which means as much as not to serve or to minister, because it prevents*

---

36 Fludd, *Utriusque cosmi maioris*: frontispiece.
37 Brancher 2015: 29.
38 Pomata and Siraisi 2005.

those who eat it from serving well and carrying out their offices"[39]. This etymology was evidently taken for granted and based on a very ancient understanding of the plant. The 'artificially induced transcendence' caused by the poppy had been used in the rituals of the Ancient Cretan cult and the Eleusinian mysteries, as Karl Kerényi has asserted in his essay[40]. One variety of poppy was called *rheas*, and another species was linked to the Goddess Demeter. These poppies were plants of the Underworld. In Greek mythology, Rhea and her daughter Demeter were the ones who originally brought the poppy to the surface from the Underworld. Kerényi explains that, "for the Greeks, Demeter was still a poppy goddess, bearing sheaves and poppies in both hands"[41]. The extraction of opium from poppies was a complex pharmacological procedure that archeologists have been able to determine thanks to excavated figurines. The Ancient Greeks probably used poppies during celebrations for Demeter, who was worshipped as the goddess of agriculture, harvest and seasons. The myth of Demeter tells how she descended into the Underworld to look for her daughter Persephone, abducted by Hades. After Zeus eventually decided that Persephone should stay six months of the year with Hades, Demeter was plunged into grief and was offered a poppy by Morpheus to ease her pain.

This myth intermixes references alluding to the plant's usage in Classical Greece, and would, in part, come to define its use during the Renaissance. In the Early Modern period, the poppy was among several plants to become symbolically associated with evil: An individual's use of the plant could be taken as evidence of witchcraft. Women were accused of making use of poppies for performing abortions, and those who used the seeds for their hallucinogenic properties were morally condemned. On psychotropic plant ointments, Daniele Piomelli and Antonino Pollio have commented that there exists "a sparse set of testimonies, scattered here and there in the writings of a few Renaissance scientists and in the transcripts of Inquisition trials [making] reference to plant ointments which 'witches' rubbed on their bodies before 'going to the Sabbat'"[42]. The use of such psychotropic substances was not only based on what Carlo Ginzburg has termed folklore belief, however, but also found its origins in the medical assumptions and scientific understanding of the early modern period.

An Anthropomorphized Root and its Subterranean Meaning

Why should subterranean flora matter so much to the early modern mythologist? Did it impart a rarity or a high value to plants like the poppy or the mandrake root? As we

---

39 Fuchs, *De historia stirpium* (Fr. transl.: *L'hystoire des plantes*, Chapitre CXCVI) [translation is mine].
40 Kerényi 1976/2020: 22-23.
41 Kerényi 1976/2020: 23.
42 Piomelli and Pollio 1994: 242.

have seen with both the *homunculus* and the small, anthropomorphized depictions of the mandrake, they both share the same property: They were made from earth and came from the earth *venter* or womb. In medical fiction, imagery of the uterine or "*venter inferior*" was often connected with the underground, as a matrix image of elementary creation. Indeed, the earth was perceived to be constantly re-fabricating itself through subterranean fires, its earthy elements, and underground rivers. This imagery is similar to an alembic[43], depicted by Kircher in his *Mundus Subterraneus*, in which he appears "to imagine the Earth itself as an alembic"[44]. The alembic was compared to the womb, where offspring are conceived and gestate.

The penetrating darkness that was believed to be the prelude to the divine creation of the world was also related to the notion or image of the womb; a mysterious symbol of the world both before and after life. Writing on primitive representations, Daniel Faivre remarks, "thus, the idea of darkness preceding the creation of the world is probably the phantasmatic representation of the unconscious memory of uterine darkness. [. . .]"[45]. This uterine cave was illustrated by Theodore de Bry (1528–1598) in the frontispiece of *Dyas chimica tripartita* (1625), attributed to Johann Grasshoff (see Figure 2 above). Thus, both the cave and the underground were two powerful literary *topoï* that explained the creation of inverted images of life on the surface. This notion and metaphor of inversion is an intrinsic component of occult and alchemical thinking. Thus, in the Latin acrostic of the word VITRIOL or VITRIOLUM, one can find this same *topos* of the lower, subterranean, uterine and cavernous stomach, for example: "Visita Interiora Terrae Rectificando Invenies Occultum Lapidem", in the *Geheime Figuren der Rosenkreuzer* and in Gérard Dorn's (ca. 1530–1584) *Congenries Paeacelsicae chemiae* (1581)[46]; just as in the *Emerald Table* or *Tabula Smaragdina* taken up by Heinrich Khunrath (ca. 1560–1605) in his *Amphitheatrum sapientiae aeternae* (1595), which can be found in the *Tabula Smaragdina*: "Quod est inferius, est sicut quod est superius. Et quod est superius est sicut quod est inferius ad perpetrada miracula rei unius" (What is inferior is similar to what is superior. And what is above is like what is beneath)[47].

In the alchemical alembic as in the womb, the elementary materials that could be used to create powerful remedies were to be found beneath the earth's crust. The physician Fracastoro explained how physicians should look for the great cure for syphilis, mercury, in the Underworld[48]. Although this metaphor represented a moral sermon, urging

---

[43] An alembic was an apparatus for distilling, consisting of a gourd-shaped container with a long tube for conveying liquids into a receiver, in order to produce alchemical substances.
[44] Nissen 2007: 26.
[45] Faivre 1995: 59–80, from Job 2:16.
[46] Dorn, *Congeries Paracelsicae chemiae*. A poem in which the letters in each line form a word or words.
[47] [Hermes Trismegistus], *Tabula smaragdina*: 363 [translation is mine].
[48] Bayle and Gauvin 2019: 111–115

the physician to administer to syphilitic patients, it was also a way to explain the disease's origins. Katharina N. Piechocki explains that for Fracastoro, the subterranean world was "a world in constant flux, where new materials are created and improved, where artifice emulates nature, and where the transformative power of the underworld exceeds and at the same time inverts the activities of the world above"[49]. This would explain why Fracastoro's fictional character, Ilceus, finds precious metals and materials in the Underworld. Fracastoro's epic poem entitled *Syphilis, sive Morbus gallicus* (1530), was written to illustrate the difficulties presented by the disease, and to explain the unknown. In his narrative, Fracastoro invents two characters, Lipare and Ilceus; the latter is a syphilitic. Lipare plays the role of explorer, showing "the underworld to Ilceus as if it were a newly discovered territory, a *mundus novus* silenced by the underworld's darkness yet eloquent in its production"[50]. For Lipare, as for the alchemists and physicians (*physici*) of the Early Modern era, the underworld represented a place of paradoxes and ambiguities, where one could find "powerful cures"[51], hidden and invisible to those living on the surface. This metaphor bears some resemblance to the image of the *homo arbor inversa*—the tree's hollow resembling in some respects the subterranean hollow of the Earth. Fracastoro's epic poem was written to convince people to seek truth in contemporary rational medicine and not in superstitious cures. The powerful image of the 'invisible cause' of the disease and the subterranean, secret cure helped Fracastoro explain his new theory of contagious particles.

In its anthropomorphized form, the mandrake root was also seen as a "demoniac anthropophyte"; a "homunculus-plant" originating in Hell[52]. It was thought to be related to witchcraft, lycanthropy and other evil spirits[53]. Thus, the connection between the mandrake and the Underworld or Hell would have been evident for many early modern naturalists and physicians. The depictions of the mandrake as Adam and Eve (see Figure 4) demonstrated its intrinsic link to original sin, and even to Satan and witches' Sabbaths[54].

Physicians and naturalists appropriated myths and medical/botanical fictions, transposing the established connection between various plants and the Underworld from a mythical realm towards a more rational explanation. This rationality was preferred to pure empiricism (represented by scientific tests, resulting in failure or success) and allowed physicians to apply Hippocratic and Galenic rules to their therapeutic and preventive medical practices. The third part of this chapter will explore how those practices that surrounded plant roots were closely connected to the subterranean—which will be used to explain the rituals involved in their harvest.

---

[49] Piechocki 2019: 169.
[50] Piechocki 2019: 170.
[51] Piechocki 2019: 153.
[52] Brancher 2015: 74.
[53] Brancher 2015: 78.
[54] See Brancher 2015: 94–101.

Mandrake Roots or the Importance of the Underground Element

Mandrake or *mandragora* refers to three species of the Solanaceae family, one found in Asia, the other two in Mediterranean regions. For early modern medical practitioners, it was its roots that were valuable, not its leaves, flowers or fruit, which underlines the importance to them of its underground provenance. Louis Irissou, writing on the mandrake in early modern pharmacopeias, declared that "the anthropomorphic aspect of their root is the key point of the fabulous legend of the mandrake which we gave human origins"[55]. As noted previously, the legend of the mandrake is as old as the Bible and is also found in various ancient sources. This had a great influence on its therapeutic use throughout the Middle Ages and Renaissance, in particular because it accorded with the theory of signatures. The mandrake also became a literary trope[56], appearing in the works of many writers, including Giovanni Boccaccio (1313–1375) to Niccolò Machiavelli (1469–1527), William Shakespeare (1564–1616), and Jean de La Fontaine (1621–1695)[57].

The first traces of the mandrake myth can be found in works dating from late Egyptian Antiquity, which describe how the plant was used for its soporific properties[58]. Further evidence of its therapeutic use in ancient Greece and the Roman Empire has also been found. In translating Dioscorides, Pietro Andrea Mattioli (1501–1578) wrote that the mandrake was sometimes named "Circae", after the enchantress of Greek mythology, because in beverage form it was believed to be a powerful aphrodisiac. Ovid suggested that Circe may have learned about medicinal remedies, herbs and potions from her mother Perse. Dioscorides goes on to describe two recipes for the preparation of somniferous substances from mandrake roots. However, some of the most prominent descriptions appear in the Bible, for example in Genesis 30:14 ("During wheat harvest, Reuben went out into the fields and found some mandrake plants, which he brought to his mother Leah. Rachel said to Leah, 'Please give me some of your son's mandrakes';")[59] and in the Song of Songs 7:13[60] ("The mandrakes give a smell, and at our gates are all manner of pleasant fruit, new and old, which I have laid up for thee, O my beloved")[61]. It is mentioned in Titus Flavius Josephus's (ca. 37-ca. 100) *De bello Iudaico* (*The Jewish War*), written in Greek during the 1st century A.D., where it is referred to as *"baaras"*: "In the valley that surrounds the city on the north side, there is a place called Baaras, which produces a root of the

---

55 Irissou 1949: 469–470.
56 Significantly, the use of the mandrake as a medicinal panacea would later be confirmed by chemical investigations that confirmed the presence of specific alkaloids (mainly scopolamine and atropine) which are responsible for pharmacological effects.
57 Irissou 1949: 469.
58 Casini 2018: 101–116.
59 Genesis 30:14.
60 See *Bible Plants*, online encyclopedia: https://ww2.odu.edu/~lmusselm/plant/bible/mandrake.php.
61 Song of Songs 7:13.

Figure 4: *The female mandrake*, in [Johannes von Cuba], *Herbarius*. Mencz [Mainz]: [Peter Schöffer], 1485: f. 208 verso. Courtesy of Wellcome Library, Latin. n 93074319.

same name", followed by a description of the plant, its dangers and methods for its extraction from the soil[62]. Flavius Josephus's passage was repeatedly studied during the 15th

---

62 Josephus, *Bellum Judaicum* 7.180 (ed. and Engl. transl. Thackeray 1961: 556–557). On this see also Céard 2013.

and 16th centuries by authors such as the Spanish historian and philosopher Pedro Mexía (1497–1551); the French writer and member of the literary circle of La Pléiade, Pontus de Tyard (1521–1605); Jean Wier (1515–1588), a physician who wrote against witch hunts; the German humanist and cosmographer, Sebastian Münster (1488–1552); and Pierre Boaistuau (also known as Seigneur de Launay, ca. 1517–1566), a French author of popular compilations. These authors all used the mandrake as a stereotypical figure in the literary genre of marvelous stories (*histoires merveilleuses*) and books of demonology. Mandrake is allegedly the poison of witche[63].

In addition to references about the traditional use of the mandrake in Antiquity, the Biblical allegories were also frequently reused in early modern natural-history treatises[64]. The mandrake was present in two plates of the frontispiece of one of the earliest illustrated copies of Dioscorides' *De materia medica*. The persistent image of the mandrake echoing through many texts of the period, remained largely the same as the one depicted in Dioscorides's frontispiece. This book is known as the *Aniciae Julianae Codex* and is conserved at the National Library of Vienna. In this frontispiece, we can find images that already testify "to the early (and frequent) anthropomorphizing of mandrake"[65] that were to be copied in many later herbals such as the so-called *Gart der Gesundheit*.

Digging and Harvesting: The Case of the Mandrake

Uprooting

The verb "to uproot" describes an action directly related to the earth, and the gesture itself is not devoid of symbolic meaning. According to Theophrastus, the individual in charge of the harvesting of plants was known as a "root cutter". They were considered healers, and occupied, like pharmacologists, an intermediate place between medicine and magic[66]. The root cutter or "*rhizotomos*" in ancient societies was the first figure in a long chain of 'medical' practitioners to be in direct contact with plants[67]. There is a whole world of imagery revolving around the act of uprooting, even in its most common, literal sense. On roots and their bonds with the subterranean, Lucien Bély quoted *Les Caractères* (1688), a Greek to French translation and adaptation of Theophrastus's work by Jean de La Bruyère (1645–1696). La Bruyère writes[68]:

---

[63] See for exemple Riddle 2010: 61–62, 71–72, and Davidson 2012: 118, 137, 160, 195.
[64] Daunay et al. 2007.
[65] Daunay et al. 2007: 60.
[66] Nutton 2004: 141, 354–355.
[67] Collard and Samama 2006: 9.
[68] La Bruyère, *Caractères*: 190 (ed. Garapon 1962: 339) (translation is mine). See also Bély 2009: 475.

> We see certain wild animals, male and female, scattered through the countryside, black, livid, and all sunburnt, chained to the earth which they dig and stir with unwavering stubbornness; they have a somewhat articulate voice, and when they get up on their feet, they show a human face; and indeed they are men. They retire at night to dens, where they live on black bread, water and roots [turnips, carrots]: they spare other men the trouble of sowing, plowing and collecting [harvesting] in order to live, and thus deserve not to lack this bread which they have themselves sown.

In this vivid and grotesque description, adjectives like "savage", "black", "livid", or verbs like "rummage", illustrate the wild and subterranean characteristics of the countryside. The text depicts men or animals retreating into dens at night and living on roots. The peasant (or the animal) is doomed only to leave his home in order to go to the fields. This fieldwork is that of someone constantly engaging with the underground world, removing roots in order to cultivate or to clear non-domesticated spaces. La Bruyère and Theophrastus are not the only authors to deploy such symbolism about roots in their writings. Roots are frequent motifs of the subterranean world: Earthy, black, somber, livid or pale; sometimes depicted as animal and wild; as stirring, crawling, or swarming. Even Galen (A.D. 129-after [?] 216) chose to place the roots at the end of the hierarchical pyramid of vegetal life and reported that the "country folk of his own day" were forced to eat roots because their supplies "had been taken away by force by the powerful men from the big city"[69]. This view of roots as the 'ignoble' part of a plant stands in contrast to many of our previous observations about the use of roots as remedies.

The act of digging up the earth is a complex one, both in terms of the early modern understanding of nature and in the picking process, which concerns the invisible part of the plant, since it was a gesture that required knowledge of what the ancient naturalists, authorities in this matter, had said; and that the act had to conform to specific preparatory rituals, as in the case of the mandrake. Digging is neither a particularly beautiful act nor an ugly one. It is an act which involves investigating or exploring the unknown and can be seen as symbolically representative of an epistemological process. Digging up soil can be related as much to the infernal imagination of Dante Alighieri's (1265–1321) *Divina Commedia* (1303–1321), or to Orpheus's descent into the Underworld, as it can be to the discovery of those "creeping" organisms, according to the *scala naturae* defined earlier in this chapter, that exist on the Earth's surface, semi-buried in caves: Snakes; swarming insects; worms; or even vegetables such as cabbages. These animals and plants are much more connected to the subterranean than to the terrestrial world, since their vegetative functions require feeding through the underground soil rather than through the air (as the early modern naturalists thought). An example of the use of "to creep" is "I Crepe as a serpent dothe"[70] in *Lesclarcissement de la Langue Francoyse* (1530) by John Palsgrave (ca. 1480/1485–1554), and "to *craule*" translated into Latin as "*repere*" or "*serpere*", in Peter Levens, *Manipulus Vocabulorum* (1570). The use

---

69 Nutton 2004: 21.
70 Palsgrave, *Lesclarcissement*: 553.

of 'to swarm' is demonstrated by "I swarme as bees do whan"[71]—a synonym of 'to flock' (or *confluere* in Latin), as explained in both Palsgrave and Levens. "Swarming" and "creeping" plants were seen as less noble beings than plants growing towards the sky, such as trees. This notion of crawling and swarming forms existing beneath the soil, could be likened to stomachal fauna or an "intestinal hell", to use Gilbert Durand's words[72]. The ambivalence manifested in the act of digging the soil originates in the opposing connotations just described: On the one hand, the subterranean physical journey of death, on the other, the thriving ecosystem nourishing itself on the terrestrial subsoil. The subterranean earth could equally be thought of in relation to the "primordial clay of the great Mesopotamian or Hebrew narratives"[73], and be "linked with burying, invagination, digging, covering, withdrawal or transformation"[74]. These symbols evoke the imagery of burrows and caves. They were fertile places, "invested by the imaginary of the womb"[75].

In the translation of the *Holy Bible* by Louis-Isaac Lemaistre de Sacy (1613–1684), "digging" is often used in reference to the image of the "well", and to the search for water. The lemma "to dig" ("*creuser*" in French) appears at least eight times in the Book of Genesis alone, for example: "The same day, Isaac's servants came to tell him that they had found water in the well they had dug"[76]. In other instances, digging the earth recalls the image of a grave or pit: "He opened a pit, and dug it; and fell into the same pit that he had made"[77]. In the Psalms, the pit is mentioned as being "intended for sinners"[78]: Their destination is not Heaven, but beneath the ground. Although the Devil is generally considered the primary figure of "evil", in the Bible he is not associated directly with the underworld, operating instead as the agent of temptation on the Earth's surface. The notion of the never-ending descent of the sinner to the infernal subterranean recesses of the world was reappropriated in Dante's *Divine Comedy*.

Additionally, for scholars in the sixteenth and seventeenth centuries, the gesture of digging the earth was considered very animalistic; an instinctive act of those creatures that dig in order to feed themselves. Pigs, for instance, as Louis Liger wrote: "are the ruin of meadows, one must never suffer them there. They plough up everything with their snouts in the search for roots"[79]. Early modern thinkers would continue to transpose

---

71 Palsgrave, *Lesclarcissement*: 797.
72 Durand 2016: 205–244.
73 Simon 2019: 229.
74 Simon 2019.
75 Simon 2019.
76 Lemaistre de Sacy, *La Sainte Bible*. *Genesis*: 32. "Le même jour les serviteurs d'Isaac vinrent lui dire, qu'ils avaient trouvé de l'eau dans le puits qu'ils avaient creusé" [translation is mine].
77 Lemaistre de Sacy, *La Sainte Bible*, *Psalms*: 16: "Il a ouvert une fosse, et l'a creusée; et il est tombé dans la même fosse qu'il avait faite".
78 Lemaistre de Sacy, *La Sainte Bible*, *Psalms*: 13: "afin que vous lui procuriez quelque adoucissement dans les jours mauvais, jusqu'à ce que soit creusée la fosse destinée au pécheur".
79 Liger, *Dictionnaire*: 276. "Les Cochons sont la perte des Prez, il ne faut jamais les y souffrir; ils labourent tout avec leur grouïn pour chercher des racines".

Biblical symbols of the divine creation into their explanations of natural phenomena. Tobias Swinden (1659–1719) exemplified this approach:

> What is more beneficial to us than the *Earth*, whose teeming womb bringeth forth such plenty of necessaries, such store of delicacies, that curious man, had he been left to his own option, could not have wishes for greater? [. . .] And yet, in how many several respects may we suppose even this our common and great benefactor to be also adverse and hurtful to us? Particularly, if it were not tilled and cultivated, how soon would it degenerate from that beauty which now delights us? How would weeds and thorns, thistles over-run the face of it? [. . .] But, as it is at present, notwithstanding all the art and industry of man to the contrary, how many unsavoury and noxious roots, how many distasteful and poisonous herbs [. . .] doth it produce? What wild and ravenous beasts, what voracious and hurtful animals, what vexatious and venomous insects doth it constantly nourish? Nay, she is apparently so free and liberal to these her productions than to the former, that she hath been rightly accounted *his mater, istis noverca*, a fond natural parent to these, but a cruel, niggardly step-dame to the other.[80]

Here, the author argues that moral reform should be used as a form of remedy. His long metaphor is simultaneously an explanation of natural phenomena and the setting out of an argument for the moralization of the human body.

In 1532, Gherardo Cibo (1512–1600), an Italian artist and herbalist, published his herbarium, *De re medica*. His drawings of plants relied on the knowledge not only of naturalists and herbalists, but also other more unusual figures such as "plant-gathering women", as Sachiko Kusakawa explains[81]. In the plate illustrating the Eryngium, he depicted an herbalist who, in the process of harvesting the plant's root with his tool, has discovered a snake he is attempting to kill with his scythe. The image shows the fears that were associated with the act of digging. In fact, the rituals that surrounded the act of digging up roots were dedicated to warding off threats related to death (represented by poisonous animals), which were also identified with defenses against evil spirits. It could also be taken as a mnemotechnic means to remember the form of the root, resembling a snake, the snake being a 'sign' of danger and poison according to the Doctrine of signatures.

Digging up Roots as Remedies?

Medical fictions, especially herbals and medical treatises, are permeated by symbols and motifs of roots. Since Antiquity, roots had largely been used as remedies and as a source of food for the poor. The former comprised several medical categories: Mandrake-root, as an aphrodisiac, was often used to solve problems relating to sexuality or fertility; belladonna (deadly nightshade) was utilized for its poisonousness and hallucinogenic effects; most roots that contain alkaloids were used to prevent madness,

---

80 Swinden, *An Enquiry*: 267–269.
81 Kusakawa 2009: 462, 464.

or to induce nightmares and hallucinations. Their use reflects a good knowledge of what we now know about the chemical composition of individual plants. The principle of *similia similibus curantur*, which translates as "let like be cured by like", also served as a theoretical explanation of such intricate processes. It was therefore not illogical to consider a beet or turnip, with their phallic and round forms, as potential remedies for problems related to sexuality. This was particularly the case in early modern societies that valued the heterosexual couple as the sexual norm, the only one to be "legitimate and capable of procreation"[82]. In his book, *A New Herball* (1551), William Turner (1509/1510–1568), writing about the *Panicaut* root (*Eryngium maritimum* L. or Sea Holly), remarked that: "The later wryters use the roote of thys herbe, to stirre up the lust of the body, and they use to gyve it bothe to men and wymen that have desyres to have chylder"[83]. Whilst other roots (or seeds, another frequent symbol of fertility) were used "to stirre up the pleasure of the body"[84].

Other authors such as Garcia da Orta (1501–1568) conveniently reused the same conceptual tools to explain why the 'China root' was an aphrodisiac. 'China root' arrived with the Portuguese from the "East Indies" to the Americas sometime after 1535, as Nicolás Monardes (ca. 1512–1588) recalled[85]. In his *Colóquios dos simples e drogas da India* published in 1563, da Orta used the narrative pattern of medical fiction in his description of 'China root'. He created dialogues between himself and a fictional character, Ruano, in which he played the role of storyteller. These dialogues represented an innovative process in the composition of a natural history encyclopedia and were intended to promote a new drug which was, according to da Orta, a more effective remedy against syphilis than guaiac wood. Like Machiavelli, da Orta used comedy as both a genre and the general tone of his work, in order to deliver a message on the potential use of the root, whilst at the time making light of it: It was thought that it would unlock the 'sexual appetite' of the consumer[86]. Like Chinese ginger and ginseng roots, 'China root' was an aphrodisiacal drug, which was believed could treat syphilis.

As for hallucinogenic roots, the night was considered the most conducive moment for potential delirium; and nightmares, which Georges Vigarello termed "disturbing perceptions", could only be cured in turn by certain plants such as the peony root. In the sixteenth and seventeenth centuries, the peony was recognized as a cure against melancholic dreams or "Night Mares" as the English botanist and practitioner of astrological medicine Nicholas Culpeper (1616–1654) claimed. The root was considered highly effective against such disorders and against those that physicians termed "Lunatickes". The English botanist John Gerard (ca. 1545–1612) remarked that peony root was particularly expensive and in demand. Its powers were believed to be enhanced

---

82 Foucault 1976: 10.
83 Turner, *A new Herball:* Part 1, 201, 311.
84 Turner, *A new Herball:* Part 1, 201, 311.
85 Winterbottom 2014: 22–44.
86 Bayle and Gauvin 2019: 203–204.

by a magic harvesting ritual: This often occurred at night, under a particular moon – the plant's human-finger shaped roots were thought to shine at night but remain invisible in sunlight[87]. For both peony and mandrake roots, a dog was required to dig the ground as it was believed that digging up these was harmful to humans. Touching the dark or pale root of the mandrake was thought to be lethal. It was also said, according to Gerard once more[88], that a woman's menstrual blood or urine, when scattered on the roots of peonies, would help to neutralize their enchantment, and the underground parts could then be dug up. John Gerard perceived these rituals as superstitions but felt he had to tell those stories to capture the imagination of his readers. The image of the dog certainly captures our attention as a frequent figure of several narratives involving plants (see Figure 5).

Why a Dog?

In the *Aconitum* genus, several species were recognized, such as monkshood (*Aconitum napellus* L.), and *Aconitum lycoctonum* L. (in French: *aconit tue-loup*). In Greek mythology, aconite was sometimes associated with Hekate, goddess of magic and herbal knowledge. Ovid described the *Aconitum* genus as among the most terrible poisons on Earth (*Lurida terribiles, miscent aconita noveres*)[89]. Aconite was the name given to various vegetal and poisonous substances by the Greeks, who invented a mythology surrounding their poisonous effects, claiming that aconite originated from the deadly saliva of the three-headed dog called Cerberus, who stood guard at the gates to the Underworld. The figure of the dog occupies an ambiguous position between the domesticated and the wild. It is a recurring feature of plant-centric myths, as is demonstrated by the story of the mandrake. It was simultaneously the beast that helped in the work of peasants or traders, or equally the wild animal which myth represented as a demonic creature in the service of evil. Indeed, the dog was a symbol of sin, and represents a diabolical reminder or tocsin in the many paintings that depict it. The dog was a frequent figure in classical paintings, as Simona Cohen describes[90]. The dog therefore symbolized the hidden vices of mankind: A metaphor of betrayal, heightened sensuality, inappropriate sexual attitudes, wild instincts; every pejorative image of a life of disorder and bestiality.

However, the dog's image in the early modern era was an equally ambivalent one. It could often symbolize "faithfulness, wisdom, vigilance, prophetic vision, appeasement and healing, fertility in magical thinking"[91]. Like the ape, a dog, often depicted lying on

---

87 Gerard, *The Herball*: 832.
88 Gerard, *The Herball*: 832.
89 See Ovidius, *Metamorphoses*: 1.3.147 (ed. Lafaye and Sers 1989: 12).
90 Cohen 2008: 137–138.
91 Cohen 2008: 137–138.

Figure 5: *Uprooting the mandrake with a dog*, in Pierre Boaistuau, *Histoires prodigieuses*. London, Wellcome Library, MS. 137: f. 138 verso.

the ground, could symbolize a man or woman's madness, representing human frailty and their mortal condition on Earth. For early modern Christians, the dog was associated with the creature who devotedly accompanied Tobias in the Bible; who licked wounds to help a human recover. On the other hand, it could be thought of as the demonic black dog – who carried disease, whose bite could cause death, who stood guard at the entrance to the Underworld. In both instances, the dog served the role of a protective guard. The figure of Cerberus and the use of dogs in the digging and harvesting of roots represents an inevitable *persona* essential in order to understand the association between roots and notions of the subterranean.

Dogs were so much associated with, and indeed also thought of as faithful to, the Underworld, that they were considered the most appropriate object of sacrifice in the ritual of harvesting the mandrake. In the mythology of the naturalists, the nymph Euresis had given the mandrake and its 'secret' cure to Dioscorides. This myth was first discovered in an ancient engraving in the *Codex medicus Graecus* (4th century A.D.), and was thereafter used by several naturalists and botanists, such as Johann Gottlieb

Gleditsch (1714–1786), who wrote an account of how the mandrake came to be a mythical cure and on how its story had been altered since Antiquity[92]; and also by Laurent Catelan (ca. 1568–1647) in his *Rare et curieux discours de la plante appellée Mandragore* (1638) in which he claimed that mandrake had been named *Cynospsatos*, meaning 'canine flesh'[93]. With this Greek etymology, he demonstrated that the ritual surrounding the uprooting of the mandrake was very old, directly linking dogs to mandrakes, and giving them the role of sacrificial animal. Mixing Biblical narratives, Ancient etymologies and contemporary testimony and practice, Laurent Catelan, an apothecary educated at the Faculty of Medicine in Montpellier, compiled all the various curious anecdotes about the mandrake, in an attempt to question the validity of such tales. His discourse on the mandrake was read aloud in front of students ("*escholiers, estudians*") of the faculty of medicine at the university of Montpellier. He deployed rhetorical arguments to study the plant as a botanist (how it grew, developed and reproduced). He then went on to challenge the myth of the mandrake, arguing, for example, that Paracelsus's experiments using mandrake root in order to create 'male sperm' led to a falsification of the results[94]. However, over the course of sixty-two pages, Catelan described two mandrakes. One was the garden mandrake, purely botanical; the other, mythic and magical. His argument then deploys all the symbols this chapter has examined, including the underground, rituals surrounding the act of uprooting, the dog, the root as a source of fertility, and the etiological myth about the mandrake. The Spanish humanist and historian of Charles V, Pedro Mexía, appeared to use the same story as Catelan in his *A Miscellany of Several Lessons*, a miscellany of humanistic knowledge. The demonic connection with the figure of the dog is emphasized here, and the story evokes, too, the requirement for warm female urine, as we have read in John Gerard's account of the magical harvest of peony roots:

> Until he experienced what the devil, or the Angel perchance discovered, he knew that by spraying it [the root] with the urine of a woman who was menstruating, he could then take it [the root] and tear it off, but the one who thus tore it off would die, unless he had another of the same root with him. And that in order to do it safely, seeing the roots, and having sprinkled them with what has been mentioned, they dug and opened up the earth around them, and tied a loop with one cord tied to the root, and with what was left of the cord, they tethered a dog, which, seeing itself tied so tightly, pulled away so hard that the root was pulled out of the ground, and then the dog died. Then any man would be able to safely take it, and use and profit from its benefits. The authors of what has been mentioned were those alleged in the last chapter, and many others.[95]

---

[92] Gleditsch, "Sur la mandragore": 61.
[93] Catelan, *Rare et curieux discours*: 11–12.
[94] Catelan, *Rare et curieux discours*: 4.
[95] Mexía, *Lección*: 321–322. "*Hasta que experimentado lo que el demonio, ò el Angel por ventura del cubrió, fe conoció que ro ciandola con orina de muger que estuuieffe con el mes, que luego te podria*

Such narratives seem to have made the tour of early modern Western Europe through medical fictions, botanical treatises, curious accounts, or folklore. The mandrake or mandragora was also sometimes called devil's apple (*Mandragora officinarum* L.). The tropane alkaloid atropine found in the mandrake root, and the botanical names *Atropa mandragora* L., and *Atropa belladonna* L., are all derived from Atropos, the name of one of the Fates or *Moirai* of Greek mythology[96]. The three Fates were believed to control the thread of life, and it was Atropos whose role it was to cut the thread—her name being a metaphor for death. In Renaissance herbals and medical treatises, Belladonna or deadly nightshade was often translated as beautiful lady or witches' fruit.

Conclusion

In conclusion, these medical or botanical fictions are not as naive in nature as they might at first seem. Indeed, as we have seen, for physicians dealing with the incongruities and novelties of new diseases such as syphilis, popularizing narratives could be a useful tool in the dissemination of new medical theories, similar to their use by their contemporaries in the discovery of new territory beyond the ocean. The process of popularizing did not, however, mean that physicians were accommodating their writing to popular culture, but rather that they were seeking for the best way to guide their readers through difficult confrontations between old and modern forms of knowledge[97]. Physicians and naturalists deployed medical and botanical myths, and metaphors of the underworld, in order to explain theories and principles that appeared complex at the time. They argued that remedies could be found with a reasonable mind; they were neither cheap nor to be found in the practices of charlatans or superstitious individuals. The opposite, however, was also true. The process of finding the Underworld's most complex cures using rituals revealed the absurdity of what some physicians considered superstitious medical practices.

This chapter has demonstrated how the theoretical constructs of naturalists and physicians concerning natural history, and specifically plants, developed over the period of the 16th and 17th centuries as a way to confront superstition. Whilst using the

---

tomar y arrancar, pero moria el que asi la arrancaua, salue et lleuaua otra de las mismas rayzes consigo. Y que para lo poder hazer seguramento, vista la rayz, y rociándola con lo dicho, cauauan y abrian la tierra al derredor, y echauan vn lazo con un cordel rezio a la rayz, y en lo que sobraua del cordel, atauan fuertemente un perro, el qual viendofe atado, tiraua tan rezio que arrancaua la rayz, y moria luego el perro. Después podia la tomar para ti qualquiera hombre seguramente, y usar y gozar de los prouechos della. Los autores de lo dicho fon los alegados en el capítulo passado, y otros muchos" [italics in the text] [translation is mine].
96 Holzman 1998: 241–249.
97 Bayle and Gauvin 2019: 177–178.

narrative structure and form of myth and folklore, they were subtly abandoning the mythological meanings of the medical and botanical fictions they were reusing, and instead increasingly relied upon a more Christian understanding of nature.

In deciphering this intricate labyrinth of references and symbols in order to comprehend the cultural dimension of the subterranean parts of plants, we first explored the nature of the symbolism surrounding the underground and Underworld in order to show that these conceptions were deployed for theoretical purposes in the philosophical and alchemical treatises, books of secrets or emblems explaining natural phenomena in which they appeared. We have seen that the symbolism of the underground was inspired by the *scala naturae* as well as passages from the Bible. The mandrake and its bonds to the subterranean implied a whole network of images and visual metaphors of the secrets of the Underworld: Grottos, caves, the alembic, the womb, and the inferno. Naturalists and physicians continued to use the ancient etymology of plant names in composing their natural histories. Through this etymology the authors recalled myths from classical Greece. In the practice of gathering plants, and in particular roots, rituals continued to take place, inspired by these myths and the new magical thought of the Renaissance, which in turn represented a cultural inheritance from the medieval era. All these metaphors were then reused through analogies that compared the human body to broader concepts of nature, specifically medical analogies (for example the womb or uterine cave), in writings termed "medical fictions". Girolamo Fracastoro made one of the most enlightening demonstrations of this in his epic poem *Syphilis, sive Morbus gallicus*. His understanding of this new disease was accompanied by an explanation of the mythological and subterranean origins of its cure[98].

Finally, another striking aspect is the presence of dogs in the magical rituals surrounding the harvesting of roots. The association between dogs and the Underworld did not seem obvious at first, but going back to the fundamental myths from which early modern doctors and naturalists were taking inspiration, we can see the various connections between the dog as a symbol and the Underworld plant. Plants which were believed to have originated in the Underworld, as described by contemporary naturalists, were the most powerful cures known at the time. The poppy, the mandrake and the deadly nightshade were depicted with such narrative *topoï* for the reasons discussed in this chapter: Firstly, in order to explain the dangers of a new and unknown disease; secondly, to emphasize the need for a powerful remedy against severe diseases, which could be extracted from both poisonous earth and plants; and thirdly, to foster devotion, faith, and morality, in the context of a Christian society, in matters specifically related to sexuality, fertility, melancholy, and madness. Magical enchantments were used as a pretense in order to unfold the mysteries and secrets of

---

[98] For more on medical fictions and syphilis, see Healy 2001.

nature. Botany was influenced by such theories, or as Dominique Brancher has designated it, *"botanique fantastique"*[99].

This transformation of thought was initially permitted and produced because of the idea that the strong therapeutic power of certain plants could draw its origin only from mythological accounts. It is hoped that this cultural study of the mandrake might be repeated for the roots of other plants, with further attention given to the ethos of the early modern naturalists.

---

[99] Brancher 2015: 69.

Aleida Offerhaus, Anastasia Stefanaki, and Tinde van Andel

Not just a Garden of Simples: Arranging the Growing Floristic Diversity in the Leiden Botanical Garden (1594–1740)

Introduction

Modern botanical gardens harbour collections that are usually a mixture of exotic and indigenous plants, frequently linked to the history of the garden. Their purpose is to awe or please their visitors, to stimulate curiosity, raise awareness about the human impact on the environment and sometimes to actively preserve species that are threatened in their natural habitats[1]. The earliest botanical gardens, however, started as places where prospective physicians learned to recognise medicinal plants. In this chapter we examine several herbaria collected in, or related to, the Leiden *Hortus botanicus*, from its origins to the mid-eighteenth century, showing how the collection, cultivation and identification of plant species intertwined with the education of physicians and pharmacists.

The First Medical Gardens in Europe

The earliest evidence of gardens comes from the Middle Ages and is rather meagre. Knowledge about plants was laid down in a classical tradition that had barely changed since the first century herbal of Dioscorides (*fl.* 1st cent. A.D.)[2] and in the relevant part of *Naturalis historia* by Pliny (A.D. 23/24–79)[3]. Around 800 A.D. the emperor Charlemagne (768–814) issued a policy document with which he tried to regulate the administration and organization of the imperial estates, covering most of West and Central Europe. His 'Capitulare de villis vel curtis imperii' (Chapter on manors and courts of the empire) listed 89 plant species whose cultivation was recommended[4]. All of these plants were useful, either medicinal, edible or used for dyeing or perfuming. In the monasteries of that period this list was most likely also adhered to[5]. While the Benedictine rule required the monks to look after the sick—which they could easily do in their well-organized communities—in the course of the twelfth century this practice led to criticism

---

[1] Johnson 2007: 76–80.
[2] Dioscorides, *De materia medica*.
[3] Pliny, *Naturalis historia*: Books 20–27.
[4] Herzog August Bibliothek, Cod. Guelf. 254/1 Helmst.: (ff. 12v–16) Captulare Caroli Magni de villis vel curtis imperii.
[5] Chavannes-Mazel and Van Uffelen 2016: 118–119.

and ultimately to a church ban on the practice of medicine by clerics[6]. From then on, the emerging universities would take on the education of physicians and the practice of medicine and pharmacy would be pursued in civilian society[7].

The birth of the modern botanical garden took place in Renaissance Italy. Ideally positioned in the Mediterranean, Italian culture was influenced by the vibrant culture of Islamic Spain, where the study of medicine was on a high level[8] and an extensive trade existed with the Eastern Mediterranean[9]. A treatise on herbs written in thirteenth-century Salerno, south of Naples, already showed a deviation from the classical description and depiction of plants. The illustrations of plants in an early manuscript of this *Tractatus de Herbis*[10] were unadorned studies, with occasional pictures to clarify the origin or the use of medicinal herbs described (see Figure 1A),

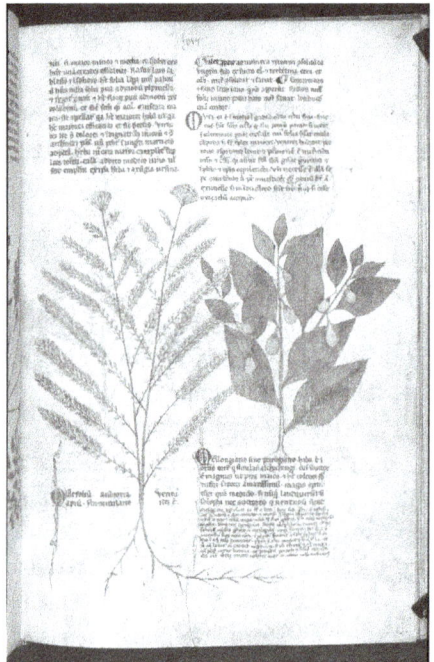 

Figure 1A: (on the left). *Yarrow* (*Achillea millefolium* L.), in Bartholomeus Mini de Senis, *Tractatus de Herbis*. London British Library, Egerton MS 747: f. 66 recto.

Figure 1B: (on the right). *Yarrow* (*Achillea millefolium* L.), in [Apuleius], *Herbarium*. Oxford, Bodleian Library, MS. Ashmole 1431: f. 23 verso.

---

6 Chavannes-Mazel 2016: 129.
7 Beukers and Bierman 2013: 7.
8 Pavord 2005: 107.
9 Pavord 2005: 212.
10 British Library, Egerton MS 747.

but without the basilisks, dragons, male and female mandragoras and inimitable representations of plants contemporary herbals were illustrated with (see Figure 1B)[11]. The identification of yarrow (*Achillea millefolium* L.) using a medieval herbal like the 11th-century manuscript of the herbal of Pseudo-Apuleius[12] (see Figure 1B) must have been quite problematic.

In order to become a physician, one had to know the 'simples' used during medical practice. The term simples derives from the Latin *simplicia* and refers to herbal, mineral or animal substances that are thought to have inherent healing properties and form the basis of a medicine, often complex recipes compiled from many separate elements[13]. Simples therefore do not only refer to plants, but also to stones, bones, skin, nails or other substances taken from nature. Here we use the term simples to refer to medicinal plant species. Educated in a culture in which observation played an important part, the physicians in Salerno were renowned and sought after throughout the Mediterranean and the rest of Europe for their skills. Their practice was facilitated by the presence of a medical garden, where exotic plants were cultivated, but the school, founded by four physicians from different cultures, primarily flourished due to its practical approach, focusing on the prevention of diseases and prescribing medicines, the efficacy of at least some of which is today pharmacologically proven[14].

What started in Salerno continued in the Northern part of Italy. Pisa was the first city where a university garden was established with plants that were used to instruct students of medicine. In 1543, Grand Duke Cosimo I de' Medici (1519–1574) commissioned Luca Ghini (1490–1556) to set up a medical garden and teach botany. Ghini, professor of botany at the University of Pisa, primarily taught his students to carefully observe plants when trying to connect the descriptions of Theophrastus (ca. 371-ca. 287 B.C.), Dioscorides and Pliny with specimens of living plants. Ghini was a committed teacher, who took his students into the garden to demonstrate what he had just described. Ghini's excursions in Tuscany resulted in a medical garden full of hitherto unknown plants. By now the interest in plants had made a decisive turn from medicinal plants based on classical descriptions to the descriptions of actual, living plants[15]. As Anna Pavord puts it, "a wider imperative emerged: the desire to make sense of the natural world"[16], facilitated by the growth of trade and the subsequent discovery of new territories[17]. Living and dried plant specimens turned out to be far better material for discussions and comparison than the revered depictions, drawn on paper or

---

11 Bodleian Libraries, MS. Ashmole 1431.
12 Collins 2000: 254.
13 De Vos 2010: 28.
14 Bifulco et al. 2020: 871–873.
15 Findlen 2017: 133.
16 Pavord 2005: 16.
17 Pavord 2005: 160.

printed from a woodcut. Paper, which was manufactured and used increasingly in the fifteenth century, turned out to be an extremely convenient and accessible way to press, dry and store the specimens of the collected plants. Simultaneously with this new-found desire to make sense of the natural plant world, a new way to preserve plants was discovered: the herbarium[18]. One of the oldest surviving botanical collections, the En Tibi Herbarium, was made around 1558 in Bologna[19]. This collection, compiled by Ghini's pupil Francesco Petrollini (*fl.* 16th cent.), possibly for the Habsburg emperor[20], is more than just an attempt to reconstruct the knowledge of antiquity[21]. Several plant species preserved in this herbarium had not yet been described previously and an effort was made to group the specimens according to habitat or medicinal properties[22].

The Leiden *Hortus Medicus*

The need to divert trade routes due to the rise of the Ottoman Empire and the subsequent establishment of the sea route to the East round the Cape of Good Hope, first by the Portuguese and then by the Dutch, led to the discovery of new territories with an unknown and quite often astonishing flora[23]. The expanding influence of the Dutch East and West India Companies (VOC and WIC respectively) resulted in trading posts in regions nowadays known as Indonesia, Malaysia, South Africa, Japan, Yemen, Iran, India, Surinam and China. Together with silk, plantation goods and spices, there was a steady flow of exotic plants from these places to the Dutch republic. The growing scientific interest in plants other than medicinal—which started in Pisa—was strengthened by this influx of new species. These novel exotic plants triggered the curiosity of both collectors of *naturalia*, physicians and botanists[24]. It ultimately led to a more fundamental study of plants, not as sources of medicine, but as plants for their own sake, followed by the desire to give them their rightful place in the existing botanical order. These attempts to categorise exotics brought the shortcomings of the existing classification systems to the fore. As a result, new, more comprehensive botanical systems had to be devised. Many exotic plants simply did not fit into the existing order.

An attempt to consolidate a political union between two of the Dutch states led to the foundation of a university in Leiden in 1575[25]. Twelve years later, in 1587, the

---

18 Findlen 2017: 135.
19 Stefanaki et al. 2019: 13.
20 Stefanaki et al. 2019: 13.
21 Stefanaki et al. 2018: 404–405.
22 Stefanaki et al. 2018: 400.
23 Van Reede, *Hortus Malabaricus*.
24 Cook 2007: 317–325.
25 Israel 1995/1996: 1.218.

curators of the university decided to ask the mayors' permission to transform the vacant site behind the university into a garden, "dienende tot leering van aller de ghene, die in der medecijnen studeren" (serving to educate all those who study medicine)[26]. It took another seven years, before the garden was finally laid out and planted[27].

The pharmacist Dirck Outgaerts Cluyt (1546–1598) was responsible for setting up the Leiden garden. He was explicitly allowed to make a profit from leftover plant material in the garden, but there is nothing to suggest that the garden officially served as a pharmacists' garden. The Leiden *Hortus botanicus* was not a medical garden in the sense that its sole purpose was to provide physicians and pharmacists with herbal simples. Already the first plant list, drawn up by Cluyt in 1595 contained more than just simples[28]. The medicinal value of peony—a plant that was recorded multiple times in this list—did most certainly not depend on its having double, red or yellow flowers (or any, come to think of it). That said, most herbal simples were found in the garden, but not all plants in the garden were simples.

Those who studied medicine at Leiden University in the first three centuries of its existence were supposed to attend the botanical demonstrations, given by the professor of Botany in the Leiden garden, just like the pupils of the Renaissance teacher Ghini. These demonstrations were primarily focused on the teaching and recognition of herbal simples, as the garden was set up to educate those who studied medicine and were to become physicians. Pharmacists, physicians and surgeons were often members of the same guild, but their activities were strictly regulated, to the extent that physicians, despite having studied botany, were prohibited to make their own medicine, not even from herbal simples. Likewise, pharmacists were forbidden to alter the recipes given to them—via the patient—by the physician. Moreover, pharmacists had to fulfil an apprenticeship and physicians needed to complete an academic study[29].

The Naming of Plants

From its creation in 1594, the names of the plants in the garden were meticulously recorded and regularly published, either in handwritten or printed form[30]. From these lists, we can gather information about the names of the plants that were cultivated in the garden, but the exact plant species to which the name refers is often unclear. More than one name was often given to one species, sometimes due to the

---

26 Molhuysen 1913: 1:140–141.
27 Veendorp and Baas Becking 1938/1990: 25.
28 Molhuysen 1913: 3.317–334.
29 Wijnands et al. 1994: 9.
30 Veendorp and Baas Becking 1938/1990: 186–187.

existence of different forms or varieties, like the above-mentioned peonies with red, yellow or double petals that were listed as separate species[31].

Garden catalogues, plant lists and pharmacopoeas usually provided quite long names that leave the modern reader guessing what species exactly is referred to. Giving a species two (Latin) names, a generic and a specific one, would ultimately turn out to be the most convenient way to classify species and communicate about them. The Swedish botanist Carolus Linnaeus (1707–1778) is credited with the invention of this binomial nomenclature system, and it is certainly through his endeavour and enthusiasm that it was accepted in a relatively short space of time by most botanists. In doing so, he turned the science of botany from a jungle into a well-ordered garden, where every plant had its place and could be found. Based on pre-Linnean names only, it is still difficult to know exactly which species are meant in historic garden catalogues.

Some effort has been made to connect pre-Linnean names with current ones[32], but even then it is hard to be conclusive about the relation between a particular plant species and its pre-Linnean name. Therefore, information from pre-Linnean plant lists, catalogues and pharmacopoeas can only be thoroughly analysed if dried plant specimens (or true-to-life depictions) are available that are linked to these names. Contemporary herbaria help us in establishing the presence of a plant species. Without the tangible evidence of dried plant specimens, statements about the presence of plant species at a certain time and place are ambiguous at the very least.

Materials and Method

The Study of Historic Herbaria

The precarious condition of timeworn vouchers often impeded extensive research of historic herbaria. If they were studied, it was mostly done by taxonomists looking for early collections or type specimens, to which the first description of a plant species could be connected[33]. The specimens that served as type specimens were often prioritised for identification and digitization, such as those of the van Royen Herbarium at Naturalis, while the non-type specimens were photographed many years later. The techniques developed to digitize documents and preserve them in an accessible way have revolutionised research on fragile historic herbaria. Finally, they can be studied in their entirety without the risk of damage[34].

---

31 Species were then understood as part of a more narrowly defined genus.
32 Jarvis 2007; Harris 2018.
33 Wijnands and Heniger 1991: 129.
34 Van Andel 2017: 7.

Based on our study of several early 18th-century herbaria related to the Leiden garden, we show how the garden flourished and expanded due to the increased supply of exotic plants through foreign trade[35]. We also show that this growing botanical diversity needed a new classification system. By identifying the plant species in several early eighteenth-century plant collections, we argue that the Leiden garden shifted further and further away from a garden of simples. We explain how the supply for medicinal plants for pharmacies was taken care of by other gardens. We show how non-medicinal plants were valued for their rare appearance and exotic origin and how this resulted in new classification systems, decades before Linnaeus proposed the binomial nomenclature[36] and his *Systema Naturae*, classifying plants based on their reproductive parts[37].

To say anything conclusive about historic herbaria it is not enough to leaf through them, mumble something about indecipherable handwritings and remark on the disappointing lack of knowledge of the person(s) who contributed to the collection. The plant specimens, the paper, the bindings, the labels added over time, and the different handwritings all contribute to an authentic, if complex, picture of the object. For the purpose of this study, we used four early eighteenth-century herbaria with proven links to their gardens of origin: the Clifford Herbarium (ca. 1720), the Zierikzee herbarium and the six-volume d'Oignies herbarium (both from the first half of the 18th century), and the Boerhaave collection, consisting of 96 specimens in the van Royen Herbarium, presumably collected by Herman Boerhaave (1668–1738), the Leiden professor of Botany and garden prefect from 1709 to 1730.

For the Clifford herbarium kept in the Natural History Museum in London we used the identifications provided by Charlie Jarvis of its 3641 specimens[38]. For the 475 Clifford specimens kept in Naturalis Biodiversity Center in Leiden we used the identifications provided by Renske Ek[39]. For the historical and physical aspects of the specimens we consulted the articles written by Wijnands and Heniger[40] and Gerard Thijsse[41]. For the d'Oignies herbarium we used the preliminary study of Alice Bertin[42], that focused on the first two volumes, and for the Zierikzee herbarium we used the recent inventory published by the first author[43]. We identified the specimens in the other four volumes

---

35 Karstens and Kleibrink 1982: 33–34.
36 Linnaeus, *Species plantarum*.
37 Linnaeus, *Hortus Cliffortianus*.
38 Jarvis 2007.
39 Ek 2011.
40 Wijnands and Heniger 1991.
41 Thijsse 2018.
42 Bertin 2016.
43 Offerhaus 2020; Offerhaus et al. 2021.

using the website of Plants of the World Online[44] and literature on the flora of the Netherlands[45], the Mediterranean[46] and South Africa[47]. Herbarium specimens of the Naturalis collection (L) were used for complementary identification[48]. Additional research on the Zierikzee Herbarium regarding the paper, watermarks, labels and ornaments was recently published[49]. All Zierikzee specimens can be viewed online[50].

All four collections were accessible and studied in digital form, even though physical contact remained indispensable when it came to research of paper, watermarks, labels and botanical details, such as hairs, glands, veins, and flower details. In all four, we looked for the ratio between herbal simples and exotic plant species. We classified species as herbal simples when they were present in the compiled list of simples by De Vos[51], the medicinal catalogue of Commelin (1698)[52], contemporary pharmacopoeas from the area, such as the *"Pharmacopoea Hagana"*[53] and the catalogue of plants from a medical garden in The Hague[54].

During this research we also established the geographical origins of the exotics. To estimate the proportion of medicinal plants in the Clifford herbarium, a random sample of 200 specimens was selected from the online specimen list to establish how many of them were simples[55].

To define the sources of the plant names used in the herbaria, we compared them to names used in contemporary botanical literature, such as the authoritative *Pinax theatri botanici* (1623) by Caspar Bauhin (1560–1624)[56], the *Institutiones rei herbariae* (1700) by Joseph Pitton de Tournefort (1656–1708)[57], the Leiden garden catalogues compiled by the prefects Paul Hermann (1646–1695)[58] and those by Herman Boerhaave[59], and the catalogues of exotic plants from the Amsterdam *Hortus Medicus* by Jan Commelin (1629–1692)[60], in 1697–1701, posthumously edited by his nephew Caspar

---

44 *POWO* 2019.
45 Duistermaat 2020; Jäger 2017.
46 Thorogood 2016 and 2019; Blamey and Grey-Wilson 2008.
47 Manning 2008.
48 Naturalis Bioportal: https://bioportal.naturalis.nl.
49 Offerhaus et al. 2021.
50 https://www.stadhuismuseumzierikzee.nl/Herbarium/nl/Herbarium-dutch.html.
51 De Vos 2010: 31–39. Only those simples were used in the list by De Vos that were identified to species level.
52 Commelin, *Plantarum usualium*.
53 *Pharmacopoea Hagana*.
54 Schwencke, *Officinalium plantarum Catalogus*.
55 Jarvis 2016.
56 Bauhin, *PINAX*.
57 Tournefort, *Institutiones rei herbariæ*.
58 Hermann, *Horti academici*.
59 Boerhaave, *Index plantarum*; Boerhaave, *Index alter plantarum*.
60 Commelin, *Horti Medici Amstelodamensis*.

Commelin (1667/1668–1731)[61]. These four historic herbaria helped us to form a picture of the Leiden garden and its living collection in the early eighteenth century.

Results: Collections, Herbaria, and Materia Medica

The Boerhaave Collection

We identified specimens as Boerhaave specimens when they were accompanied by a label in his handwriting, when it was explicitly stated by a later author that this particular specimen came from Boerhaave's collection, herbarium or even mind (*ex mente*) or when the text on the label directly referred to the seed register maintained by Boerhaave. On 25 of the labels concerned a year was indicated, ranging from 1712 to 1721. We assumed that these were collected by Boerhaave himself or one of his gardeners. The labels were originally unattached tags, kept with the herbarium specimens. This we deduced from the fact that some labels were glued on top of the labels of later prefects, which of necessity must have been done at a later stage. The specimens belong to 76 different species distributed over 19 families. Most of these species originate from the Mediterranean (36), 34 species have a wide geographical distribution (Temperate Eurasia), two are from the Americas, two from the Indian subcontinent, one from South Africa, and one is a green alga with a global distribution (*POWO* 2021). Between 1718, the deadline for the second edition of Boerhaave's catalogue[62], and 1740, the publication of the catalogue of his successor Adriaan van Royen (1704–1779)[63], no catalogue was composed. Until 1727, often rudimentary information can still be gathered from Boerhaave's seed register, the *"Index Seminum Satorum"*[64], but from 1728 till 1740, when van Royen published his catalogue, we have limited knowledge of what entered the garden. 50 specimens were decorated with vases and ribbons, 38 were attached to the herbarium sheet with paper strips[65].

The Clifford Herbarium

The Clifford Herbarium was primarily compiled from specimens in the extensive gardens and greenhouses that were once part of the manor De Hartekamp near Heemstede, which belonged to the Clifford family. George Clifford III (1685–1760) was a

---

61 Commelin, *Rariores et Exoticae*.
62 Boerhaave, *Index alter plantarum*.
63 van Royen, *Florae Leydensis Prodromus*.
64 Leiden, Universiteit Bibliotheek, BPL 3654.
65 Offerhaus et al. 2022.

wealthy banker, but he had much interest in botany and specifically in exotic plants. He had close contacts with both Boerhaave and Linnaeus, who wrote the authoritative *Hortus Cliffortianus* (1737) based on the Clifford herbarium and the living collection from De Hartekamp. The Clifford Herbarium contains 4116 specimens but up till now it was particularly famous for its 546 Linnean type specimens. The herbarium was originally grouped according to the botanical system of Boerhaave, which was outlined in the first edition of Boerhaave's garden catalogue (1710) and applied to the Leiden garden. The author of the labels on the Clifford specimens started using the first catalogue and afterwards switched to the second, published in 1720, as shown by the corrections he made on some of the labels. Species were exchanged with the Leiden garden[66]: on 11th March 1715, Boerhaave confirmed that he sent Clifford seeds of cucurbits: *"misi ad d. Clifford semina plantarum Cucumeriferum varias species"*[67]. The Clifford herbarium contains predominantly exotic plants, cultivated in his garden and partly acquired via Boerhaave and his many contacts abroad. Browsing through his herbarium, however, the indigenous, the medicinal and the exotic species rise proudly and indiscriminately from their vases, side by side.

The d'Oignies Herbarium

The d'Oignies Herbarium was acquired by the Dutch National Library in 1818 and transferred to the collection of the National Herbarium in Leiden in 1869. This collection originally consisted of stacks of loose herbarium leaves with dried plant specimens, from which later six books were assembled. The specimens are carefully mounted and decorated with vases, bows and ribbons. There are 766 specimens, containing 663 species, of which 250 can be classified as simples. A total of 305 species originates from the Mediterranean, South Africa, tropical and temperate Asia, the Americas, East Europe or mountainous parts of Europe. The remaining 108 species are indigenous species with no known medicinal properties.

Simon Joseph d'Oignies, after whom the herbarium is named, was born in Leuven in 1738 in the Austrian Netherlands and joined the army at the age of 17. He had no verifiable education when in 1780 he became a surgeon in the battalion of General De Murray (1718–1802), commander-in-chief of the Austrian army. Shortly before or in 1780 he acquired the herbarium. The names on the pages and in the separate register, written by d'Oignies himself, were copied from unattached labels with plant names on them, that came with the herbarium, as is shown by the mix-up of the names of several unrelated specimens in different books. D'Oignies was not a trained botanist, because he frequently misspelled names (e.g., Aenauster for Oenanthe) and attributed

---

66 Wijnands and Heniger 1991: 130.
67 Leiden, Universiteit Bibliotheek, BPL 3654.

incorrect names to several plants. From those names, copied by d'Oignies, half are either incorrect identifications or only made on genus level. A total of 89 explicit references to the garden catalogue by van Royen[68] are written next to the specimens, usually followed by the abbreviation "v:v:R:" (vide van Royen) and a number corresponding with the number in the catalogue, e.g., Knautia 1:v:v:R:, referring to the first species of the genus Knautia described in the catalogue. 30% of those plant species were not identified correctly. Some of the often-garbled names written on the herbarium sheets, could be retraced to a variety of botanical authors, from Rembert Dodoens (1517–1585), Matthias de L'Obel (or Mathias de Lobel, 1538–1616), Carolus Clusius (1526–1609), Basilius Besler (1561–1629), Hermann, Boerhaave and Linnaeus. Some exotic, non-medicinal plants were given names of well-known exotic simples, often tropical species, that were not cultivated commercially in Europe, e.g., "Sarsaparilla" for *Smilax aspera* L., "Acmella" for *Bidens pilosa* L. and "Ipecaquana" for *Gillenia trifoliata* (L.) Moench. Finally, 80 labels are present with often unique, quite long plant names that suggest that the author described the plant in his own words with Bauhin's *Pinax* as a starting point, only partly succeeding in identifying the plants. Judging by the alphabetic order interspersed with Linnean names, the herbarium was rearranged at a later date, bound and provided with covers.

The Zierikzee Herbarium

The Zierikzee herbarium is virtually identical in appearance to the d'Oignies Herbarium. The dried plant specimens are mounted in a similar way and the same decorations are used[69]. Both herbaria contain duplicate species, but within a particular plant genus they also complement each other. This loose-leafed collection contains 348 specimens belonging to 311 species. The majority (235 species) represents simples, while 76 non-medicinal species originate from the Mediterranean, South Africa, Tropical Asia, the Americas and Eastern and mountainous parts of Europe. It contains three types of labels, written by different hands in different time periods. The oldest labels reveal pre-Linnean plant names, which show a high overlap with the plant descriptions in Boerhaave's catalogues (1710, 1720), but also have a distinct, as yet unspecified relation with a pharmacopoea from The Hague[70]. This particular publication was the first manual to list the herbal simples with extensive references, using the works of amongst others Bauhin and his brother Johann (1541–1613), Dodoens, Robert Morison (1620–1683) and Joseph Pitton de Tournefort (1656–1708).

---

68 van Royen, *Florae Leydensis Prodromus*.
69 Offerhaus et al. 2021: 4.
70 *Pharmacopoea Hagana*.

Ratio between Simples and Exotics

The ratio between simples and exotics in the analysed herbaria is as follows: of the identified Boerhaave species 80% is exotic and 20% simple of the Clifford Herbarium species 80% is exotic and 20% is simple, of the d'Oignies species 65% is exotic and 35% is simple. The Zierikzee Herbarium reveals the opposite ratio, 68% simples vs. 32 % exotics. Those exotic plant species came from all over the then known world. Boerhaave, as prefects before him, held up a steady correspondence with botanists from Venice, Montpellier, Florence, Chelsea, Paris, Surinam and many other places, people with whom he exchanged seeds and cuttings[71]. Clifford also had many connections due to his prominent position in Dutch society at the time[72]. In his catalogues, Boerhaave indicated the medicinal plants by underscoring their numbers, like the medicinal *Gentiana major* L. (**1**. Gentiana; major; lutea) vs. the non-medicinal *Gentiana acaulis* L. (4. Gentiana; Alpina; flore magno.)[73]. Both species are present in the analysed herbaria. In between the publication of the first (1710) and the second (1720) edition of Boerhaave's catalogue the percentage of plants in general grew (+37%), but that of medicinal plants declined (-20%: from 511 to 405).

Paper

The paper sheets of the Zierikzee and the d'Oignies herbaria have the same anomalies, characteristic double chain lines and numbering on the lower right side in the same handwriting.

The paper used in the Clifford Herbarium is from a different manufacturer than the paper in the Zierikzee and the d'Oignies herbaria. Labels of a similar design are found in both the Clifford and the Zierikzee herbaria: an ornamental frame adorned with naturalistic elements such as leaves and shells, a style popular at the end of the 17th century and the beginning of the 18th century[74].

Decorations

The common feature in these four early 18th-century plant collections is the use of vases and ribbons to embellish the plant specimens (Figure 2). Half of the decorative elements in the Clifford herbarium, five of the 12 vases used in the d'Oignies herbarium and all those present on Boerhaave's own specimens have been designed by two Leiden artists:

---

[71] Leiden, Universiteit Bibliotheek, BPL 3654; Baas Becking and Veendorp: 106–111.
[72] Wijnands and Heniger 1991: 140.
[73] Boerhaave, *Index alter plantarum*: 1.205.
[74] Offerhaus et al. 2021: 5.

the painter Hieronymus van der Mij (1687–1761) and the print-artist Johannes van der Spijk, active in Leiden from 1716 till 1761[75]. Van der Mij set up business in 1710, and Van der Spijk started out in 1716. Evidence of their cooperation lies in the archive of Naturalis, where several printed pages with decorative vases, a label and several ribbons are kept[76]. At the bottom it reads, "*H:v:d:Mij invent.*" and "*J:v:d:Spyk fecit*" (designed by H. van der Mij and executed by J. van der Spyk). The start of their cooperation could have been as early as 1716. This is supported by Boerhaave specimens, dating from the period 1712–1721, which were decorated with their vases and ribbons. The use of these decorations supports the connection of these herbaria with Leiden.

The remaining seven vases in d'Oignies are identical to the ones in the Zierikzee herbarium, but it is not known by whom these decorations were designed and engraved. The specimens in all four collections have been mounted with the utmost care, so that botanical details stand out (Figure 2). The way the plants are displayed shows that the aim was not only to reveal their beauty, but also their identity. Based on the paper, the numbering and the decorations we believe that the Zierikzee and the d'Oignies herbaria once formed a single herbarium collection, produced in the Leiden botanical garden under Boerhaave's management[77].

Sale of the Herbaria

Following the death of Boerhaave, his library was auctioned off in July 1739. Number 523 in the section *Manuscripta et Icones* of the auction catalogue listed a herbarium in six volumes or stacks (*voll.*)[78]. The auctioneers' catalogue reveals the price for which it was sold (18 guilders), but it is not known who bought it. Following the death of Boerhaave's head gardener, Jakob Ligtvoet (1685–1752), a herbarium in 13 parts was sold with a description that could apply to both the Zierikzee and the D'Oignies Herbarium[79]. This supports our claim that both herbaria formed a single herbarium, originating in the Leiden garden, with the larger herbarium with more exotics held by Boerhaave and the smaller herbarium with more simples held by the head gardener. Their origin however, is still a matter of debate. According to Thijsse and Wesseling, evidence for the existence of the herbaria of Boerhaave is found in a report by a travelling book collector (Johan Meerman, 1753–1815) at the end of the 18th century, who mentions having seen herbaria in the Hermitage in Saint Petersburg he simply describes as consisting of several parts and having "eenige geschreeven aantekeningen van onzen onsterflijken Landsman"

---

75 Waller 1938: 310.
76 Naturalis Archive; Thijsse 2018: 142.
77 Offerhaus et al. 2021.
78 Luchtmans, *Bibliotheca Boerhaaviana*: 27.
79 Haak, *Catalogus*: 19.

("some written annotations by our immortal compatriot"). Born 15 years after Boerhaave died it is difficult to imagine how he was able to recognise his handwriting[80].

The Leiden Garden under Boerhaave

The *commercium botanicum*, the correspondence botanists undertook in order to exchange plants and seeds, was fuelled by the foreign trade that led to a growing supply of exotic plants from all over the world. Trade, correspondence and subsequent exchange led to a spectacular growth of the garden collection in a relatively short space of time[81]. In 1594, the year that the garden was created, Cluyt compiled a list with over 1000 plant names[82], of which only a third was medicinal and the others non-medicinal, both indigenous and exotic[83]. Some 125 years later, in 1720, Boerhaave's catalogue contained 5846 plant names. With the surface area of the garden scarcely doubled in a century, the number of species seems to have increased sixfold: a huge amount for an area smaller than a modern-day football field, even though it is likely that a considerable part of the collection was cultivated in the garden of Boerhaave's manor, 'Oud-Poelgeest'. This would explain the substantial drop in the number of species after Boerhaave's death, as shown in the garden catalogue published by his successor Adriaan van Royen (1704–1779)[84].

Boerhaave studied medicine on his own and never attended the lectures of the professor of botany and garden prefect Hermann. He graduated in Harderwijk, because it was the cheaper alternative, but also because he had no goodwill left for having failed to attend any lectures during his study in Leiden[85]. Boerhaave had to write a new catalogue for the garden, since the one written by his predecessor Petrus Hotton (1648–1709) was denied him by Hotton's heirs, who did not want Boerhaave as his successor.

Although the botanical system that Boerhaave devised was applied in the Utrecht botanical garden, its diffusion was limited. However, it was one of the first botanical systems that was based on extensive and original descriptions of the morphology of flowers, fruits and seeds[86]. Paul Hermann, prefect of the Leiden garden from 1680 till his death, also devised a botanical system that accommodated the many exotic plants that had entered the garden during his prefecture[87]. He used different characteristics, primarily focusing on the size, number and form of the seeds, but also on the form of

---

80 Thijsse and Wesseling 2021: 178–179.
81 Karstens and Kleibrink 1982: 33–34.
82 Molhuysen 1913: 1.317–334.
83 Tjon Sie Fat 1991: 7.
84 Karstens and Kleibrink 1982: 34; van Royen, *Florae Leydensis Prodromus*.
85 Lindeboom 1979: 45.
86 Sprague 1939: 93.
87 Hermann, *Horti academici*.

the flower, but the descriptions Boerhaave gave of the plant genera were far more elaborate and precise than those of Hermann[88].

From the botanical gardens in Padua and Montpellier we know that numbers were written on stones to ensure the plants could be located, even when dormant[89]. From Albrecht Haller (1708–1777), a young Swiss student of Boerhaave, we know that the garden was in fact arranged according to the order of Boerhaave's system: "Das schönste daran ist, daß alles nach einer künstlichen Ordnung eingerichtet ist, und ein Fremder mit dem Boerhaavischen Verzeichnüß in der Hand alle Pflanzen an ihrer Ordnung und den bezeichneten Stäben erkennen kann" (The best thing is that the whole is arranged in an artificial manner so that even a stranger—with the register of Boerhaave at hand—can recognise the different plants by their location and the inscribed markers)[90].

Canary Sage

One of the exotic species that appears in three of the four analysed herbaria is *Salvia canariensis* L., a sage species endemic to the Canary Islands (see Figure 2). Leonard Plukenet (1642–1706), botanist in service of Queen Mary II, describes the species as coming from the island of La Gomera[91]. It is a whitish, woolly, aromatic shrub with large pink-purplish flowers, that can grow up to 2.5 m height and 1.5 m width[92]. In the seed register of 1718, we find a reference to this plant, the seed of which Boerhaave received from his correspondent Francesco Cornaro (1670–1734), a Venetian nobleman and former ambassador to England[93]. A matching description is found in the catalogue of 1720[94]. It must have found its way soon afterwards to the garden of De Hartekamp where it was cultivated, dried and preserved in the Clifford herbarium (see Figure 2A). Both the Clifford and Boerhaave specimens are decorated with an identical type of vase (Figure 2A and 2B). The d'Oignies herbarium has a specimen with a different vase, but mounted with the same attention to detail (see Figure 2C).

There are several medicinal species of the genus "Salvia" present in the analysed herbaria (*S. officinalis* L., *S. rosmarinus* L., *S. sclarea* L. and *S. viridis* L.), but "Salvia canariensis" was included because it was rare and exotic, not because it was known for its medicinal properties. Plukenet introduced the plant in England, since he is the only one who mentions a specific place of origin—the isle of La Gomera—in his description. He listed the species under the genus "Horminum" in his *Almagestum botanicum* (1696), as

---

88 Sprague 1939: 93.
89 Grämiger 2016: 245.
90 Lindeboom 1979: 45.
91 Plukenet, *Almagestum botanicum*: 185.
92 Clebsch 2003: 57.
93 Leiden, Universiteit Bibliotheek, BPL 3654.
94 Boerhaave, *Index alter plantarum*: 1.164.

 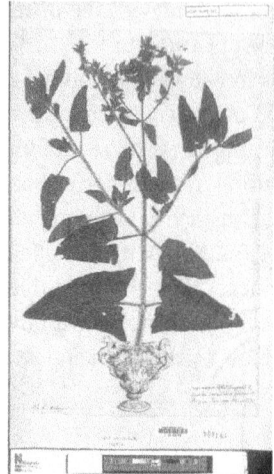 

Figure 2A: (left). *Specimen of Salvia canariensis* L., described as *'Sclarea'*, in *Clifford Herbarium*. London, BM000557604, now kept at London, Natural History Museum.
Figure 2B: (middle). *Specimen of Salvia canariensis* L., described as *'Horminum'*, in *Boerhaave specimens*. Leiden, Naturalis Biodiversity Center, L 0142243.
Figure 2C: (right). *Specimen of Salvia canariensis* L., described as *'Salvia'*, in *D'Oignies Herbarium*. Leiden, Naturalis Biodiversity Center, Book 6: f. 25.

Morison did in his *Plantarum Historiae*[95]. In Boerhaave's second edition of his catalogue, it was listed under the genus "Sclarea"[96], and it ended up in the genus 'Salvia' where Linnaeus, followed by van Royen[97], lumped together all plant species hitherto covered by the genera "Salvia", "Horminum" and "Sclarea"[98].

Lightning Bush

The presence of altogether four specimens of the South African species *Clutia pulchella* L., known as "Lightning bush", suggests the Leiden garden as the origin of the Zierikzee and d'Oignies herbaria (Figure 3A–D). Boerhaave named this man-sized bush after the son of the first supervisor of the Leiden garden, Auger Cluyt (1577/1578–1636), a competent botanist and physician[99]. It was cultivated in the Amsterdam Hortus Botanicus according to Jan Commelin[100], who reportedly received the seeds from Simon van der Stel (1639–1712),

---

95 Morison, *Plantarum Historiae*: 3.394.
96 Boerhaave, *Index alter plantarum*: 1.164.
97 van Royen, *Florae Leydensis Prodromus*: 308.
98 Linnaeus, *Species plantarum*: 1.23–27.
99 Boerhaave, *Index alter plantarum*.
100 Commelin, *Horti Medici Amstelodamensis*: 1.177–178.

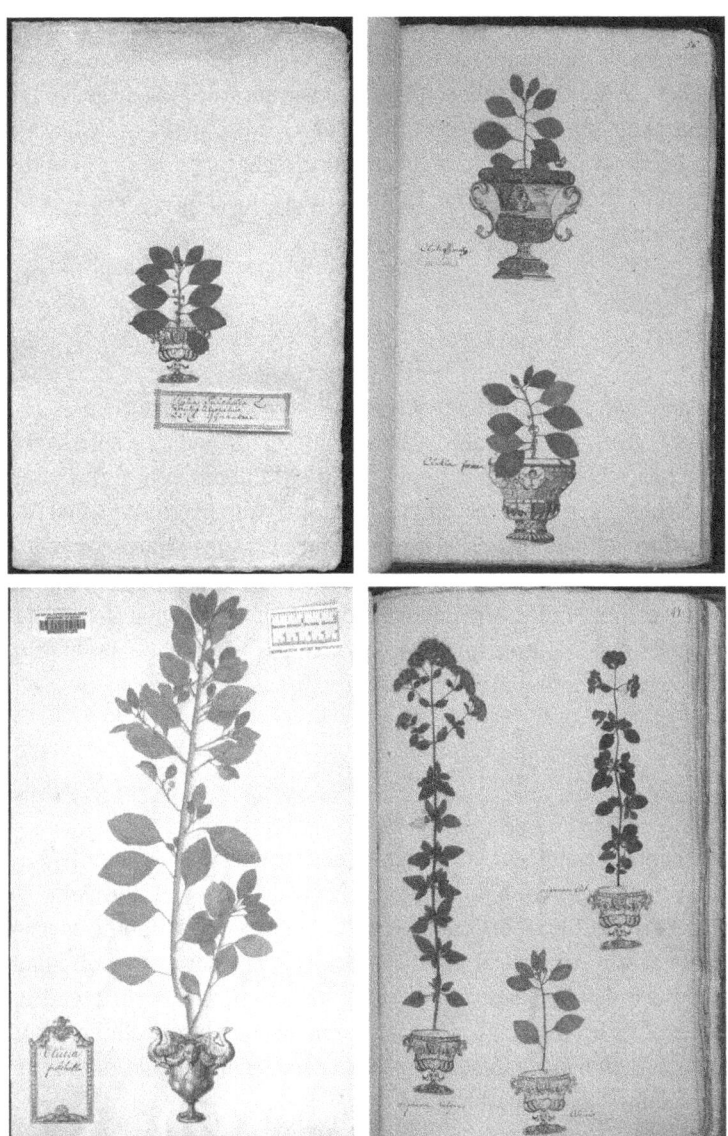

Figure 3A: (top left) *Specimen of Clutia pulchella* L., in *Zierikzee Herbarium*. Zierikzee, the Stadhuismuseum: No 327.
Figure 3B: (top right). *Specimen of Clutia pulchella* L., in *D'Oignies Herbarium*. Leiden, Naturalis Biodiversity Center, L.3961055 and L.3961056, Book 2: f. 56.
Figure 3C: (bottom left). *Specimen of Clutia pulchella* L., in *Clifford Herbarium*. London, British Museum, BM000647328, now kept at London, Natural History Museum.
Figure 3D: (bottom right). *Specimen of Clutia pulchella L*., in *D'Oignies Herbarium*. Leiden, Naturalis Biodiversity Centre, Book 5: f. 47.

governor of the Cape of Good Hope. It is possible that Boerhaave received the seeds or seedlings from Amsterdam, as the input of Amsterdam is prominent in his botanical correspondence[101]. Commelin writes he cannot confirm whether the species belonged to the genus Solanum, as Hermann, the then prefect of the Leiden garden, states, as he has not seen the flowers yet. Boerhaave was not sure either and thought that it belonged to the genus *Ricinus*, based on its three-locular fruit. In the nineteenth century, *Clutia* was included in the Peraceae family[102].

Hart's Tongue Fern

Hart's tongue fern (*Asplenium scolopendrium* L.) was used as a simple, but the presence of seven forms in the d'Oignies and Zierikzee herbaria, with corresponding – if garbled – descriptions from Boerhaave's catalogue[103], shows the interest in different forms of an indigenous plant, then considered as distinct species (Figure 4)[104]. In Boerhaave's first catalogue there are eight forms listed for this fern species, and 15 forms in his second edition. The Clifford herbarium contains 10 forms. Linnaeus[105] and van Royen[106] reduced these forms to one species, later called Asplenium scolopendrium[107] and our physical examination of the d'Oignies and Zierikzee specimens allowed us to draw the same conclusion, namely that the seven forms displayed and described belong to a single species.

Glory lily

There are dozens of plant specimens in the d'Oignies herbarium that are identified by names that could only have come from the catalogues of Boerhaave and Hermann.

One example is the glory lily, *Gloriosa superba* L., (Figure 5) a climbing species from the Colchicaceae family, a tropical plant species, that is ferociously poisonous and at the same time exquisitely beautiful, working its way upwards with tendrils coming from the apex of its leaves. Here it carries the name *Metonica malbarorum*, a corruption of the name *Methonica malabarorum*. This name is only found in the garden catalogue of Hermann[108] and of Boerhaave[109].

---

101 Leiden, Universiteit Bibliotheek, BPL 3654.
102 Klotzsch 1860.
103 Boerhaave, *Index plantarum*: 234; Boerhaave, *Index alter plantarum*: 1.23–24.
104 Boerhaave, *Index alter plantarum*: 1.24, 11: "Lingua Cervina; quae Phyllitis; major; ex uno pedunculo, quandoque bifolia".
105 Linnaeus, *Hortus Cliffortianus*: 474.
106 van Royen, *Florae Leydensis Prodromus*: 498.
107 Linnaeus, *Species plantarum*: 1.1079.
108 Hermann, *Horti academici Lugduno-Batavi*: 688.
109 Boerhaave, *Index alter plantarum*.

Figure 4: (on the left). *Asplenium scolopendium* L., in *D'Oignies Herbarium*. Leiden, Naturalis Biodiversity Centre, Book 4: f. 68 (accompanied by an inexpertly copied text of Boerhaave's catalogue [1720, I.24]: "Lingua Cervina qua pilltris Major: in uno pediculo quadroqua bifida").

Figure 5: (on the right). *Gloriosa superba* L., described as '*Metonica malbarorum*', in *D'Oignies Herbarium*. Leiden, Naturalis Biodiversity Center, Book 5: f. 16.

The Pharmacist's Training Tool: The Simplicia Cabinet

For pharmacists it was important to be able to recognise medicinal plants, but for the deliverance of the simples they needed in order to prepare a recipe, they often had to rely on other professionals, like growers, grocers, chemists and even brokers of simples[110]. The simples mentioned in the *Pharmacopoea Hagana* were only available '*in officinis nostris*' (in our shops)[111]. Many simples came from overseas and were delivered in a pulverized, dried or ground state that made identification and the assessment of their quality a challenge[112]. For this, pharmacists took recourse to cupboards filled with samples of simples to compare and determine the quality of their merchandise. Together with their on-the-job training

---

110 Wittop Koning 1954: 1–5.
111 *Pharmacopoea Hagana*.
112 Swart et al. 2019.

as pharmacists and the information from pharmacopoeas, these simplicia cabinets were the tools of their trade, providing physical samples of the materia medica listed in the pharmacopoeas, and much more useful than a herbarium with entire plants. Pharmacists learned to recognise living plants, either cultivated in the Netherlands or collected in the wild, but exotic simples were taught via simplicia cabinets[113].

The bark of Cinchona or the root of the tropical Smilax (either *Smilax china* L. or American species, generally known as 'Sarsaparilla') were considered important medicines in curing venereal diseases and combating fever[114]. Tropical Smilax was cultivated in greenhouses of botanical gardens[115], but as an interesting exotic and not commercially, for the benefit of pharmacists, who needed the root for their recipes.

A Pharmacists' Herbarium?

A pharmacists' herbarium would contain plant species that were present in the pharmacopoeas and cultivated in the Low Countries. Our research shows herbaria with an overwhelming presence of exotic, non-medicinal plant species. Combined with the absence of many plant species from contemporary pharmacopoeas in these herbaria, it becomes clear that for a pharmacist such a herbarium had no added value.

"True" Gardens of Simples

Although most surviving historical information concerns academic botanical gardens, "true" gardens of simples that provided pharmacies with medicinal plants are of no minor historical importance. Pharmacists often had their own gardens, ranging from a garden the size and quality of the early Leiden garden to relatively simple gardens at the outskirts of cities[116]. In the wake of the Dutch botanical gardens, guilds of physicians, pharmacists and surgeons founded medical gardens that came close in quality and content to their academic counterparts. In Haarlem a garden was founded in 1696 and abandoned ca. 150 years later.

Next to educating pharmacists, the garden was meant to provide its members with medicinal plants they were not able to find elsewhere. It also boasted a greenhouse, where orange and laurel trees were housed and melons were grown. Additionally, the gardener provided the members of the medical guild with attractive flowers[117]. In The

---

113 Van der Ham 2006: 28.
114 Winterbottom 2014: 25. On Cinchona bark, see Federica Rotelli in this volume.
115 Boerhaave, *Index alter plantarum*; van Royen, *Florae Leydensis Prodromus*: 228; Schwencke, *Officinalium plantarum catalogus*: 34.
116 Cohen 1927: 84.
117 Bitter 1914: 5, 23–24.

Hague, a house was bought in 1675 for the benefit of a medical garden, but no other information about this garden could be found[118]. In 1750, professor of botany Martinus Schwencke (1706–1785) was appointed for the surgeons' guild. He was responsible for relocating this garden of simples to the outskirts of The Hague[119]. Schwencke's catalogue listed 596 medicinal species, but he mentions that he did not list all the plants in the garden[120]. Because he did not do so, we have no way of knowing how many and which plant species grew in this medical garden. The auction catalogue of his possessions[121] at the time of his death in 1785 mentioned 58 bound volumes with altogether more than 5000 dried specimens, of which – to this day – the fate is unknown. This shows that herbaria were also made in true gardens of simples.

The immediate aim of these gardens of simples was to educate surgeons and pharmacists and provide them with the necessary simples, but exotic plants were also present. At the end of the 18th century, exotics could even be purchased from growers and their presence no longer depended on exchange[122].

Conclusion

A number of medical gardens in the beginning of the 18th century are recorded, but no real survey has been made so far, and little is known about their living collection. The way in which pharmacy gardens and commercial plant growers copied the academic gardens in their interest in and pursuit of exotic species warrants further research. Boerhaave's seed registers, the *Index Seminum Satorum*[123], are not yet digitized, but are key documents to understanding how he obtained and exchanged seeds within his extensive network, and when exactly the exotics that appeared in the Dutch garden herbaria entered the country. The separate notes written by Boerhaave and digitized by Erfgoed Leiden en Omstreken will complement the information from the *Index Seminum Satorum* and may provide an interesting insight in the way the garden was organised[124].

Cultivating exotic species would be inconceivable without the framework of a botanical garden with its skilled personnel, its greenhouses, its knowledge, its exchange of plants and preferably the dynamics of an affiliated educational institute.

Prefects and professors of botany were well-known figures, who appropriated the contribution of their gardeners. Gardeners were well-informed, capable men, who,

---

118 Krul 1891: 26.
119 Endtz 1972: 99–109.
120 Schwencke, *Officinalium plantarum Catalogus*, IV.
121 Plaat and Scheurleer, *Bibliotheca*: 110–113.
122 Kuijlen et al. 1983: 63.
123 Leiden, Universiteit Bibliotheek, BPL 3654.
124 Erfgoed Leiden en Omstreken, Boerhaave, Herman, NL-LdnRAL-1658/I Herman Boerhaave/I.B.5,27.

while not academically educated, played a vital role in the upkeep of the gardens under their supervision. When in 1739 Dirck Nietzel (1703–1756), Clifford's head gardener, departed for Uppsala with Linnaeus, he wrote a farewell letter in fluent Latin. His departure brought about the onset of the decline of the garden. Nowadays only the buildings are testimony to the once world-famous garden of George Clifford.

As intermediaries of medical and scientific knowledge, gardeners were part of the scientific and educational venture of an institution[125]. To paraphrase Steven Shapin in his article on invisible technicians: "In the case of *horticultural* practice, the price of *gardeners'* continued invisibility is an impoverished understanding of the science of *botany*"[126]. In a recently published monograph, the contribution made by the gardeners in service of the princes of Orange was made visible[127], but gardeners working for other institutions, like Universities, certainly deserve a better testimony.

The attribution of an anonymous herbarium is, by its nature, a precarious enterprise. For example, the specimens in the analysed herbaria are nearly all mounted with attention to botanical detail and the skills and effort to obtain and cultivate the plant specimens are visible in the species richness of the herbarium. However, this appears to be at odds with the obvious gap in knowledge displayed by the author of some of the names in the d'Oignies Herbarium. The sometimes totally inadequate identifications we're presented with suggests that the person who made them was not the maker of the herbarium. Furthermore, cultivating exotic species presupposes a botanical garden, where facilities and people with skills and knowledge were present to do so. Together with the Leiden provenance of some of the vases this leads us to believe that the d'Oignies herbarium (along with the Zierikzee Herbarium) started as a loose-leafed herbarium in the Leiden garden with separate name tags accompanying the herbarium sheets, which, as shown by the Boerhaave specimens, sometimes survived[128]. The original herbarium changed hands only after the death of Boerhaave in 1738 (d'Oignies Herbarium) and the death of Ligtvoet in 1752 (Zierikzee Herbarium). The subsequent owners might have been botanists—not all identifications are incorrect—but their identification skills were no match for the plants they were presented with.

Finally, the survey we have given in this chapter uncovered the limitations and problems of interpreting early modern botanical sources, but it also showed the richness of the collections discussed. Together with the Boerhaave collection and the specimens of the Clifford Herbarium, the Zierikzee and the d'Oignies herbarium present us with a plausible picture of what the Leiden garden and its living collection might have looked like in the early eighteenth century.

---

125 Hickman 2019: 560–561.
126 Shapin 1989: 563 [quotation modified].
127 Berkhout 2020.
128 Offerhaus et al. 2022.

Fabrizio Baldassarri

From the Analogy with Animals to the Anatomy of Plants in Medicine: The Physiology of Living Processes from Harvey to Malpighi

> The anatomy of plants is so diffuse a subject and so recondite
> that it demands enormous labor and great patience. I might therefore
> spend the rest of my life pursuing it.
> (Malpighi to Oldenburg, 23rd July 1672.)

Introduction

Throughout the ages, the analogy between vegetal and animal forms has importantly shaped medical knowledge. While an anthropomorphic understanding of plants developed in the Greek philosophers Plato (428/427 or 424/423–348/347 B.C.) and Aristotle (384–322 B.C.),[1] the knowledge of resemblances and analogies between plants and animals flourished in diverse frameworks and contexts, especially characterizing the medical understanding of a few animal processes, namely the basic activities of nutrition, growth, and reproduction. Still, since scholars primarily devoted their attention to the animal world, this analogy appeared engulfed in the effort of anthropomorphising plants, revealing a restricted focus and generally resulting in absurdities and monstrosities.[2] In contrast to this, it is especially with the anatomical studies of vegetation performed by Nehemiah Grew (1641–1712) and Marcello Malpighi (1628–1694) that the science of plants surfaced and helped shape a new approach in early modern medicine.

Besides the metaphorical uses in Plato and the Platonic strand, the attempts to uncover analogies between plants and animals characterized the medical knowledge that goes back to Hippocrates (ca. 460–between 375 and 351 B.C.), Aristotle, and Galen

---

1 Repici 2000. See also the contributions in Baldassarri and Blank 2021.
2 Brancher 2015. See, for example, Aldrovandi, *Dendrologiae* (ed. Ovidis Montalbani, 1668): 3–6.

Note: Research for this article has been carried out with the support by a grant of the Romanian National Authority for Scientific Research and Innovation (CNCS-UEFISCDI), project number PN-III-P1-1.1-PD-2016-1496 and has received funding from the European Union's Horizon 2020 research and innovation programme under the Marie Sklodovska-Curie grant agreement No 890770, "VegSciLif". I presented part of this work at Ca' Foscari Venice, at HSS Annual Conference 2018 in Seattle, and at HSS Annual Conference 2019 in Utrecht in a panel I co-organized with Alain Touwaide. I would like to thank the participants of these meetings for their questions and comments.

(A.D. 129-after [?] 216). Such analogies furthered the knowledge of the lower functions of the body and, in some cases, revealed conceptions of the human being as a plant, or claims that organs like the liver functions as plants. The medical analogy especially acquired momentum in the Renaissance. From Jean Fernel (1497–1558) to André du Laurens (1558–1609), Hieronymus Fabricius ab Aquapendente (1533–1619), and Jean Riolan the Younger (1577 or 1580–1657), among others, Renaissance and early modern physicians generally reappraised the Ancient tenets in their embryological works, claiming that the fetus lives the life of a plant at the beginning, as it does not need a heart and simply draws nutrition from the mother.

Scholars used the analogy to delve into two main physiological activities: (1) generation, as eggs and fetus were compared to seeds, an early modern commonplace with roots in ancient medical knowledge; and (2) nutrition, therefore entailing a similarity in the movements of animal and vegetable juices. Albert the Great (ca. 1200–1280), for example, importantly demonstrated in *De Vegetabilibus* that radical moisture, digestion, and the rise of nutrients characterize plant nutrition in the Aristotelian way, making these bodies physiologically comparable to animals. Accordingly, the rise of nutrients in plants corresponds to the flow of blood in animals, making this activity a benchmark for understanding living bodies and life in general.[3]

Yet, questions remain, as these positions appear scarcely grounded on the anatomical observations of plants. It is well-known that the studies of plants rarely concerned the nature and structure of vegetal bodies, or the reason for their growth and decay, which were however the subject of the pseudo-Aristotelian *De plantis*, a well-known but difficult-to-grasp text.[4] Basically, established on the metaphor of plants as inverted humans, i.e., as bodies sharing similar activities,[5] this analogy took for granted that both plants and animals live through vegetative activities, whose principle is the vegetative soul—that is an Aristotelian claim, but this concept was already present in Plato and in several pre-Socratic traditions, which incorporated desiderative and sensible activity in vegetal bodies. As a result, throughout the centuries, the comparison between animals and plants appears more a theoretical-philosophical framework to understand bodies, even on its medical and natural philosophical sides, while the anatomical and physiological investigation of vegetation per se lagged behind.

---

3 See [Aristotle], *De plantis* (ed. Drossart Lulofs and Poortman 1989). On it, see Stannard 1980/1999c; Crisciani 2005; Repici 2009.
4 On medieval restrictions in the study of the structure and nature of plants, see Ventura 2016. On [Aristotle], *De plantis*, see Giglioni and Ferrini 2020.
5 It is to be noted that this comparison present in both Plato and Aristotle is absent in the pseudo-Aristotelian *De plantis*, whose authorship was uncertain until the Renaissance. See Repici 2000: ix–x, and 2009: 92–93.

This grows as a problematic issue. If Albert the Great had already solicited for an anatomy of plants, this investigation failed to develop completely for centuries.[6] In the Renaissance period, natural history and collecting prevailed, while scholars restricted the study of plants per se to outward qualities and differences. Yet, a more structural study of plants created the conditions for a more precise medical knowledge. For example, a system to apprehend the therapeutic virtues of plants appeared mandatory to contrast the role of occult doctrines, magic, or celestial influences in explaining the healing powers of plants, such as the Doctrine of signatures, which acquired momentum during the Renaissance and thrived in the seventeenth century. Grounded on the microcosm-macrocosm analogy, this doctrine aimed to unearth analogical correspondences between bodies as a way to reveal signs of plant healing powers.[7] In contrast to this, scholars attempted to find alternative views to explain pharmacological powers as resulting from the combinations of elementary qualities or parts of plants. In this sense, scholars started dealing more precisely with the form of plants, a section of the natural historical description of plants, as it is in texts such as Leonhart Fuchs' (1501–1566) *De historia stirpium commentarii insignes* (1542). However, this mostly consists of a representation of outward qualities, and the therapeutic powers of plants remain a metaphysical combination of qualities rather than a material exploration of their structure.[8] As Andreas Blank has recently shown, different approaches to pharmacological knowledge arose at the time, as reductionism and emergentism outlined two efforts to find within plants the explanation of the origins of their powers.[9] Sixteenth-century physicians such as Jacob Schegk (1511–1587), Thomas Erastus (1524–1583), and Girolamo Mercuriale (1530–1606), among others, suggested that pharmacological powers pertain to the structure of plants, somehow anticipating Gottfried Wilhelm Leibniz's (1646–1716) claim that one must investigate the inner structure of natural objects (such as plants) to rebut the doctrine of signatures and other occult doctrines. The anatomy of bodies helped understand their qualities and therapeutic virtues. Studies in the field were, however, slow to develop. In 1671 Johann Ferdinand Hertodt von Todtenfeld's (1647–1714) *Crocologia, seu curiosa croci*, an outstanding early modern summary of all the knowledge of the medical uses of saffron, the plant acquires a central role in describing all the parts of human bodies and listing a never-ending number of diseases. Yet, no exploration of the structure and nature of the plant in relation to its healing powers ever surfaces in this text.[10]

---

6 On Albert the Great, see Moulinier 1993: 93; Draelants 2011: 99. On the Renaissance, see Brancher 2015: 180–187; Findlen 1994: 260–261. Also, Lenoble 1969: 49.
7 Bianchi 1987; Shackelford 2004; Woolley 2004; Hirai 2014; Kikuchihara and Hirai 2022; Baldassarri 2022b.
8 Brancher 2015: 186.
9 Blank 2018.
10 Francis and Ramandi 2020.

Nevertheless, the anatomical observation of vegetal bodies started to become a crucial issue in medical knowledge, as the speculations over the common activities of plants and animals found a more observational ground. This slowly appeared towards the end of the sixteenth century. If in Giovanni Costeo's (1528–1603) *De Universali stirpium natura libri duo* (1578), an Aristotelian account of plants, the comparison between plants and animals arises to explain nutrition and generation, his exploration of vegetation appears restricted to superficial qualities, as he mostly followed Theophrastus's (ca. 371 – ca. 287 B.C.) work. In contrast to this theoretical approach to botany, a deeper attention in the structure of plants is in 1603 *De venarum ostiolis*, where Fabricius ab Aquapendente paralleled veins with plants (see Figure 1). He indeed compared the valves in the vein to the knots of verbena, which he showed both as a whole (*Figura iiii*) and with the side branches cut off (*Figura iii*), as he likely observed the structure of the plant itself.[11] In 1600 *De formato foetu*, where he discussed the umbilical veins, Fabricius probed into the similarity between veins and the roots of plants, something he borrowed from Aristotle and Galen, as these bodies are comparable and the fetus lives like a plant.[12] Yet, how much he observed the structure of plants is unclear, although at this point of the text he repeats Galen's suggestion that knowing plants is important for medical knowledge, specifically referring to anatomical knowledge.[13] Moreover, Fabricius claims a similarity develops between the movement of blood and nutrients in animal veins with the movement of humor within plants.

While a close connection between anatomy and botany gradually flourished in the late sixteenth century, and especially at Padua, where professors of anatomy were in contact with people at the Hortus botanicus, as Florike Egmond has shown,[14] and Fabricius's texts reveal several analogies that connect the anatomical study of animal bodies with the understanding of the structure and functioning of plants, this second field remains underdetermined at this stage. Insofar as plants attracted new attention from scholars who unearthed the physiology of vegetal bodies, the animal-plant analogy slowly acquired a new condition in medical knowledge.

In this chapter, I aim to disclose the ways the anatomical study of plants in the second half of the seventeenth century played a role in furthering the knowledge of the physiological processes that are in common between plants and animals. Moving from the presence of such an analogy in European culture that developed in (1) embryology and nutrition, and (2) the analogy between the flowing of blood and nutrients in veins

---

11 Fabricius, *Tractatus Quatuor: IV. De venarum ostiolis*: 149, 152. See Bertoloni Meli 2019: 46.
12 Fabricius, *De formato foetu*: 112: "Sparguntur igitur, & ad uterum propagantur umbilicalia vasa, non dissiimili ratione, ac radices plantarum in terram. Unde Aristoteles dicevad, quid animalia umbilicum, quasi radicem agunt in uterum; et iure quidem, quoniam foetus eodem iam modo, ac planta gubernatur".
13 Fabricius, *De formato foetu*: 112: "Unde ex ijs que necessaria plantis sunt, dicebat Galenus, licet cognoscere . . .". See Favaro 1911–1912; Ekholm 2010.
14 Egmond 2021.

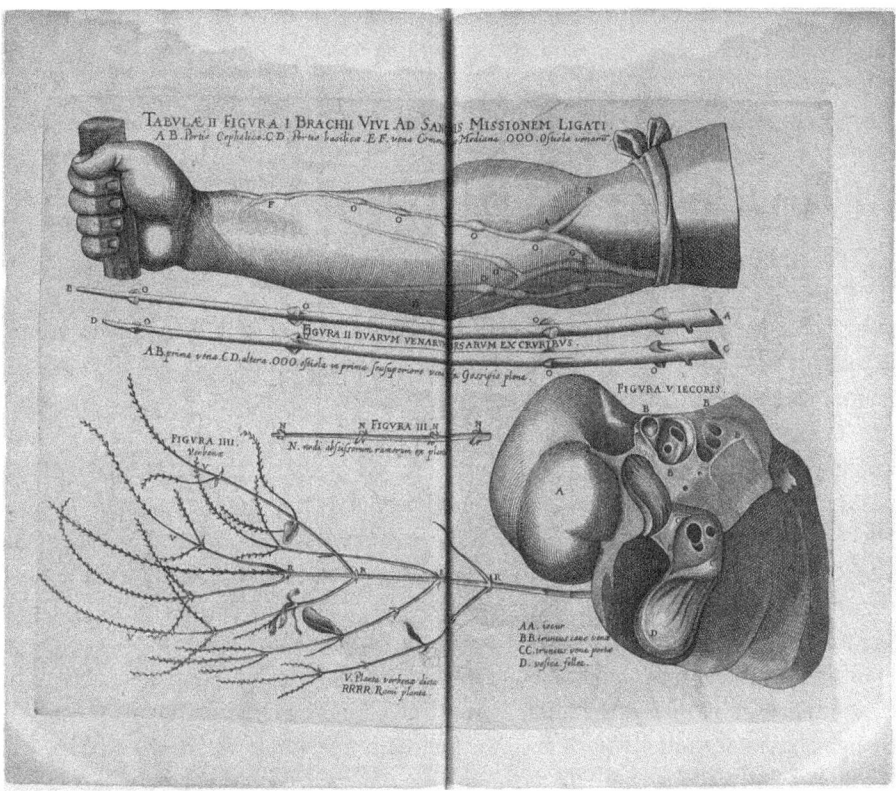

Figure 1: *Branch of verbena* (as a whole, *Figura iiii*, and cut off, *Figura iii*), in Hieronymus Fabricius ab Aquapendente, *Tractatus Quatuor, IV. De venarum ostiolis*. Frankfurt: Harman Palthenij, 1624: plate II, pp. 152ff.

and arteries with the movement of humors in the channels or vessels of plants, in this chapter I explore how much these studies ultimately benefited from Grew's and Malpighi's anatomies of plants.

Before concentrating on their work, I present the case of William Harvey (1578–1657), who studied in Padua with Fabricius, and who liaised between this context and the physicians of the second half of the century. Then, I focus on a few scholars of the Royal Society, and especially on Francis Glisson (1599–1677), who repeated the plant-animal analogy, therefore uncovering the context in which Grew and Malpighi developed their research. Throughout these pages, I show the nuances shaping the animal-plant analogy from a theoretical tenet into a subject of plant anatomical studies, as the works of Grew and Malpighi began to emerge. From their texts, this analogy developed as an acquired subject of the anatomical and physiological explanation of living bodies, as it finally surfaces in the eighteenth century, and for example in the static and chemical experimentation with plants of Stephen Hales (1677–1761) and Herman Boerhaave (1668–1738). These cases finally show how far the analogy between plants and animals grounded the medical studies of living processes, ultimately benefiting from the anatomical and physiological study of plants.

William Harvey: *Primordium vegetale* or the Vital Principle in Bodies

In his work, Harvey repeated the analogy between plants and animals he found in Fabricius (and also Galen and Aristotle), especially developing this claim from an experimental perspective, as he observed seeds, eggs, as well as animal bodies.[15] While his 1628 *De motu cordis* concentrates on the circulation of blood, Harvey also describes exsanguine (or bloodless) animals he conceives "to live after the manner of vegetables or of those other productions which are therefore designated plant-animals",[16] or zoophytes, whose entire body is conceived as the heart, or the source of heat, and have the vitality of a vegetable, as they live through nutritive functions and manifest "a kind of vegetative existence".[17]

According to Harvey, insofar as they live a vegetative existence, the processes of plants and animals are comparable. Interestingly enough, this comparison surfaces in a text entirely devoted to the circulation of blood, an activity missing in plants, although Harvey meaningfully claims that the entire body of zoophytes is a heart and the movement of fluids within plants thus corresponds to the circulation of blood.

This point constitutes the ground for Harvey's further work on embryology, where he moves from speculations on blood's purpose to a theory of the role of blood in the body, as Benjamin Goldberg has shown. In *Exercitationes de Generatione Animalium* (Exercises on the Generation of Animals), published in 1651, Harvey focused the analogy between plants and animals on generation, as similar processes concern these bodies. Accordingly, eggs and seeds reveal similar activities, and "the hen [. . .] produces egg without a male [. . .] in the same way that trees bear their fruits and grasses their seeds without any distinction of sex".[18] Additionally, Harvey claims the seed of plants as a suitable example to understand the implantation of the soul in the fetus, to define a similarity between the nutrition of the fetus and the vegetation of plants, as the former "when enclosed within its mother first draws its nourishment from her womb like a plant from the earth",[19] and to claim a correspondence in the life of these bodies. Yet, while Harvey

---

15 Baldassarri 2022a.
16 Harvey, *De motu cordis*: 28 (Engl. transl.: Willis and Guyton 1965: 30).
17 Harvey, *De motu cordis*: 65 (Engl. transl.: Willis and Guyton 1965: 76).
18 Harvey, *Exercitationes de Generatione Animalium*: exc. 62, 215 (Engl. transl.: Willis and Guyton 1981: 463). See Adelmann 1966: 3.1096–1097.
19 Harvey, *Exercitationes de Generatione Animalium*: exc. 28, 89 (Engl. transl.: Willis and Guyton 1981: 282).

goes deep into the continuity between animals and plants, unfolding a study of seeds to resolve questions concerning animal generation, he also specifies that a vegetative power inheres in the body, as life is specifically related to the vegetal activities. This power is the *primordium vegetale* or the original vital principle in the initial stages of life.[20] This vegetative power lies in animal eggs and in plant seeds.[21] As a result, the analogy goes beyond the mere similarity of activities, that is, the observation of the physiology and anatomy of bodies, and concerns the philosophical definition of life.

In developing the animal-plant analogy, Harvey provides it with a broader aim. While this analogy continues to help explain similar processes occurring in both bodies, it also provides a theoretical framework to any explanation of life. As Guido Giglioni points out, Harvey suggests that if one wants to perform an anatomy of the vegetative soul, one should anatomize plants, for the latter hold the secret of animal life.[22] However, despite the experimental approach on seeds and fetuses, and besides revealing that animals grow like plants, Harvey's understanding of the animal-plant analogy reveals a persisting philosophical inclination.

Francis Glisson and the Royal Society: Analogies and Circulations

Fellow of the College of Physicians and Regius Professor at Cambridge University, and then a fellow of the Royal Society, Glisson played a non-secondary role in the history of medicine and the sciences. If I reduce him to a link between Harvey on the one hand and Grew on the other hand it is not because I consider Glisson a minor figure. He was a student of Harvey at Cambridge, and from "the 1640s Glisson was involved in an anatomical research program based on Harvey's discovery of the circulation of blood" concerning "the organs of nutrition and their related apparatus".[23] In his work

---

20 Harvey, *Excercitationes de Generatione Animalium*: exc. 61, 210: "Liceat hoc nobis *primordium vegetale* nominare: nempe, substantiam quandam corpoream, vitam habentem potentia, vel quoddam per se existens, quod aptum sit, in vegetativam formam, ab interno principio operante, mutari" (Engl. transl.: Willis and Guyton 1981: 457).
21 Harvey, *Excercitationes de Generatione Animalium*: exc. 61, 211: "praesertim cum talis partus, sit primordium vegetale, plantarum seminibus analogum, quale etiam ovum habetur" (Engl. transl.: Willis and Guyton 1981: 457).
22 Harvey, *Excercitationes de Generatione Animalium*: exc. 61, 51: "Hoc certo constat futuri foetus [. . .] et punctum saliens venasque illas quae omnes ab uno trunco (quemadmodum iste a puncto saliente) propagantur et per totum colliquamentum plurimis fibrarum ramificationibus sparguntur; quae postmodum vasa umbilicalia constituunt [. . .]. Harum venarum earumque propaginum vivum exemplar videas in arborum foliis, quorum fibrae omnes a pedunculo oriuntur et ab uno trunco per totum folium diffunduntur" (Engl. transl.: Willis and Guyton 1981: 238). See also Giglioni 2021.
23 Giglioni 2022: 763.

on the anatomy of the liver, the *Anatomia hepatis* (1654), Glisson suggested that plants might usefully be studied for the light that they might throw on animal anatomy.[24] As reported by Anna Marie Roos, among others, this passage importantly influenced Nehemiah Grew's comparative anatomy.[25]

If, on the one hand Glisson's work on animal irritability appears simultaneously with the study on plant sensitivity, as claimed by Charles Webster,[26] therefore confirming a continuity between the anatomical and physiological investigations of living bodies, on the other hand, Glisson claimed a similarity between the circulation of blood and the movement of sap through the tree and the movement of spirits through animal nerves. In his text, comparisons between plants and animals are indeed frequent. For example, when speaking of the veins of the liver in chapter 25, Glisson quotes Galen's *Liber de venis*, where he compared the Portal veins to the roots and branches of plants.[27] Besides the structural analogy, Glisson provides a functional comparison between the movement of chylum within these veins and plants' attraction of food from the ground.[28] And again in chapter 26, entitled "Of radical, or ramose distribution of the vena cava in the liver" (*De radicum, sive ramorum venae cavae in hepate distributio*), Glisson stretches the comparison a bit further.[29] Then, when he speaks of the principles regulating the functioning of the liver, he draws a *"principium radicationis"*, insofar as roots nurture and sustain plants.[30] Then, in chapter 45, Glisson compares the structure of nerves to the fibers of plants.[31]

In the Royal Society, the exploration of the physiological analogies between plants and animals acquired momentum. In Robert Hooke's (1635–1703) *Micrographia: or Some Physiological Descriptions of Minute Bodies Made by Magnifying Glasses* (1665), these issues surface. It is to be noted that, in these years, Hooke was the curator of experiments within the Royal Society. In *Micrographia*, several drawings show a comparison between vegetables and other bodies. A famous case is the one of the offensive parts in stinging nettle and bees (see Figure 2), as Hooke specifies the parallels and differences in these bodies.[32] At the same time, questions concerning whether the movement of sap could be compared to the circulation of blood were analyzed by the

---

24 Glisson, *Anatomia hepatis*: 4. See Roos 2007: 92–93.
25 See Delaporte 1982.
26 Webster 1966: 22.
27 Glisson, *Anatomia hepatis*: 286: "*Imaginare*, inquit [Galenus], *arboris quendam truncum, qui parte inferiore in multas radices, superiore in numerosam ramorum [. . .] findantur. Hoc pacto quae in ventrem & intestina deferuntur venae, proportione radicibus respondent; truncus vero est in hepate*" [italics in the text].
28 Glisson, *Anatomia hepatis*: 286: "chylum omnem in hepar per venam Portae deferri, eundemque fibrarum hujus capillarium ope è ventriculo atque intestinis exfugi: perinde atque radices arboris è terra succum attrahunt, qui postea toti plantae nutriendae inserviat".
29 Glisson, *Anatomia hepatis*: 297.
30 Glisson, *Anatomia hepatis*: 397.
31 Glisson, *Anatomia hepatis*: 505, 507.
32 Bertoloni Meli 2019: 70. See Hooke, *Micrographia*: 142–147.

means of microscopes, generally with not entirely satisfying answers—as Hooke writes, "though I have with great diligence endeavoured to find whether there be any such thing in those *Microscopical* pores of Wood or Piths, as the *Valves* in the heart, veins, and other passages of Animals, that open and find a passage to the contained fluid juices one way, and shut themselves, and impede the passage of such liquors back again, yet have I not hitherto been able to say any thing positive in it; though, me thinks, it seems very probable, that Nature has in these passages, as well as in those of Animals bodies, very many appropriated Instruments and contrivances, whereby to bring her designs and end to pass . . .".[33]

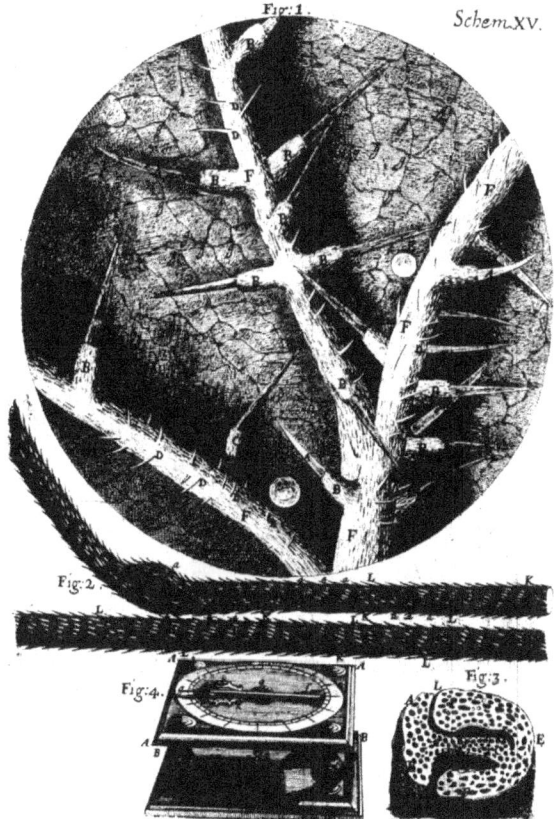

Figure 2: *Stinging nettle*, in Robert Hooke, *Micrographia*. London: John Martyn and James Allestry, 1665: scheme XV.

A few pages later, Hooke speaks of the metamorphosis of plants into animals, unearthing a similar principle of life between bodies, despite the huge differences occurring between

---

[33] Hooke, *Micrographia*: 116.

plants and animals. Indeed, he writes that "there are many *Zoophytes*, and sensitive Plants [. . .] so have we, in some Authors, Instances of Plants turning into Animals, and Animals into Plants".[34] In these cases, Hooke's attempt to explore the similarities in the physiology of plants and animals clearly surfaces. More importantly, he grounded, and restricted these comparisons on the microscopic observation of bodies.

In the years after the publication of the *Micrographia*, fellows of the Royal Society discussed vegetal bodies at large, frequently investigating the issue of the physiological similarities between plants and animals.[35] Indeed, the *Philosophical Transactions* of the time contains queries comparing vegetation with animal physiology. Observations performed in 1666 by Christopher Merrett (1614/1615–1695) reveal how much he was trying to find out the similarities between plants and animals, while *Philosophical Transaction* number 43 (January 1668) and number 46 (April 1669) suggest an analogy between the movement of sap and blood circulation. It is to be noted that these texts contain experiments contributing to knowledge, and contain queries that were not completely elucidated or resolved. Contemporary with these queries, and encouraged by Harvey's work, the German professor of theoretical medicine and naturalist Johann Daniel Major (1634–1693) of Kiel outlined a circulatory theory of plants in his *Dissertatio botanica de Planta Monstrosa Gottorpiensi* (1665), ultimately claiming that the movement of juices vivifies plants in the same way as the circulation of blood in animals.

In a series of letters published in the *Philosophical Transactions*, Martin Lister (1638–1712) claimed that plants and animals shared a similar system of vessels: "I find some new Observations concerning the *Veins* of Plants or such *Ductus*'s, as seem to contain and carry in them the noblest juices of Plants [. . .] will probe vessels Analogous to *Human Veins*, and not meer *Pores*".[36] Additionally, Lister postulated that the "*primary* use of these *Veins* [. . .] is to carry the *Succus nutritivus* of *Plants*, because, where they are not, there is no Vegetation".[37] Physiological and anatomical similarities seem to definitely surface in the context of the Royal Society. As Lister suggests, several experiments on the circulation of sap were performed, confirming that (1) "Vessels in Plants [are] analogous to Veins in Animals", which is clear as diverse plants have differently colored juices, and (2) in "many Plants [the] Juice seems never to be at rest, but will spring at all times freely, as the Blood of Animals, upon Incision".[38] In a 1672 letter, Lister explores this issue further, accounting for the existence of veins in plants and for the similarities between the juice of plants and blood, and between a few botanical activities and animal sensation. The letters detail the experiments performed to visualize such veins in cutting the body of a plant or a leaf. Indeed, by means of microscopic observations he claimed that the veins held fluids as animal veins do blood, and the primary use of

---

34 Hooke, *Micrographia*: 124.
35 Thomas 1955: 19. See Roos 2011: 152.
36 Lister, "An Ingenious account of Veins": 3052.
37 Lister, "An Ingenious account of Veins": 3054.
38 Lister, "Circulation of Sap": 2122.

these veins was to carry the *succus nutritivus*, insofar as without the vessels there is no vegetation, nor life.

From Glisson to Lister, a more precise understanding of plants surfaced as increasingly grounded on the microscopic and anatomical observation of vegetal bodies. While the physiology of plants attracted the attention of scholars, the similarities between plants and animals became a subject of anatomical and physiological investigation. Speculations over these analogies focused on the similarities between blood circulation and the movement of juices in plants, and on plant sensation, somehow transferring to plants Harvey's claim that blood is the living part of the body. As in animals blood circulation vivifies the body, scholars applied the same framework to the motion of juices and sap in plants, ultimately claiming, as Lister did, to expect that "the analogies betwixt plants and animals be in all things else, as well as the motion of their juice, fully cleared".[39] In this sense, when the experimentalist approach began, the analogical tradition slowly acquired a more precise condition, and the anatomical work on plants appears to play a central role in the physiological understanding of living processes.

Grew and Malpighi: The Anatomy and Physiology of Plants Developed

In a letter published in the *Philosophical Transactions*, the astronomer Giovanni Cassini (1625–1712) reported that

> an ingenuous person [sought] for finding out, whether there be a Circulation of Sap in Trees, resembling that of Bloud in Animals; Offers it to the consideration of Naturalists, Whether it be likely to find a place in Vegetables, whence the Sap may part, and whither it may return, such as in the Heart of Animals; adding that, whereas Vegetables are always to put forth new branches, leaves, &c. it seems to be sufficient for them, that there be a continual and plentiful course and supply of Juyce, to thrust out every way, without any necessity of such a Circulation.[40]

This person is Marcello Malpighi. In these same years, Malpighi and Grew each worked on their anatomical observation of plants, trying to detail the similar processes in plants and animals. Despite having been inspired by different sources, both of them laid the groundwork of comparative anatomy between plants and animals. As Agnes Arber wrote, "Malpighi's project [consisted] of understanding animals through plants", although it was "a forlorn hope, so far as morphology was concerned, but physiologically it was an anticipation of a modern viewpoint".[41] At the same time, even Grew discussed

---

39 Lister, "Juice in Vegetables": 5136–5137; Barker 2002.
40 Cassini, "Philosophical Particulars": 2218.
41 Arber 1942: 10; Arber 1950.

the analogy between plants and animals as drew from Glisson's text, Hooke's work, and the Royal Society context, and his work played a crucial role in the definition of a modern science of plants.[42] Although in November and December 1671 they reported their botanical discoveries to the Royal Society, inaugurating the anatomical and physiological science of plants,[43] Malpighi's and Grew's works are independent and different. In this chapter, I am not entering the debate between them, which attracted great attention from scholars,[44] but mostly focus on their contribution to the investigations of the physiological similarities between plants and animals in the medical studies of the body.

Let us turn to Grew. As Bertoloni Meli writes, "Grew did draw significant comparisons with animals, but in no way was this a key feature of his work".[45] In the dedicatory letter, Grew claimed a similarity between plants and animals: "a Plant, as well as an Animal, is composed of several Organic Parts [. . .]. So that a Plant is, as it were, an Animal in Quires; as an Animal is a Plant, or rather several Plants bound up into one Volume".[46] Yet, Grew does not merely repeat such a claim, but explores this similarity by means of microscopic observations, sometimes rejecting a few claims of his predecessors or shedding new light on other aspects.

In writing to Henry Oldenburg (1619–1677), Grew expressed doubts concerning Lister's complete analogy between the juice of plants and blood.[47] As Anna Marie Roos has perfectly reconstructed, this doubt unfolded a debate between the two, as Grew clarified this point in *An Account of the Vegetation of Trunks* (1675), where he demonstrated that the circulation of sap in plants was not the same as blood in animals. In chapter 3 of *The Anatomy of Trunks*, that is the first part of this text, Grew claims that "the Analogy betwixt the *Vessels* of an *Animal* and a *Plant*, is the more clear and proper. For as the *Sanguineous Vessels* in an *Animal* are composed of a number of *Fibres* [. . .] so are these *Lymphaeducts* of a *Plant*".[48] Then, he repeats that nutrition is as necessary in plants as in animals, and this should occur throughout vessels. In chapter 1 of the *Vegetation of Trunks*, Grew started describing the movement of sap and the role of vessels, ultimately revealing the differences between veins and the vessels of plants. First, when cut, "the Trunk or Branch of any Plant [. . .] always bleeds at both ends [. . .], as well as divers other Experiments plainly shews [. . .] in the *Sap-Vessels* of a Plant, there are no Valves",[49] which in contrast are within the veins of animals. Similarly, Grew's explanation of the movement of sap develops from the appeal to the parenchyma quality: "the Parenchyma being filled and swelled with sap, hath thereby a continual Conatus to

---

42 Bolam 1973: 227; Morton 1981: 194; Garrett 2003: 72.
43 Le Fanu 1990: 8.
44 For example, Bertoloni Meli 2011: 262.
45 Bertoloni Meli 2011: 263.
46 Grew, *Anatomy of Plants*: unpaged ff 2. See Basse Eriksen 2018: 117–158.
47 Hall and Hall 1965–1986: 9, 357.
48 Grew, *Anatomy of Plants*: 112 [italics in the text].
49 Grew, *Anatomy of Plants*: 125–126 [italics in the text.]

dilate itself [. . .]. And the said vessels being cut, their actual Contraction and the Eruption of the sap, do both immediately follow".[50] In explaining the movement of sap within plants, Grew did not rely upon the analogy with animals, but on direct observations of the vegetal body (see Figure 3). As a result, the cause of movement is the pressure and a conatus to dilate vessels, and not a heart, and the movement of sap does not correspond to a true circulation, as it occurs in animals. The analogy with animals therefore fails to define the movement of sap, as the observation of plants reveals. A broader epistemological shift results, as the microscopic observation of the analogy of plants replaces the analogical study of similarities and resemblances, as Catherine Wilson pointed out: "the microscope takes away the privilege of surface. What the object looks like on the outside is no guide to what it is in the sense of what it can do; the key to its powers is to be found in its inner *invisible* structure. And in the interior of things there is no resemblance: here is indeed a new world".[51]

External or internal similarities, from signatures and resemblances to more precise analogies, are slowly abandoned as the knowledge of the internal structure (the anatomy) and functioning (the physiology) of plants develops.

Yet, this does not mean that Grew abandons the analogy altogether. In other parts of the text, Grew compares animals and plants as physical objects. In describing the liquor within plants, Grew speaks of animals; accordingly, "the same thing is the cause of the whiteness of *Vegetable*, as of *Animal-Milk*: that is to say, a more copious mixture of *Watery* and *Oily Pars per minima*, or into one Body".[52] In this case, Grew adds the chemical investigation of liquors to the microscopic observation of plants. Then Grew compares the texture of leaves to the skin of animals, which share a similar function, namely that of perspiration. Indeed, Grew writes that "as the *Skins* of *Animals* [. . .] are made with certain open *Pores* or *Orifices*, either for the Reception, or the Elimination of something for the benefit of the *Body*, so likewise the *Skins*, of at least many *Plants*, are formed with several *Orifices* or *Pass-ports*, either for the better *Avolation* of *Superfluous Sap*, or the Admission of *Aer*".[53] Yet, even in this case, Grew takes the move from a similarity between bodies, but then investigates the structure and functioning of plants in themselves. Then, he compares seeds to fetus, as in principle generation follows the same mechanics in both bodies, or compares the membranes of the seeds to placenta.[54] In *Musaeum Regalis Societatis* (1681), that is a catalogue of rarities and curious bodies, Grew claims that "the Seeds of all Plants whatsoever, which are not merely Metaphorically, but really so many Eggs (like those of many Animals) without a Yelk",[55] therefore suggesting that, besides the analogy, the microscopic observation of living nature specifies similarities. In

---

50 Grew, *Anatomy of Plants*: 126.
51 Wilson 1995: 62 [italics is mine].
52 Grew, *Anatomy of Plants*: 134 [italics in the text].
53 Grew, *Anatomy of Plants*: 153 [italics in the text].
54 Grew, *Anatomy of Plants*: 206, 210.
55 Grew, *Musaeum regalis societatis*: 198.

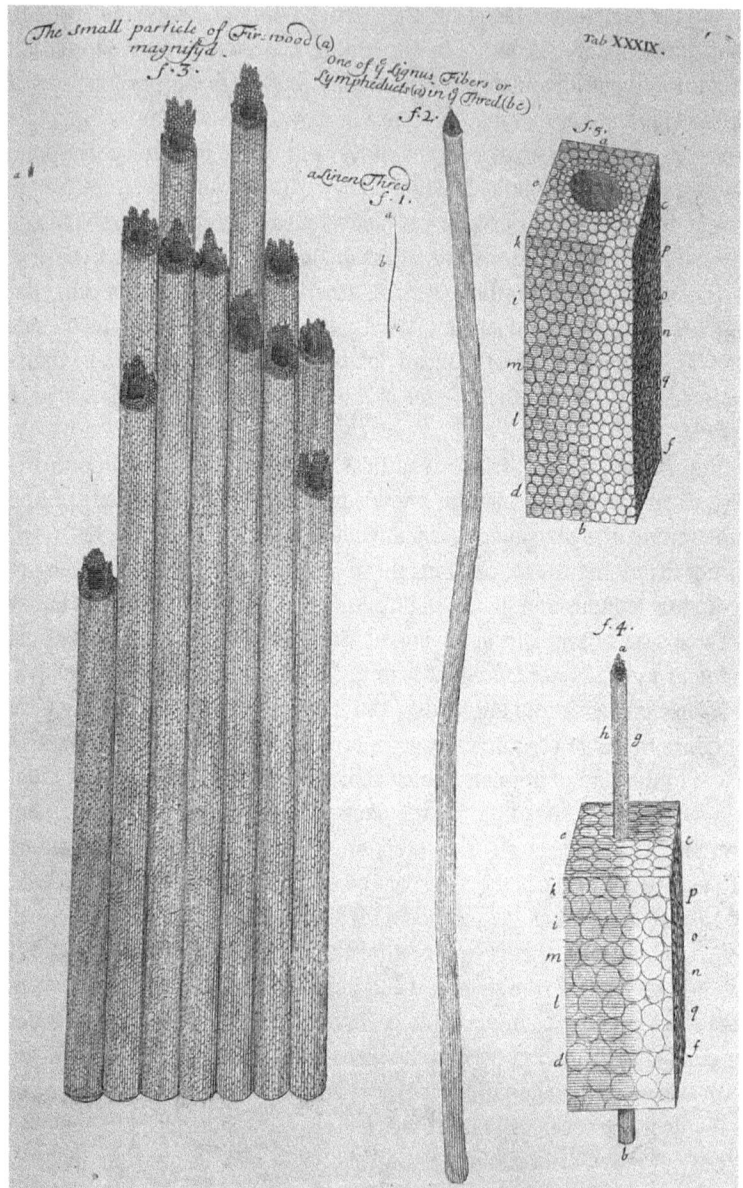

Figure 3: *Representation of channels and threads within woods*, in Nehemiah Grew, *The Anatomy of Plants*. London: W. Rawlings, 1682: tab. xxxix.

this sense, while analogies and similarities surface in Grew, plants are not just a segment of a comparison, but have their own anatomy and physiology, which in the end reveals an autonomous science. As he writes a few pages later, "if any one shall require the Similitude to hold in every Thing; he would not have a *Plant* to resemble, but to be, an

Animal.⁵⁶ According to Grew, plants have their own independence from animals, while stretching the comparison too far should result in conceiving plants to be animals, which cannot be said.

Malpighi provided a different understanding to this analogy, as he provided a more systematic exploitation of the analogy in both directions.⁵⁷ As Domenico Bertoloni Meli reports, Malpighi "used animal anatomy as a guide to understand plants and, conversely, studied processes such as growth in plants in order to tackle the same problem in animals".⁵⁸ In *Anatome plantarum* (1675–1679), Malpighi's analogy between animals and plants reveals several layers. Indeed, such analogies helped him in uncovering the *œconomy* of living nature, insofar as studying plants provides him with a simple path for understanding living activities such as growth and generation. Let us look at a few examples.

When discussing growth in *Anatome plantarum*, Malpighi delineates a comparison with bones, "a similar way of growing does Nature operate to achieve the growth of bones",⁵⁹ and more specifically of teeth (depicted in figure 37, in the appendix to the text), as both bodies reveal the arrangement of different layers and the presence of roots, as in plants⁶⁰ According to Malpighi, the intimate structure of bones is that of a net of fibers or filaments, composing bodies similar to intertwining branches⁶¹ These fibers are structurally analogous to the structure of plants. In illustrating the composition of the bark of cherry and plum trees, Malpighi shows the intertwining arrangement of fibers.

In the text, Malpighi compares the bodies of plants with insects, as he describes the structure of plants and that of insects with air vessels surrounded by vessels carrying nutrients, and especially in animals for respiration. Malpighi claims that *sicut* in animals' nutrients ferment and circulate within the body, *ita* in plants nutritive juices perform similar activities;⁶² and *sicut* in animals' trachea is composed of cartilaginous circles, *ita* in plants and insects Nature fabricates this part arranging particles in different ways to allow air to penetrate within the body (see Figure 4). While Malpighi describes superior animals that have lungs, he specifies how plants breathe from the roots or other parts, as the disposition of fibers make clear.⁶³

Additionally, in describing buds, leaves, and flowers, Malpighi systematically compares plants with animals. Indeed, buds are like fetuses.⁶⁴ As the latter are fastened to

---

56 Grew, *Anatomy of Plants*: 173 [italics in the text].
57 See Adelmann 1966, 3, *Excursus* VI, "The Analogy between Plants and Animals": 1092–1103.
58 Bertoloni Meli 2011: 269.
59 Malpighi, *Anatome plantarum*: 1.19: "Pari incremento procedit Natura in *Ossium* augmento" [italics in the text]. [translation is mine].
60 Malpighi, *Anatome plantarum*: 1.21–22.
61 Malpighi, *Anatome plantarum*: 2.47–50.
62 Malpighi, *Anatome plantarum*: 1.13.
63 Malpighi, *Anatome plantarum*: 1.15.
64 Malpighi, *Anatome plantarum*: 1.22: "*Gemmae* itaque sunt veluti infans seu foetus ita custoditus, ut suo tempore auctus, in surculum excrescens, tandem *ova* promanat".

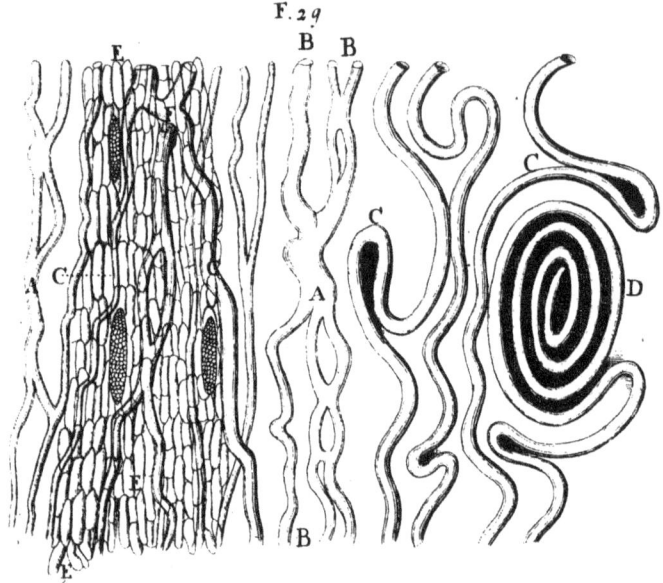

Figure 4: *Trachea*, in Marcello Malpighi, *Anatome plantarum*. London: John Martyn, 1675–1679, vol. 1: tab. vii, figura 29.

the womb of the mother, the former are fastened to the branch of the plant. He then describes several buds of diverse plants, whose illustrations are collected from Tabula ix to Tabula xiii. In early seventeenth-century embryological texts, physicians generally described the analogy between fetuses and plants, while Malpighi used the same analogy as a point of departure to explore the growth of leaves and flowers in detail. In Figure 63, he depicts the growth of the leaves of lemons and oranges (see Figure 5),[65] revealing their specific conformation as they extend from the stalk in Figure 77 (see Figure 6). Additionally, on the surface of leaves Malpighi identifies utricles or receptacles, analogous to glands in Figure 88 (see Figure 7).[66] In comparing leaves to the skin, Malpighi claims that the utricles allow the expulsion via transpiration of humor, "*quasi sudor*" [like sweat].[67] Like sweat expelled from animal bodies, the utricles in leaves work to expel the waste of sap concoction, as Malpighi acknowledges leaves with such a function.

Yet, Malpighi speaks of the movement of sap "*quasi peculiarem circulationem*" [like a peculiar circulation], as the juice concocted in the leaves reaches to all parts of

---

65 Malpighi, *Anatome plantarum*: 1.26–27.
66 Malpighi, *Anatome plantarum*: 1.33: "In *mali medicae & limoniae* 88 foliis [. . .] seu utriculus C; hujus consimiles D quamplurimi, folii latitudinem occupant".
67 Malpighi, *Anatome plantarum*: 1.37.

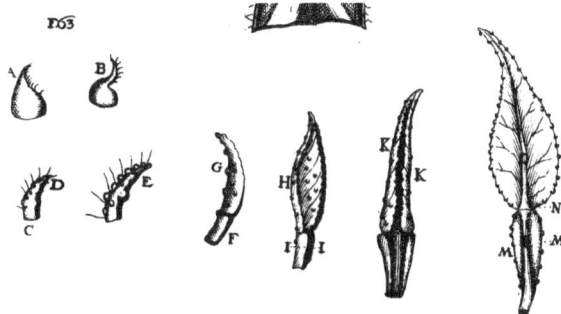

Figure 5: *The growth of the leaves of citrus*, in Marcello Malpighi, *Anatome plantarum*. London: John Martyn, 1675–1679, vol. 1: tab. xiii, figura 63.

Figure 6: *The stalk of lemon*, in Marcello Malpighi, *Anatome plantarum*. London: John Martyn, 1675–1679, vol. 1: tab. xv, figura 77.

the plants and does not return to the leaves.[68] In the two pages concluding the section on leaves, Malpighi delves into the analogy between plants and animals, as the former are deprived of animal muscles and organs, while animal flesh [*animalium viscera*] corresponds to the roots, which carry out the same activities of elaboration.[69] Still, Malpighi suggests that leaves are the analogous to the system of nerves, veins, vessels, glands, and

---

68 Malpighi, *Anatome plantarum*: 1.39: "Quoniam tamen sensim vegetante radicis trunco, & plantulae germine, contabescere incipiunt hujusmodi folia; hinc constat, à foliis in caulem & caudicem regressum esse concocti succi, & quasi peculiarem circulationem".
69 Malpighi, *Anatome plantarum*: 1.38.

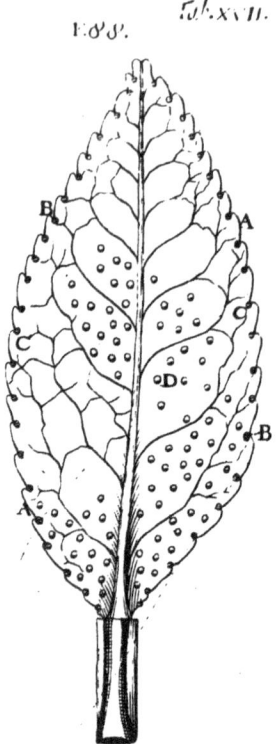

Figure 7: *The surface of a leaf with utricles*, in Marcello Malpighi, *Anatome plantarum*. London: John Martyn, 1675–1679, vol. 1: tab. xvii, figura 88.

channels wherein nutrient juices move and arrange animal bodies, as in leaves sap is concocted, distributed within the body, and new buds (both flowers and fruits) grow.

In the following sections, Malpighi deals with the processes of generation. Yet, he draws a sharp distinction between animals and plants: Accordingly, the latter do not have sexual reproduction, such as animals show.[70] Plants grow from a seed or a detached branch. Yet, he calls *uterus* the part of the body or the fruit where the seeds grow. A lavish image depicts the *uterus* of opium in Figure 262 (see Figure 8), in which he represents the membrane containing the seed of opium as a womb with fibers departing from it similar to veins. In these sections, Malpighi generally compares seeds to fetuses, ultimately claiming that as nutrients reach the fetus from the umbilical veins and through the bones, something similar occurs in plants that (1) are nurtured "*quasi per umbilicum*", and (2) analogously "*per os* [. . .] *in foetu*".[71] In the last two pages of the

---

[70] Malpighi, *Anatome plantarum*: 62–63.
[71] Malpighi, *Anatome plantarum*: 1.80. Malpighi constructs the paragraph as a comparison between animals and plants: "*Sicut* enim in *Animalibus* foetum non tantum per umbilicum nutritur, [. . .] sed per os [. . .] ita probabiliter contingere reor in *seminalibus plantulis* [. . .]" [italics in the text].

first volume of *Anatome plantarum*, Malpighi reassesses the analogy between plant seeds and animal eggs—"The old dogma from Empedocles is: the seeds of plants are [like] eggs".[72]

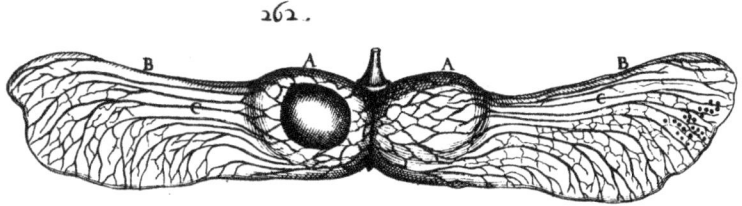

Figure 8: *A seed of opium in the plant*, in Marcello Malpighi, *Anatome plantarum*. London: John Martyn, 1675–1679, vol. 1: tab. xlvi, figura 262.

Despite highlighting the differences between animals and plants, Malpighi extended the preformation model to plants: In describing a plantlet in the seed, which could be traced back to Giuseppe degli Aromatari's (1587–1660) *Epistola de plantarum ex seminibus generatione* (1626), a text Malpighi probably knew, the latter drew a common paradigm for plants and animals. Yet, the theory of the *plantula seminalis* was challenged by scholars, insofar as difficulties surfaced (especially for the case of sea plants).[73]

In response to such criticism, Malpighi continued working on plants in the years after the publication of the *Anatome plantarum*, and a few important comparisons between plants and animals surfaced again. This is especially true in the letter to the Lyonnais physician and archeologist Jacob Spon (1647–1685). Dated 1st November 1681, the letter appeared in *Philosophical Transactions* 1684–1685.[74] In a section of this letter, Malpighi describes the uterus and uterus vessels by means of the analogy with plants: As he did in *Anatome plantarum* he repeated the comparison between the animal uterus and parts of the plants, namely the uterine tubes in animals and the style and ovary of pistil in plants, and ultimately claimed that these parts are more noticeable in plants rather than animals.[75] In describing the same uterine vessels in the *Autobiography* published in the *Opera posthuma*, Malpighi included an image that resembles a plant. The figure (see Figure 9) helps understanding the explanation of the uterine vessels, an obscure section of the letter to Spon.

Besides the visual similarity, in the letter to Spon Malpighi repeated the general claim of an analogy between plants and animals. It is to be noted that in the first part of the letter Malpighi describes the vegetation of animal horns, in the section on ovaries

---

72 Malpighi, *Anatome plantarum*: 1.81.
73 Ottaviani 2002.
74 Malpighi, "Letter to Spon". See Bertoloni Meli 2011, 2016 and 2019.
75 Malpighi, "Letter to Spon": 637: "Illud tamen perpetuum videtur, tubis ditari, quae in plantarum uteris prae reliquis luxuriant".

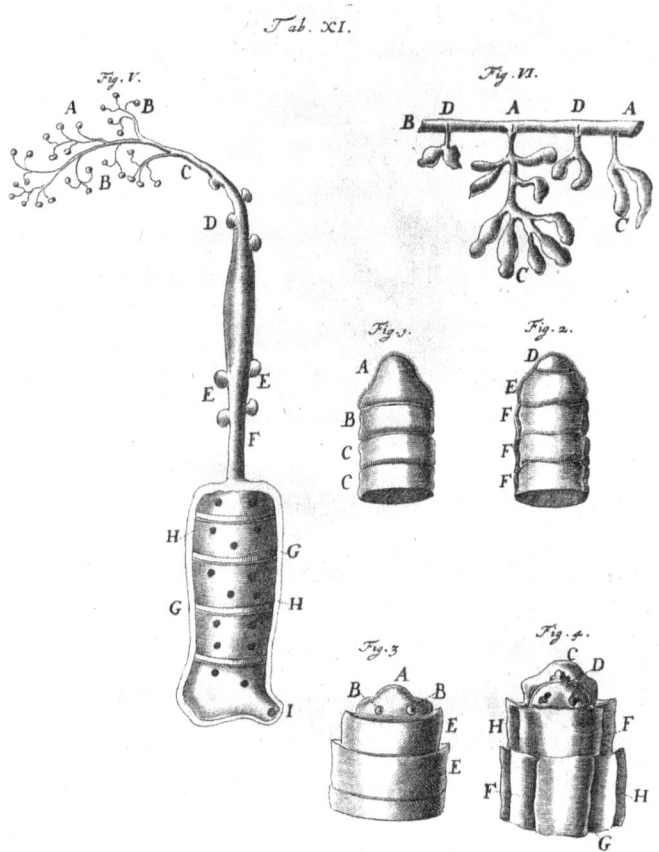

Figure 9: *Uterine vessels*, in Marcello Malpighi, *Opera posthuma*. London: A.&J. Churchill, 1697: tab. xi, figura v.

Malpighi's description differs from the *Anatome plantarum*, but similarities persist: "eggs are produced in an analogous fashion in plants [as it] reveals the plantlet";[76] while in the last section he presents a few experiments of botanical ligatures to show the movement of juices within plants, or the growth of plants from other bodies, which was the topic of a section of the second volume of *Anatome plantarum*. In the posthumous writings, then, he describes bones, teeth, vessels, hair and nails generally suggesting similarities with vegetal bodies. Indeed, the animal-plant analogy concerns different areas, such as the movement of fluids within the body, the structure of their vessels, the identity of seeds and eggs, the similarities of various parts and organs, and the idea that basic living

---

76 Adelmann 1966: 2.860. See Malpighi, "Letter to Spon": 640, "Hujus analogam productionem in plantarum ovis intuemur, in quibus umbilicale vas primo emergit, [. . .] quod tandem plantulam manifestat".

activities could be reduced to vegetation. Malpighi used the analogy as a conceptual framework for his anatomical investigation of plants and his study of animal physiology, as a tool to visualize several processes of life. Yet, while reconstructing the "oeconomy" of living nature, his microscopic study of vegetation reveals the differences between bodies and helps isolate a science of plants in their own right. The case of the analogy between sap and blood is a crucial example of this: Despite repeating the widespread comparison in his work, Malpighi cautiously spelled out the differences, laying bare the specific characteristics of the movement of sap within plants.[77]

The German botanist and physician Rudolf Jakob Camerarius (1665–1721) relied on Malpighi's systematic analogy between plants and animals, as well as on the experimental investigation of plants, for establishing his argument for sexual reproduction in plants, notoriously collected in *De sexu plantarum epistola* (1694). In these years, the microscopic observations of plants opened a new field of scientific investigation, framing (and circumscribing) the analogy between plants and animals to a specific role in the exploration of the organization of living nature and grounding the anatomical and physiological study of vegetation.

The Aftermath: Boerhaave and Hales

In the late seventeenth and eighteenth centuries, the representation of the animal-plant analogy recurred in the study of animal anatomy and saw the birth of the experimental science of plants. Within this framework, scholars studied plants as an important addition to understand anatomy and physiology. For example, the Dutch physician and anatomist Govard Bidloo (1649–1713) continued representing blood circulatory vessels shaped like plants, generally using botanical vocabulary, as he spoke of ramifications, trunks, and so on.[78] Yet, while naturalists delved into questioning the separation of realms, physicians unearthed the connections between animals and plants as a way to describe the structure and functioning of living nature in general.[79] In the *Dissertation on the sexes of plants* (1759, translated into English in 1786), Carolus Linnaeus (1707–1778) insisted that an analogy could be drawn between plants and animals—wood is equivalent to bone, plants have a system of channels transporting fluids similar to animal vessels, and generation too reveals similarities—as a way to clarify the field, while the anatomy of plants clearly emerged as an autonomous science.

Before Linnaeus, two more cases appear significant to understanding the contribution of the anatomy of plants to medicine. The first is the case of the Dutch physician Herman Boerhaave (1668–1738), whose chemical theory of plants contained in the landmark

---

77 Bertoloni Meli 2011: 248.
78 Bidloo, *Anatomia humani corporis*: tab. 38, fig. 5, fig. 2. See Fend 2019.
79 Gibson 2015: 85. See Delaporte 1982: 91–148.

textbook on chemistry, *Elementa chemiae* (1732), is consistent with animal physiology; the second is the English clergyman and physiologist Stephen Hales (1677–1761), whose *Vegetable Staticks* (1727) collects a number of experimentations on fluid pressure in plants and animals.

As Ursula Klein has pointed out, Boerhaave paralleled plants to animals under various respects: The juice in the stem is called the chyle of the plant, the roots and the body of the plants are the stomach and intestines of an animal, the transformation of such a chyle in nutriment is operated by the lungs of the plant, namely the leaves, the channels of plants are like arteries and veins, where juices circulate, the seed (*plantula*, as an echo of Malpighi's work) is like the animal fetus, and juices bleed from the bark and branches.[80] In unearthing the chemical composition of nature, Boerhaave ultimately claims that "we may discover a near resemblance between the elements of plants and animals, which latter are made up of the same matter with vegetables: the chief difference consists in the variety of their structure, and the quicker passage of the aliment thro' the bodies of animals, than of plants . . .".[81] Yet, in the chemical and anatomical study of plants, differences surface as well: Boerhaave asks how one should name the "golden yellow and very bitter [humor]" in aloe, since this is "neither arterial, nor venal . . .".[82] Grounded on his chemical studies, similarities between the physiology of animals and plants constituted a crucial point of departure, as Boerhaave's study of the movement of juices and transpiration in plants ran parallel to that of fluids in animals.[83] Yet, within the chemical study of bodily compositions and processes, he provided the analogy with a more precise condition, and plants with their own field of investigation.

In these same years, Hales published a study of the movement of juices within plants that paralleled a work on animals, *Haemastaticks* (1733). In his work, Hales collected a number of experiments performed with plants, starting from the Santorian investigation of the quantities of matter perspired by plants to the study of the force of the movement of sap, concluding with the study of the role of air in composing plants (and animals and minerals). Yet, Hales embedded these observations within a precise framework, namely that "since in vegetables, their growth and the preservation of their vegetable life is promoted and maintained, *as in animals*, by the very plentiful and regular motion of their fluids [. . .]; it is therefore reasonable to hope [. . .] there being, in many respects, a great analogy between plants and animals".[84] In conceiving both plants and animals as hydraulic apparatus and subjects of static, Hales, however, highlighted differences (such as circulation), ultimately making plants constitute an

---

80 Boerhaave, *Institutiones et experimenta chemiae*: 1.119–127; Boerhaave, *Elementa Chemiae*: 1.57–63 (English translation Shaw, 1741: 1.134–147). See Klein 2003. On Boerhaave and plants, see Offerhaus, Stefanaki, and van Andel in this volume.
81 Boerhaave, *Elementa Chemiae*: 1.69–70 (Shaw, *A new Method of Chemistry*: 1.154).
82 Boerhaave, *Institutiones et experimenta chemiae*: 1.126.
83 Verwaal 2020: 179–182.
84 Hales, *Vegetable Staticks*: 3.

autonomous field of scientific investigation,[85] if not even a laboratory to shed new light on living activities.[86]

These positions show how much the anatomical observations of plants shaped a science of vegetation that nuanced the analogy between animals and plants in medicine. While this latter has been a commonplace in medical traditions since ancient times, and acquired special momentum during the Renaissance and early modern time, the microscopic study of plants in the second half of the seventeenth century secured a new perspective on the medical understanding of the physiological processes in common with animals. In this chapter I have tracked this change from the work of Fabrici and Harvey, in which the analogy disclosed a more theoretical constitution. In the first half of the seventeenth century, physicians slowly referred to plants to understand vegetative activities such as nutrition, growth, and generation, but also the internal movement of juices. Developing from the embryological observations of eggs and seeds, Harvey brought this analogical view to a strong consequence, claiming that the embryonic development is *botanical* and that to understand the beginning of life one should start from the plant. Still, Harvey's *primordium vegetale* outlines a theoretical understanding of living nature, encompassing embryology within a philosophical framework.

Then, scholars and physicians of the Royal Society welcomed Harvey's suggestion to delve into plant anatomy, as they started observing the structure and functioning of plants, confirming the analogy with animals. This especially surfaces in Glisson's and Hooke's works but appears as a common procedure at the time. Yet, only with Grew and Malpighi did plant anatomy acquire a methodological stance, bearing important consequences in the understanding of the animal-plant analogy, and on medical knowledge. On the one hand, as in Grew, a science of plants developed as an autonomous field of knowledge, namely a philosophy of botany, in which the activities of plants are studied in their own respect. On the other hand, as in Malpighi, grasping the functioning of plants shed light on the study of the living processes shared by animals and plants. As Malpighi wrote in response to Girolamo Sbaraglia (1641–1710), such studies on "the anatomy of plants [. . .] have not labored over [. . .] for the sake of practical medicine and to discover remedies from it, but to throw light on that part of Nature [. . .] and to add to knowledge for the sake of natural history".[87] In their observations of plants, Grew and Malpighi uncovered the similarities between animals and vegetal bodies, and circumscribed the analogy to a few physiological processes, ultimately revealing the anatomy of plants as a crucial addition to medical understanding. Within their work on chemistry and static, Boerhaave and Hales repeated the analogy between animal and plant, now grounded on the anatomical study of vegetation, to enhance medical knowledge.

---

85 Hales, *Vegetable Staticks*: 358–376. See Begley 2022.
86 Bandinelli 2013: 23.
87 Malpighi, *Opera posthuma*: 89; Adelmann 1966: 3.1093.

In sum, grounded on the anatomical observations of bodies, the animal-plant continuity developed as a fruitful methodology to boost the medical understanding of living processes. This is something Malpighi confirmed to Sbaraglia, as he wrote that "if Galen", together with the Renaissance and early modern physicians who reappraised the latter's medical interpretation of the animal-plant analogy, such as Fabricius and Harvey, had known "the structure of plants responsible for the propagation of the aliment [. . .] [they] would surely *not* have thought that this sort of philosophy (namely, Malpighi's and Grew's anatomical and physiological studies of plants) is useless to medicine".[88] Developing from the animal-plant analogy, the anatomy of plants emerged as a crucial field to complement the physiological knowledge of living nature.

---

[88] Malpighi, *Opera posthuma*: 89; Adelmann 1966: 3.1093. [Italics is mine.]

Edoardo Pierini
Opium Taking: Blurring Experimentation and Pharmaceutical Theories

> What is the greatest Dose, any men are known to have taken of Opium?
> And how prepared?
> ("Inquiries for Turky": 360.)

Introduction

In 1665, the Royal Society of London made a call for enquiries in its *Philosophical Transactions*, asking "whether the Turks do not only take Opium themselves for strength and courage, but also give it to their Horses, Camels and Dromedaries, for the same purpose, when they find them tired and faint in their travelling?" and, "What is the greatest Dose, any men are known to have taken of Opium? And how prepared?".[1]

These enquiries reflect a specific trend in the pharmacological investigations of the time. The Royal Society was interested in broadening knowledge of the effects of opium, and especially in examining the effects of its administration to animals and measuring the maximum dosage the human body could tolerate. The reference to the Turks related to their undisputed reputation for opium addiction.[2] In answering these queries, scholars and physicians performed a number of experiments involving opium taking. Known since antiquity, opium fitted the Hippocratic definition of *pharmakon*, as it could have either a curative or a toxic effect on the body, depending on its dosage.[3]

Reappraising ancient sources, Renaissance scholars and pharmacists gave renewed attention to opium, drawing especially on the work of Theophrastus (ca. 371-ca. 287 B.C.),[4] Aulus Cornelius Celsus (*fl.* 1st cent. A.D.), Dioscorides (*fl.* 1st cent. A.D.),[5] Scribonius Largus (A.D. ca. 1-ca. 50),[6] and Galen (A.D. 129-after [?] 216).[7] As a result, sixteenth-century

---
1 "Inquiries for Turky": 360–362.
2 Rycaut, *The Ottoman Empire*: 135: "they drink Wine not to appear Cynical or unsociable, but more generally addict themselves to Electuaries composed of Opium, which tends to augment their natural stupefaction". See also Pierini 2021.
3 See Touwaide 1996; Jouanna 1999; Nutton 2005; Totelin 2008; Totelin 2018.
4 Theophrastus, *Historia plantarum*.
5 Dioscorides, *De materia medica*.
6 Scribonius, *Compositiones*.
7 Galen, *De simplicium medicamentorum*.

physicians administered the opium drug not only as an analgesic and a narcotic – that is, making use of its most evident properties – but also as a diaphoretic and as a remedy for diarrhoea, vomiting, and coughs. Along with the therapeutic uses of the drug, however, these physicians were interested in its theoretical status and its *supposed* virtues, which they found in the qualitative medical system elaborated by Galen. In this context, however, the problematic assessment of opium's quality represented a dangerous issue that Galenic physicians had to address.

Recently, Saskia Klerk has analysed the extent to which opium was a Lydian stone for the reappraisal or rejection of Galenic pharmacology. The studies of opium appearing in Pietro Andrea Mattioli's (1501–1578) *Di Pedacio Dioscoride Anazarbeo Libri Cinque della historia, et materia medicinale tradotti in lingua volgare italiana*,[8] Rembert Dodoens's (1517–1585) *Stirpium historiae pemptades sex*,[9] and Adrianus Spigelius' (1578–1625) *Isagoges in rem herbariam*[10] show that these authors challenged Galenic tenets, despite endorsing a Galenic pharmacology.[11] An ambiguity ultimately emerged in the investigation of opium. This conflict developed around the experimentation with opium to establish its uses. From the late sixteenth century onwards, learned physicians across Europe performed a diverse array of trials, indicating the vigorous debates taking place in this period over old and new theories of drug action, as Evan Ragland has shown.[12] Alisha Rankin has recently examined how sixteenth-century European physicians devised specific trials for testing poisons and antidotes on condemned criminals, such as the case of opium administration discussed by Gabriele Falloppio (1523–1562).[13] Experimentation with drugs, and with opium in particular, characterized seventeenth-century studies in medicine and pharmacology. Melvin P. Earles has highlighted the role of seventeenth-century pharmacological experiments in which physicians explored whether opium and other drugs acted directly on the nervous system or indirectly via absorption into the blood;[14] Andreas-Holger Maehle has studied the new experimental basis of opium investigation, stimulated by the discovery of the circulation of blood and the advent of iatrochemistry;[15] and Claire Salomon-Bayet and, more recently, Emma Spary have written about the auto-experiments with the drug conducted by the pharmacist Moyse Charas (1619–1698).[16] At the same time, institutions such as the Royal Society and the Paris

---

8 Mattioli, *Di Pedacio Dioscoride Anazarbeo Libri cinque*.
9 Dodonaeus, *Stirpium historiae*.
10 Spigelius, *Isagoges*.
11 Klerk 2014: 287–316.
12 Ragland 2017a: 503–528, and 2017b: 331–361.
13 Falloppio, *de Tumoribus*. See also Corradi 1886; Rankin 2017: 274–302; Leong and Rankin 2017: 157–182.
14 Earles 1961: 97–110, and 1963: 241–254.
15 Maehle 1998 and 1999.
16 Salomon-Bayet 1972: 125–150; Spary 2022: 49–67.

Academy of Sciences provided new motivations for virtuosi and learned scholars to investigate and experiment with pharmacology.[17]

In this article, I focus on the description and the background of pharmacological experiments with opium from the second half of the sixteenth century to the end of the seventeenth century. First, I deal with the theoretical debate over the virtues of opium, and I discuss the first experiments of this kind, describing the first trials involving condemned criminals conducted in the sixteenth century. Second, I highlight the gradual transformation in medical theory and in the experimental background that helped establish the practice of administering opium to living animals, both orally and by injection, in the seventeenth century. Third, I show how these experiments with living animals contributed to the development of new theories of opium action. Finally, I examine the factors that led to a rise in auto-experimentation with opiates among physicians in the last decades of the seventeenth century.

Early Trials with a Problematic Drug

Physicians in sixteenth-century Europe had not yet established a clear methodology for investigating the virtues of medicaments, and the first experiments of this sort that we find involved the oral administration of high doses of opium to test subjects in order to evaluate its effects. The opium poppy was certainly one of the most intensely debated medical plants, as the reassessment of the works of Galen and Dioscorides posed a number of obstacles to the understanding of its properties.

The Italian physician Pietro Andrea Mattioli seems to have been the first to recognize the problematic contradiction in classical theory concerning the opium poppy, whose bitter taste did not tally with the cold quality attributed to it by Galen and Dioscorides.

In the Galenic medical system, drugs were composed of combinations of the four elements and the four qualities, and they interacted with diseases by the logic of opposition: a disease caused by an excess of a certain element could be counteracted with a drug of the opposite quality, so that a physician would cure with a cold drug a patient with an excess of heat.[18] In addition to the qualities of drugs, physicians were expected to know their quantitative degrees on a scale of one (the smallest presence of a certain quality) to four (the greatest). When discussing the resin extracted from the opium poppy, Galen placed it in the fourth category of coldness, the strongest, claiming that "est enim ex quarto et ultimo refrigerantium ordine".[19] This attribution

---

[17] Hall 1974: 421–452; Porter 1989: 272–293; Cook 1990: 37–43.
[18] Teigen 1987: 60–64.
[19] Galen, *De simplicium medicamentorum temperamentis ac facultatibus*: 7.13 (ed. Kühn 12.73.18–74.1).

to opium of the fourth degree of coldness reflected the danger that was then associated with its use, with Galen writing in his *De simplicium medicamentorum temperamentis et facultatibus* that "valide enim refrigerat, ad stuporem usque et mortem perducens".[20]

In his work, while commenting on Dioscorides, *De materia medica*, Mattioli discussed the virtues of the opium poppy and noted that he had detected in its taste a "manifestly not mediocre amount of heat", but, probably aware of the theoretical consequences of this observation, he avoided any further comment.[21] He simply remarked that he did not want to be "cheeky", preferring to accept "the judgement of those who considered the qualities or faculties of opium very diligently before us".[22] Mattioli did not wish to challenge the ancient medical authorities, but his concrete critique nonetheless had important consequences and stimulated a lively debate among physicians and natural philosophers.

We find the consequence of this debate in the increasing relevance some physicians ascribed to the so-called 'experience' of Falloppio, one of the most celebrated physicians of the sixteenth century, who adopted a concrete approach to the problem of opium experimentation in medical practice. As a leading medical theoretician, Falloppio knew very well the debate surrounding the qualities of opium, and he seems to have truly believed in the Galenic idea of its coldness. In his work *De Tumoribus*, he advocated the use of warm medicaments like wine and theriac against a cold poison like opium. Following this pattern, he believed that the cold effect of opium could be countered by the heat produced by a fever, as he had demonstrated in an experiment "*in homine anatomizando*" on a condemned criminal.[23]

The first experiment involving the use of opiates emerged within the framework of Galenic medicine, probably because Galen's own insistence on the *qualified* use of experience, intended as a way of judging whether a particular view on the power of a drug was true, could be read as advocating the value of experimentation.[24] The doubts of a Galenic physician such as Mattioli probably originated from his practical experience in

---

20 Galen, *De simplicium medicamentorum temperamentis ac facultatibus*: 7.13 (ed. Kühn 12.73.15–16).
21 Mattioli, *Discorsi*: 493: "l'Opio, come benissimo, & diligentemente insegna Dioscoride, il quale quantunque sta tenuto da tutti frigido nel quarto ordine; nondimeno se dal sapore si conosce il temperamento delle cose, & parimente da gli effetti, ritrovo io, che l'Opio al gusto è amaro, & che tenuto in bocca vescica la lingua. Il che dimostra manifestamente, che sta in lui calidità non mediocre" [translation is mine].
22 Mattioli, *Di Pedacio Dioscoride Anazarbeo Libri Cinque*: 493: "Pure per non essere tenuto sfacciato, & contrario à tutta la caterva de i medici, me ne rimetto al giudicio di coloro, che avanti à me hanno benissimo essaminato i temperamenti suoi" [translation is mine].
23 Falloppio, *de Tumoribus*: 47.
24 van der Eijk 1997: 35–57.

Figure 1: *Domestic poppy*, in Pietro Andrea Mattioli, *I discorsi di M. Pietro Andrea Matthioli sanese . . . nelli sei libri di Pedacio Dioscoride Anazarbeo della materia medicinale*. Venezia: Appresso gli Heredi di Vincenzo Valgrisi, 1559: p. 539.

testing opium on himself or on his patients, and showed how experimental medicine could also emerge within the authoritative framework of Galenic medicine.[25]

The tradition of carrying out experiments "in facinorosis hominibus morte damnatis" was historically considered to have been started by the legendary king of Pontus Mithridates VI Eupator (b. 135, king 120, d. 63 B.C.), and was discussed extensively in classical medical sources.[26] The Roman physician Aulus Cornelius Celsus encouraged this kind of experimentation, especially vivisection, on condemned criminals, so that over the centuries good people would benefit from their suffering. In the Galenic treatise *De Antidotis*, the author outlined the practice of testing the antitoxic properties of theriac, a medical compound that included opium among its ingredients, on condemned criminals. Falloppio, conversely, used opium as a poison when a prince allowed him to administer a dose of two drachms to a condemned criminal who was suffering from a quartan fever. This high dose should have killed a healthy man, but in this case the heat generated by the fever mitigated the cold poisoning effect of the opium, allowing the man to survive. Unfortunately for him, Falloppio reported, the man then asked for more opium and a promise of amnesty if he survived the dose, but this additional administration of two drachms of opium eventually killed him.[27]

From a theoretical point of view, this experiment was a clear success for supporters of the Galenic system, as it confirmed that the qualitative value of medicaments, and in this case of coldness in particular, could influence medical practice. Physicians generally considered just half a drachm of opium to be a dose that could harm a patient, while Falloppio showed that even a higher dosage could be tolerated in certain situations, where the coldness of opium was balanced by a warmer counterpart. In this way, the medical audience would have been convinced to believe in the effectiveness of compounded medicaments like theriac and laudanum, which, even if they contained a good amount of opium, were considered qualitatively warm because of the presence of high quantities of warmer substances such as cinnamon and viper's flesh. The spread of these kinds of medicament from the sixteenth century onwards clearly demonstrates the success of this view, and again reflects the ambivalent role of opium, which in the experiment discussed by Galen acted as an antidote, while Falloppio used it as a poison. This ambivalence, however, did not pose a problem in the Galenic system of medicine, as these compounded medicaments were believed to preserve the curative properties of opium while mitigating its dangerous power.

---

25 Klerk 2014: 287–299.
26 Totelin 2004: 1–19.
27 Falloppio, *de Tumoribus*: 47–48: "Nam Princeps iubet ut nobis dent hominem, quem nostro modo interficimus, & illum anatomizamus: cui exhinui 3.ij.opii & adveniens paroximus (nam hic patiebatur quartanam) prohibuit opii actionem. Hic gloriabudus rogavit, ut bis adhuc exhiberemus, quod si non more retur, ut procuraremus pros eius salute Apud Principem rursus illi exhibuimus extra paroxysmum 3.ij. opii, & mortuus est".

At the time, the fact that Christian men could be used to test poisons apparently did not cause any ethical concern. Even Pope Clement VII (1478–1534), in August 1524, commanded his physicians to 'carry out an experiment' testing a promising new poison antidote on two Corsican assassins who had been condemned to death. Interestingly enough, this anecdote was reported by Mattioli and involved his master of surgery, Gregorio Caravita (*fl.* 16th cent.). Mattioli's acquaintance with this kind of practice again indicates his reluctance to challenge the authority of ancient authors on the opium problem, as he never reported on any experiments that might have supported his own convictions.

We do not know how often these kinds of opium experiments were tried, though there is a statement by the Dutch physician Levinus Lemnius (1505–1568) in *Les Occultes merveilles et secretz de Nature* (1567) that seems to suggest that they spread to the Italian medical context. In Lemnius's view, however, the administration of opium to criminals was not aimed at testing its cold quality, but rather at trying to preserve the humours and the animal spirits of test subjects for further dissection.[28] From the Renaissance onwards, we find a renewed interest in the study of anatomy, which included the specific desire to uncover the hidden mechanisms of the human body. This attitude had significant consequences for the role of opiates, as vivisection and pharmacological experimentation were closely connected.[29] According to Lemnius, a solution of wine and three drachms of black poppy juice would first make a man merry, and then induce a deep sleep that would eventually lead to death. He claimed that the power of opium penetrated the vital parts so quickly that an observer might see its effect on the heart of the dissected man.[30] In the sixteenth century, it also seems that physicians carrying out vivisections and dissections quite often administered opiates and other hallucinogenic drugs to make tissues and organs more distended and relaxed for their examinations.[31]

Other reports of experiments on condemned criminals in sixteenth-century Italy suggest that this kind of practice was aimed mainly at testing poisons and their alleged antidotes on human beings.[32] The role of opium would in this case have been more

---

28 [Lemnius], *Les Occultes merveilles*: 421–422: "Les medecins en Italie, en certain temps de l'annés demandét aux magistrats & gouverneurs des villes, les malfaiteurs qui sont condannes à mourir par exécution de Justice, pour les ouvrir & déchiqueter, à celle fin que ceux qui etudiét en medecine se puissent exercer au fait de Anatomie. Et pour obvier qu'aucunes humeurs point ne soient dissipées en eux, ou que les plus gros espris ne se perdent, & que tout se demontre plus manifestement, ils leur donnent à boire en bò vin pur, deux ou trois drachmes de jus de pavot noir".
29 Camporesi 1985: 115–126.
30 [Lemnius], *Les Occultes merveilles*: 421–422: ". . . après avoir beu lequel bruage, ils commencent premièrement à se refiouir & à rire tant qu'ils peuvent comme fouls, puis soudain eprins d'un profond sonneil, ils meurent tout endormis, un tel bruage ayant si vite penetré ès veines & aux parties vitales, on voit à l'œil comme un tel ius leur saisi le cœur".
31 Camporesi 1985: 129.
32 Rankin 2017: 277.

ambivalent, and often probably less important for the main goal of the experiment. Apart from the experiment of Falloppio, which reflected a theoretical controversy, other reports of similar experiments did not mention any administration of opiates. The statement by Lemnius nonetheless suggests that opiates were used both as a poison, whose value as an inducer of death had been highlighted by Theophrastus, and to condense the spirits in order to facilitate the subsequent dissection.[33] We also cannot exclude the possibility that opiates were also used as antidotes, as Falloppio showed that an appropriate means of counteracting a cold poison would be to administer a warm solution of theriac, which included among its ingredients a considerable quantity of opium.

A New Experimental Approach: The Rise of Animal Experimentation

From the beginning of the seventeenth century onwards, the theoretical debate surrounding the qualitative nature of opium within the Galenic framework seems to have gradually subsided. The proliferation of new theories about its action, based mainly on its iatrochemical or occult properties, polarized the medical debate, with Galenic physicians tending to accept its cold quality as an orthodoxy and avoiding any experiment that could have helped their opponents. This trend, however, did not stop the conduct of experiments, which, helped by the spread of these new medical systems and by several important discoveries, started to be carried out more often.

Particularly important were the consequences of the emergence of the chemicomedical doctrines developed by Philippus Aureolus Theophrastus Bombastus von Hohenheim, known as Paracelsus (ca. 1493–1541) and Jean Baptiste van Helmont (1579–1644), who considered traditional book learning worthless and demolished the Galenic doctrine of qualities. Paracelsus revolutionized the Galenic concept of the quantitative degree of elements, with the result that the fourth degree of coldness attributed to opium was no longer linked to its effects.[34] Van Helmont went further, asking provocatively how opium could be cold and hot at the same time and claiming that its effects were caused by a hypnotic "sulphur" and an "acrid, sudorific salt", which calmed the vital principle *archeus*.[35] The discrepancy between opium's taste and its qualities again became a central argument in this harsh critic of Galenism:

---

33 [Lemnius], *Les Occultes merveilles*: 420–421; Theophrastus, *Historia plantarum*: IX.303.
34 Paracelsus, *De gradibus*: 1.4–7 (ed. Sudhoff, 1922–1933, IV.9–12). See also Debus 1977: 387.
35 Van Helmont, *Ortus medicinae* [Chandles, *Van Helmont's works*, 1664]. See also Maehle 1999: 131–132.

Galen according to his manner, transcribing Dioscorides word for word, and being willing to measure the Elementary Degrees of Simples, he hath not attempted it by the discretion of his Tongue: and so he divined, that more of the fire had concurred to a mixture, where he found the more sharpness and bitterness. Which thing, the Schooles even till now hold as authenticall; although Opium being bitter, hinders it.[36]

Van Helmont used the same argument as Mattioli, in highlighting the fact that the Galenic theory lacked empirical validation. Unlike Mattioli, however, Van Helmont did not shy away from controversy, and used this argument as a foundation for a new system of medicine. While his new explanation for opium's effects was still vague, and was unproven by the results of any experiment, these critiques stimulated demand for a new chemical understanding of the drug.[37] Also, within this new theoretical context, opium medicaments gradually started to be considered less dangerous and appear to have been used more frequently in the European pharmacopoeia.

The other watershed development that contributed to changing the experimental approach towards opium followed the discovery of blood circulation, when physicians were investigating the transportation of drugs throughout the body. In this instance, opium was not the subject of discussion but rather a tool to test a particular thesis, as in the examination of the effects produced by certain substances if applied through intravenous injection. This new practice of drug injection seemed best suited to being tested on animals, especially dogs, to avoid any moral issues arising from its cruelty.

Experimentation on living animals was also a good method of observing the action of poisons and antidotes. The problem of the discrepancy in the results of such experiments between humans and animals was well known, and the (pseudo-)Galenic treatise *De Theriaca ad Pisonem* had already pointed out that experimentation on animals was inferior to that on human subjects, but that in some cases it could nonetheless be useful, especially in the search for antidotes. The author of this tract recorded that he had tested the antidotal properties of theriac on animals that he had put in contact with poisonous snakes. The animals who had not been given the theriac had died immediately after being bitten, whereas those who had been given the theriac had survived.[38]

Robert Boyle (1627–1691), probably the most influential British empiricist, observed that animal experimentation was most suitable for the investigation of antidotes and wound remedies, with an eye to finding treatments for human conditions that could be artificially induced in animals.[39] There is no evidence of the persistence of the administration of poisons and antidotes to condemned criminals in the seventeenth century, as Boyle observed that "I likewise propos'd, That if it could be done

---

36 Van Helmont, *Ortus Medicinae* [Chandles, *Van Helmont's works*, 1664]: XI, 69.
37 Maehle 1999: 131–132.
38 Galen, *De Theriaca*: 2 (ed. Kühn 14.215.8).
39 Earles 1963: 242.

without either too much danger or cruelty, trial might be made upon some human Bodies, especially those of Malefactors".[40] Given this, the derivatives of the opium poppy seemed a perfect choice for animal testing. They could work both as poisons and as antidotes, and their narcotic and 'stupefiant' effects were easily detectable in animals too.

The first experiment involving opium and animals was recorded by Robert Boyle, whose medical and chemical knowledge certainly gave him a deep understanding of the drug's effects. He stated, "That opium is reckon'd by Physitians among Poisons, I need not tell you; and yet such powerful Remedies may be made with it for many desperate Cases".[41] The idea that a poison could act as a medicine had long been established via the Greek notion of *pharmakon*, and Boyle, like the majority of the medical community, believed that a poison's noxiousness "is not incorrigible, but that by Mans Art and Chymical Preparations, they may be made, not onely innocent and harmless, but useful too".[42]

Following this introduction, Boyle explained that he had discussed this matter with John Wilkins (1614–1672) and Christopher Wren (1632–1723), and that "the latter of those *Virtuosi* told us, That he thought he could easily contrive a way to convey any liquid Poison immediately into the Mass of Blood".[43] The experiment, which was conducted in 1656, was carried out as follows: "with the assistance of some eminent Physicians, and other learned Men a large dog" was bound to a table and an incision was made in "the hind leg, where the larger Vessels that carry the Blood are more easie to taken hold of".[44] Boyle and his colleagues then "[made] a Ligature upon those vessels, and to apply a certain small Plate of Brass . . . made a Slit along the Veins, from the Ligature towards the Heart, great enough to put in at it the slender Pipe of a Syringe. By which I had propos'd to have injected a warm solution of Opium in Sack, that the effect of our Experiment might be more quick and manifest".[45]

The structure of the experiment shows that the vivisection of the dog was necessary to ensure that the poison directly reached the bloodstream and the internal organs. Boyle noted that he had decided to inject a very popular electuary made of opium and fortified Spanish wine, whose effects were straightforward and well known, as it provided the base of one of the most famous medicaments of the time, Sydenham's laudanum. This choice of infusion stemmed from the hope that "the effect of our Experiment might be more quick and manifest", and indeed this worked well, as "a small Dose of the Solution or Tincture into the open'd Vessel, whereby, getting into the mass of Blood it was quickly, by the circular motion of that, carry'd to

---

40 Boyle, *Of the Usefulness of Experimental Naturall Philosophy*: sect I 64/5, 329.
41 Boyle, *Of the Usefulness of Experimental Naturall Philosophy*: sect I 64/5: 231–232.
42 Boyle, *Of the Usefulness of Experimental Naturall Philosophy*: sect I 64/5: 231–232.
43 Boyle, *Of the Usefulness of Experimental Naturall Philosophy*: sect I 64/5: 327.
44 Boyle, *Of the Usefulness of Experimental Naturall Philosophy*: sect I 64/5: 327.
45 Boyle, *Of the Usefulness of Experimental Naturall Philosophy*: sect I 64/5: 328.

the Brain, and other parts of the Body".[46] The *tortured* dog lost some blood and struggled violently "before the Opium began to disclose its Narcotick Quality, and almost assoon as he was upon his feet, he began to nod with his head, and faulter and reel in his pace, and presently after appear'd so stupifi'd, that were Wagers offer'd his Life could not be sav'd".[47] Luckily, Boyle reported that the poor dog "not onely recover'd, but began to grow fat so manifestly that 'twas admired", and "having made him famous, he was soon after stoln away from me".[48]

The experiment was a success: the opium's narcotic and anodyne properties revealed themselves quickly and confirmed Wren's intuition about the possibility for certain substances to be conveyed directly into the bloodstream. Boyle and his colleagues also repeated the experiment, which demonstrated "that the Plate was not necessary, if the Finger were skilfully employ'd to support the Vessel to be opened; and that a slender Quill, fasten'd to a Bladder, containing the matter to be injected, was somewhat more convenient then a Syringe; as also that this notwithstanding, unless the Dog were pretty big, and lean, that the Vessels might be large enough and easily accessible, the Experiment would not well succeed".[49] Opium was particularly suitable for this test because of its powerful effects, and Boyle acknowledged that he "wish'd, that not onely some vehemently working Drugs, but their appropriated Antidotes (or else powerful liquid Cordials) and also some altering Medicines, might be in a plentiful Dose injected".[50] These comments suggest that these early injection experiments were not directed specifically towards the study of the pharmacology of opium. Boyle and his colleagues were much more concerned with this new method of application, whose effects promised to be basically the same as those of oral administration.[51]

Following Boyle and his colleagues, other physicians tried similar experiments to provide new evidence on the effectiveness of drug infusion. Johann Sigismund Elsholtz (1623–1688) was court botanist, alchemist, and physician of Friedrich Wilhelm of Brandenburg's botanical gardens, and in 1667 he published his *Clysmatica nova* in Latin,[52] adding more content and illustrations to the German edition of 1665. This work was particularly important as it included not only an illustrated account of the author's experimental study on intravenous injection, but also a discussion of the state of the field in the seventh decade of the seventeenth century.

Elsholtz described in considerable detail his series of experiments that had followed the discovery of blood circulation, which had had a very practical goal: the demonstration of the effectiveness of injecting drugs directly into the blood vessels. In order

---

46 Boyle, *Of the Usefulness of Experimental Naturall Philosophy*: sect I 64/5: 328.
47 Boyle, *Of the Usefulness of Experimental Naturall Philosophy*: sect I 64/5: 328.
48 Boyle, *Of the Usefulness of Experimental Naturall Philosophy*: sect I 64/5: 328.
49 Boyle, *Of the Usefulness of Experimental Naturall Philosophy*: sect I 64/5: 328.
50 Boyle, *Of the Usefulness of Experimental Naturall Philosophy*: sect I 64/5: 328.
51 Maehle 1999: 133.
52 Elsholtz, *Clysmatica nova* 1667 (tr. Gladstone, 1933).

Figure 2: *Injection into crural vein of a dog*, in Johann Sigismund Elsholtz, *Clysmatica nova sive ratio qua in venam rectam medicamenta*. Coloniae Brandenburgicae: Reichelius, 1665: p. 13.

to convince his readers, he provided a highly accurate and precise description of his experiments, supplemented by useful technical details and beautiful illustrations.

In the first chapter of his work, Elsholtz described an experiment involving drug infusion in a dead woman, but explained that then, unsure of the consequences for living humans, he had decided to begin experimenting with dogs.[53] He had managed to obtain a large number of dogs in order to make multiple attempts, and was able to

---

53 Elsholtz, *Clysmatica Nova*: 43.

describe in great detail how to make an incision in the crural vein and inject drugs with a syringe. As a trial to test how the animal would react to this procedure, he had first injected some water, and had noticed that the dog had survived without too much concern.[54] He had then tried to inject another dog with a solution of Spanish wine, but had not noticed any sign of drunkenness in it. Remaining convinced of his initial assumption, Elsholtz had moved on, injecting a new dog with an ounce of liquid "*extractum opii*". At this point, he provided a vivid description of the dog's reaction over the course of the experiment. The dog had been "very strong and ferocious", but after the "infusion" had become "astonishingly tame", and then "so sleepy that did not arouse himself when he was struck hard on the nape of the neck with a finger".[55] Elsholtz had convinced himself of the dog's "incipient stupor" when he had noticed that only touching the dog's injured leg or stabbing a needle deep into its tongue provoked a reaction.[56] The goal of his experiment had been to determine how a drug would work if injected directly into the blood vessels, and his choice of a tincture of opium in this case had probably been based on the extent to which its effects on the animal would be easily discernible. The well-known anodyne and narcotic properties of the opium poppy had seriously affected the dog and confirmed Elsholtz's initial assumption.[57] It must be emphasized, however, that in this case too the experiments were not aimed specifically at the investigation of the pharmacology of opium, but rather at the study of drug infusion directly into the bloodstream.

---

54 Elsholtz, *Clysmatica Nova*: 45–47.
55 Elsholtz, *Clysmatica Nova*: 46: "This dog was very strong and ferocious, and all through the infusion made an uproar that nearly exhausted us; but as soon as the operation was over he became astonishingly tame. During the next hour his eyes grew dull and half-closed, and he became so sleepy that he did not arouse himself when he was struck hard on the nape of the neck with a finger".
56 Elsholtz, *Clysmatica Nova:* 46: "It is true that as often I touched his injured foot he would raise his head and then immediately lie back as if overcome by sleep. About noon I noticed that his tongue was hanging out, and pressed it with my finger. He did not move in response to this pressure, and so I pricked it with a needle. Since he did not notice this, I thrust the needle in deeply; then he raised his head a little, but almost instantly let it fall back. Convinced in this manner of his incipient stupor, I began to prick his uninjured foot, and observed that the dog made no movement when the needle was thrust through the skin in the usual way; but when it was continually plunged up to the head into the flesh itself, he gave signs of feeling".
57 Elsholtz, *Clysmatica Nova*: 46: "Convinced in this manner of his incipient stupor, I began to prick his uninjured foot, and observed that the dog made no movement when the needle was thrust through the skin in the usual way; but when it was continually plunged up to the head into the flesh itself, he gave signs of feeling. At about one o'clock in the afternoon he got up, turned around in a circle several times, and then fell down and went to sleep again, like a drunkard. He also refused bread and a pan of water when it was put to his mouth, so that he seemed to have hydrophobia. Then, since I opportunely recollected that he was a hunting dog, we aroused him to the chase by fictitious shouting. He struggled to come to himself and get up; but the attempt was useless because of his stupor and the condition of his legs, which were as if paralyzed. So he slept the rest of that day and all the next night and day before he came to himself and was able to stand steadily on his feet. Nevertheless, he finally became perfectly well".

In the sixth decade of the seventeenth century, these kinds of experiment were performed and discussed by members of the various newly established European scientific academies and societies, including the Royal Society and the Académie des Sciences de Paris. This new scientific context provided a solid foundation for collective discussion and an increasing focus on experimental science, encouraging the transmission of medical and pharmacological research. Recent historiography has in particular highlighted the importance of physicians in the early Royal Society, attested by an increase in medical communication and written records of clinical practice, and also by a focus on medical curiosities and wonder drugs.[58] As a consequence, the growing number of accounts of drug testing became more detailed and precise, as physicians and practitioners began to invoke 'public interest' as a new justification for their work.[59] This new experimental approach to drugs kept opium, with its peculiar status as both a dangerous poison and a formidable antidote, centre stage.

The importance of experimentation with opiates is attested by the discussions that took place at the meetings of the Royal Society that resulted in the publication of the call for enquiries in the *Philosophical Transactions* in 1666, where interest was expressed in the alleged Turkish administration of opiates to horses, camels, and dromedaries, which seemed to mirror the European experimentation with opium and dogs. The idea that the Turks took opium to make themselves stronger and braver first appeared in an early account by Pierre Belon (1517–1564) in 1555,[60] but its dissemination throughout Europe can be attributed to the influence of Dutch botanist Carolus Clusius (1526–1609), who translated Belon's work into Latin for an international readership in 1589,[61] and then reprinted it as an appendix to his popular *Exoticorum libri decem* in 1605.[62] In the mid seventeenth century, the idea that opium could work in the same way in Turkish animals was appealing and worthy of attention. The Royal Society's call for enquiries apparently did not receive any direct response, but interest in the administration of opiates to animals continued in the following decades. The administration of opium to animals was repeated in other experiments under slightly different conditions, and seems to have been quite popular in the second half of the seventeenth century.

As Roy Porter has noted, this central role of the Royal Society in the exchange of medical information not only provided an important stimulus for the Society's fellows to "initiat[e] longer-term chains and networks of correspondents", but also contributed to establishing specific topics as objects of medical enquiry.[63] The case of opium injection is revealing in this respect, as the first injection trials by Boyle and Wren inspired

---

[58] Hall 1974: 421–450; Cook 1990: 397–436; Leong and Rankin 2017: 175–176.
[59] Leong and Rankin 2017: 175–176.
[60] Belon, *Les Observations de plusieurs singularitez*.
[61] Belon, *Les Observations de plusieurs singularitez* (Latin translation Clusius 1589).
[62] Clusius, *Exoticorum*.
[63] Porter 1989: 275.

further experiments across Europe. These were communicated to the Society's secretary, Henry Oldenburg (1619–1677), who seems quite optimistic in his comments on the account of a series of experiments with infusions sent by a Baltic physician: "As far as I see, both those Experiments, that met wth so much difficulty and contradiction at first (I mean yt of Infusion and Transfusion) may at last prove very beneficiall to the Health of Man".[64]

In 1678, William Courten (1642–1702) carried out a number of similar experiments relating to the toxicology of opium in dogs and cats, an account of which was published in the *Philosophical Transactions* posthumously in 1712,[65] when Hans Sloane (1660–1753) provided an English translation of the original Latin manuscript to the Society.[66] Courten used opium both as a poison and as an antidote to test its properties, and in his manuscript, he recounted how in each experiment, after intravenous injection, symptoms of stupor had appeared more rapidly, accompanied by convulsions that had killed the animal. Compared with the experiments of Boyle and Elsholtz, Courten's tests appear to have been conducted with higher dosages, with different animals, and with greater attention to the dissection of their bodies. He also repeated the experiment discussed by Galen involving the administration of theriac to a dog that had been poisoned by a viper bite.[67] Of particular interest was his injection of a huge dose of fifty grains of opium into the crural vein of a cat, which eventually died after "Tremblings of her Limbs, Convulsive Motions of her Eyes, Earss, Lips and almost of her Breast".[68] In the subsequent dissection, Courten seems to have been interested mostly in the condition of the animal's blood; he stated that he "did not find the Blood much altered from its Natural State",[69] but apparently did not use his observations to

---

64 Porter 1989: 279; Oldenburg, *Letter to Boyle*, 3 December 1667, IV.6–7 [Hall and Hall 1965–1986: 4.6–7].
65 Courten and Sloane, "Experiments and Observations": 485–500.
66 Maehle 1999: 54.
67 Courten and Sloane, "Experiments and Observations": 486: "Being willing to save this Dog . . . we thought fit to have recourse to several Remedies; and therefore, cupped and scarified the part that was wounded, and applied Treacle (Theriaca). After this we let him alone for about two Hours: But his Sleepiness increasing more and more, and his Vital and Animal Functions sinking, we were forced to have recourse to another Method of Cure".
68 Courten and Sloane, "Experiments and Observations": 493: ". . . we injected warm into the Crural Vein of a Cat, 50 Grains of Opium, dissolved in an Ounce of Water. The Cat presently after the Injection seemed very much dejected, but did not cry, only made a low, interrupted, complaining Noise. After this Tremblings of her Limbs, Convulsive Motions of her Eyes, Earss, Lips and almost of her Breast: Sometimes she vould raise up her Head, and seem to look about her, but her Eyes were very dull and deadish, and tho' she was let loose, and had nothing tied about her Head or Neck, yet her mouth was so filled with Foam and Froth, that she was like to be strangled. At last, her Convulsive Motions continuing, and being seized with a stretching of her Limbs, she dy'd within a quarter of an hour".
69 Courten and Sloane, "Experiments and Observations": 494: "Upon opening her Body, we did not find the Blood much altered from its Natural State".

formulate any kind of theory on the mode of action of the drug, confining himself to producing and describing the characteristic symptoms of intoxication.[70]

The increasing focus on the state of the blood shows how this pattern of experimentation through infusion gradually shifted attention from external symptoms to the internal effects of opiates on the blood. This attitude indicates a growing interest in the mechanico-chemical explanation of opium action, which in the final decades of the seventeenth century led to the birth of new theories.

A New Pharmacological Theory of Opium Action

Thomas Willis (1621–1675), a physician and a very influential member of the Royal Society, was probably one of the most important supporters of iatrochemistry in England. In his celebrated work *Pharmaceuticae Rationalis* (1674), he described the medicines then available and provided anatomical, pathological, and clinical observations, dedicating considerable attention to opiate medicaments and their effects. Following Van Helmont's explanation of opium action, he elaborated his own theory based on the action of "sulphurous particles", which extinguished the animal spirits in the brain and caused a "hipnotick" effect. In the section dedicated to "the Kinds, Preparations, and Forms of Opiates", he suggested that opiates "may work more strongly and virulently" on Europeans than on Middle Eastern peoples because of the "divers texture of various bodies", and provided a description of animal experimentation to support this assertion.[71] Just as the Turks, unlike Europeans, could consume opium without risk of death, he claimed that "Dogs devour Opium in a great quantity without any sleepiness or death", but that "a little dose of it presently kils a Cat".[72] These differences in the effects of the drug were explained by the Helmontian theory of opium action, "for that the spirits of this beast, so very saline volatile cannot at all endure the narcotic Sulphur".[73] Following this introduction, Willis recounted an experience where he had witnessed the oral administration of two drachms of opium, placed in a piece of bread, to an ordinary dog. At the very beginning, the opium had seemed to have its usual dangerous effect, as the dog, "after which growing very sick he became torpid or rather stupid, and drawing his Breath difficultly and laboriously seemed as if he was about to dye".[74] This was not surprising, as two drachms of opium was considered highly dangerous to human life. In this case, however, after half an hour the dog had "eased his belly

---

70 Maehle 1999: 54.
71 Willis, *Pharmaceutice rationalis*: 149.
72 Willis, *Pharmaceutice rationalis*: 149.
73 Willis, *Pharmaceutice rationalis*: 149.
74 Willis, *Pharmaceutice rationalis*: 149.

plentifully with the most horrid and troublesome stink that ever I smelt in my life",[75] and had managed to survive without any other complication. Willis then tried to explain why the drug had affected the dog differently:

> But that such a quantity of Opium did the less hurt to this Animal, perhaps the cause may in part be the notably acid ferment of his Ventricle, by which the Particles of the narcotic Sulphur might be somewhat broken and overcome; and so by reason of this Sulphur being very much dissolved by the acid Menstruum of the Stomach, the highly stinking dejection of the Belly proceeded.[76]

In this way, Willis explained how the dog's "acid ferment" had counteracted the narcotic power of the opium, which had been caused by its "Sulphur". The strong and acidic stomach of the dog had saved it by breaking up and purging the narcotic and dangerous sulphur. This reaction was not limited to dogs, but could also be observed in other bodies, as whenever "an acid body acts on a sulphureous, there is excited an ingrateful smell".[77] This experiment revealed how the "Particles of Opium" exerted their action "on the dogs spirits" located in the "Encephalon", causing "the torpor and difficult and disturbed Respiration", but in contrast to "the spirits of a Man, or of other Beasts", the "dogs spirits" were able to better "resist the narcotic Venom" and finally overcome it through purging.[78] To support his theory, Willis provided an account of another "anatomical experiment" he participated in some years earlier.[79] This experiment appears to have been very similar to the injection of opium carried out by his friends Robert Boyle and Christopher Wren:

> Many years ago we saw about three ounces of the tincture of Opium done in Canary tranfused into the Jugulary Vein of a little Dog, the vein being closed the dog ran about after his wonted manner and seemed little or nothing affected or changed, after a quarter of an hour he began to be a little stupefied, to nod his head, and at length began to fall into a Sleep, but not permitting him when we had hindred him a while by beating, terrifying, and coursing him about, at length the soporous Affection being so done away he became whole and lively enough.[80]

On this occasion, three ounces of a tincture of opium had been injected into the jugular vein of a little dog, though no other details were specified. Unlike in the earlier

---

75 Willis, *Pharmaceutice rationalis*: 149.
76 Willis, *Pharmaceutice rationalis*: 149.
77 Willis, *Pharmaceutice rationalis*: 149: "For it is to be observed of other Bodies whilst that an acid body acts on a sulphureous, there is excited an ingrateful smell".
78 Willis, *Pharmaceutice rationalis*: 149: "But indeed 'tis not to be denied that the Particles of Opium do work on the dogs spirits which notwithstanding do more strongly resist the narcotic Venom than the spirits of a Man, or of other Beasts, as it is clear by the History but now mentioned; for as much as when the torpor and difficult and disturbed Respiration by reason of the spirits inhabiting the Encephalon being affected by the narcotic began to arise, those symptoms as soon as the narcotick matter was overcome by purging presently vanished".
79 Willis, *Pharmaceutice rationalis*: 149: "Besides I have known the truth of this opinion confirmed more by an anatomical experiment".
80 Willis, *Pharmaceutice rationalis*: 149–150.

experiment, which was not directed specifically towards the study of the toxicology of opium, this time there seems to have been a clear desire to investigate the effects of opium on dogs. For this reason, Willis did not linger on the description of the injection, but instead focused his observations on the dog's reaction, which had been quite similar to that in the experiment carried out by Boyle. The dog had initially appeared to be "stupefied", but in this case it had not been allowed to succumb to the opium's narcotic power and had become "lively enough" again. This reaction was again attributed both to the "Viscera of concoction by their ferments" and to the resistance of the animal spirits, and persuaded Willis that opium "is not always, nor alike in all, either an Hypnotick or deadly".[81] Willis referred to this procedure as an anatomical experiment, reinforcing the clear link between pharmacological investigation and animal experimentation aimed at the examination of the toxicology of opium. This claim also indicates how this kind of opium experimentation contributed to a shift in the status of the drug. Throughout the seventeenth century, physicians began to consider opium medicaments less and less dangerous, and increasingly rejected the traditional Galenic recommendation to administer it only in desperate cases and in very small dosages. This attitude was central to the emergence of a new pharmacology that progressively abandoned all ties to the Galenic system of medicine.

To summarize the situation in the second half of the seventeenth century, it appears that experiments on living animals were finally being recognized as a legitimate and objective means of studying the effects of opiates. The results of these experiments then contributed to the development of new pharmacological theories.

A New Way of Taking Opium: Auto-Experimentation

In the second half of the seventeenth century, the increasing attention given to the properties and uses of opiates in medicine required more specific experiments that could not be performed on animals. Even the idea of poisoning condemned criminals seems to have fallen completely out of fashion, and even the mere remembrance of this kind of practice shocked Christian physicians like Jean Astruc (1684–1766).[82] Probably because of the lack of serious alternatives, a number of physicians and learned men tried to shed light on this ambiguous drug by testing it on themselves.

The medical practice of testing drugs on oneself probably originated in ancient times, but there is no evidence of a generalized study of the properties of opiates conducted in this way before the end of the seventeenth century. In the sixteenth century,

---

81 Willis, *Pharmaceutice rationalis*: 150: ". . . as it appears from hence, the opiatick Poyson is so overcome either with the Viscera of concoction by their ferments, or else resisted by the animal spirits themselves, as that it is not always, nor alike in all, either an Hypnotick or deadly".
82 Astruc, *De morbis venereis*: II.203.

Pierre Belon briefly recounted taking opium during his journey to the Middle East, but he never tried to integrate his experience into a theory of any sort. Other physicians probably tried the drug, either as scholars seeking to determine its taste or simply as patients, but they did not publish any accounts of experiments in the typical sense.

The first clear report of systematized self-experimentation with opiates was published by Moyse Charas, a learned French apothecary, who gained fortune with his pharmaceutical boutique Vipères d'Or, where he became famous for his public preparation of theriac.[83] In his first published treatise, *Histoire naturelle des animaux, des plantes, et des mineraux qui entrent dans la composition de la thériaque d'Andromachus* (1668), he followed the ancient theriac recipe of Andromachus and Galen, but integrated it with his practical chemical skills. He described all the ingredients of this celebrated antidote and paid particular attention to opium, which he considered the most important. In the chapter dedicated to the opium poppy,[84] he displayed his acquaintance with the derivatives of this plant, and described how he had personally collected the latex from the plant's capsule to analyse its qualities.[85] In addition, he confessed that he had tried, "just out of curiosity", six grains of an extract of opium, claiming that he had not felt the narcotic power of the medicament at all, but had in fact felt absolutely fine, and even stronger than before.[86] This new attitude towards the danger associated with the drug differed consistently from the classical exhortation to take great care in dealing with it, and allowed Charas to claim that taking a much higher dose was perfectly safe.[87] This liberal use of opiates probably led Charas to develop some sort of dependency, as he seems to have taken opium very frequently. In his celebrated *Pharmacopée royale galénique et chymyque* (1676), published in Paris eight years after the treatise on theriac, he argued that he was certain of the effectiveness of his "Extraction d'Opium" because of the "innumerable experiences" he had had with the drug for a long time, and which he still had every day.[88] These

---

83 Salomon-Bayet 1972: 131.
84 Charas, *Thériaque d'Andromachus*: 74.
85 Charas, *Thériaque d'Andromachus*: 75: "Cette liqueur est en sortant en forme de laict, & venant peu à peu à se coaguler, elle change insensiblement de couleur, & s'obscurcit par succession de temps. Ce que j'ay verifié autresfois moy-mesme, ayant incisé des testes vertes de Pavot dans le Languedoc".
86 Charas, *Thériaque d'Andromachus*: 86: "Et je puis dire d'avoir pris moy-mesme une fois par curiosité, le poids de six grains de mon extrait d'Opium, & d'avoir bien remarqué que mon sommeil ne fut pas plus long qu'à mon ordinaire, à sçavoir de cinq ou six heures, mais je reconnus principalement en moy une tranquilité interne si douce & si agreable que rien plus, & je me sentis en mesme temps tres sensiblement fortifié, & fort en estat de faire toute sorte de fonctions, sans qu'il me restat aucune envie de dormir".
87 Charas, *Thériaque d'Andromachus*: 87: "Je suis pour tant asseuré que j'en pouvois bien prendre une dose beaucoup plus grande, & qu'il ne m'en fut arrivé aucun accident . . .".
88 Charas, *Pharmacopée royale*: 724: "C'est icy, selon mon sens, la plus pure & la meilleure Extraction d'Opium, qu'on puisse inventer, dont je puis hautement assûrer le bon succez, tant de la preparation, que de l'usage, pour les experiences innombrables que j'en ay faites depuis long tems, & que j'en fais encore tous les jours".

experiences confirmed his belief that opium was neither dangerous nor a narcotic, and he reinforced his assertion by describing the most important experiment he had conducted on himself.[89]

Charas related that he had wanted to convince another physician that the narcotic properties attributed to opium were not especially powerful, and that in this physician's presence he had taken the sizeable dose of twelve grains of opium, despite the fact that the physician had been very worried and had begged him not to do so.[90] He had taken the drug in the morning and had continued to work as usual in his pharmacy with no effect aside from some "embarrassment by the vapours of opium", which did not prevent him from sitting down and working until nine o'clock in the evening.[91] Determined to understand the effects of opium more deeply, Charas had gone to bed and had managed not to sleep the whole night, and as a result had experienced a state of "restoring wakefulness".[92] The only effect he noted that could have prevented him from sleeping was the itching caused by the diaphoretic properties of opium, which nonetheless had not overly concerned him.[93] In commenting on the experience, the French apothecary

---

[89] Charas, *Pharmacopée royale*: 725: "Je crois néanmoins fort à propos de joindre aux experiences que j'avois faites alors, une bien plus considerable que je fis encore quelque année aprés sur moy-mesme, dont voicy la veritable Histoire".

[90] Charas, *Pharmacopée royale*: 726: "Raisonnant dans ma boutique un jour d'Esté vers les huit heures du matin, avec un Medecin de Provence fort curieux, sur la nature & les effets de l'Opium, & luy voulant démontrer que la qualité somnifere; qu'on avoir jusques-là attribuée à l'Opium n'étoit pas telle qu'on se l'estoit imaginé, je coupay en sa presence par le milieu une grosse piece d'Opium, qui se trouva au-dedans fort belle & pure t & en ayant séparé du milieu un petit morceau du plus pur, qui pesa douze grains à bon poids, je l'arrondis avec mes doigts, & en ayant fait une pilule, je l'avallay devant luy, nonobstant les instances qu'il me faisoit-pour m'obliger de m'en abstenir, craignant quelque mauvais succez".

[91] Charas, *Pharmacopée royale*: 726: "Je continûay après cela d'agir dans ma boutique de mesme que si je n'eusse rien pris, jusqu'à l'heure du disner, qui fut un peu aprés midy, auquel tems, je me mis à table, mais aprés avoir mangé de la soupe à mon ordinaire, je me trouvay rassasié, & je recommençay d'agir dans ma boutique jusques sur les deux heures, qu'estant monté dans ma chambre je me couchay sur des chezes, à dessein d'y dormir une heure on deux; Mais y estant je n'eus aucune envie de dormir, quoy que je m'y trouvasse fort tranquille, & si satisfait de ce repos que j'y demeuray jusques vers les six à sept heures de soir- auquel tems quelqu'un estant venu me demander pour quelque malade, je me levay d'abord, mais me sentant en quelque sorte embarrassé des vapeurs de l'Opium, lors que je fus debout, je me remis sur les mesmes chezes, où je fus avec la mesme tranquillité jusques sur les neuf heures".

[92] Charas, *Pharmacopée royale*: 726: "Auquel tems n'ayant point d'envie de manger, je resolus de ne pas souper, afin de mieux connoître jusques où pourroient s'étendre les effets de l'Opium. Et pour le sçavoir, m'estant mis tout-a- fait au lit, je passay la nuit dans un fort agreable repos; mais ce qui est le plus remarquable, c'est queje n'eus jamais envie de fermer l'œil, & que je fus toujours éveillé de mesme que si j'euue esté debout".

[93] Charas, *Pharmacopée royale*: 726: "Il est bien vray que pen-dant tout ce tems-la, tant sur le lit, que sur les chezes, je sen-tois de tems à autre des demangeaisons par tout le corps, quiestoient des marques sensibles de la vertu diaphoretique de l'O-pium, & qui pouvoient mesme contribuer quelque chose à mes veilles, quoy qu'elles fussent tout- à fait sans inquiétude".

claimed that his certainty when talking about the effects of opium stemmed from his experimental knowledge, but also noted that he was a healthy man, and that the effects might be different for sick people.[94] These experiences convinced him that the strong diaphoretic properties of the drug derived from its volatile salt, which reacted with its sulphur and pushed the "sooty humours" (which caused the pain) towards the skin, thus causing the itching.[95]

Twenty years after the publication of the extremely successful *Pharmacopée*, Charas made another series of observations on the nature of the drug in a manuscript titled *Nouvelles observations sur l'Opium*, which he discussed at the Académie des Sciences in 1696. At that time, the apothecary had returned to France after his conversion to Catholicism and was working successfully as an experimenter at the Jardin des Plantes. Presumably encouraged by his academic success, he vehemently expressed his arguments against those who still considered opium too dangerous. He believed that those who subscribed to this view lacked "intimate knowledge" of the drug and avoided using it because of the fear generated by their false assumption.[96] Charas was aware that his position as a self-experimenter was quite rare, and expressed his regret that there were no other men as curious, strong, or determined to learn more in this field.[97] This suggests that he believed a collective effort by learned auto-experimenters could have finally resolved the debate surrounding the virtues of opium.

---

94 Charas, *Pharmacopée royale*: 727: "Car me trouvant d'une fort bonne constitution, & n'ayant d'ordinaire point de mauvaise humeur qui prédomine en moy, & qui eust esté capable dalterer les effets naturels de l'Opium, on doit plustost avoir égard, & se fonder sur ce qui m'est arrivé, & que je proteste estre fort veritable, que sur plusieurs autres experiences qu'on pourroit avoir faites sur des personnes malades, dont le divers estat, & la diverse constitution, ne peu- vent que diversifier beaucoup tous les effets de l'Opium ; Outre qu'on n'a pas accoustumé de le leur bailler en si grande dose, velt qu'on ne le donne gueres que depuis demy grain, jusqu'à un ou deux, Et si on y prend bien garde, à peine peut on remarquer qu'ils procurent aux malades aucun sommeil excessif, quelque crainte qu'on puisse avoir de sa qualité somnifere".
95 Charas, *Pharmacopée royale*: 727: "Le resultat de toutes les reflexions que j'ay faites en divers tems sur mes experiences, & surtout ce que j'ay veu arriver dans l'exhibition de l'Opium; est, que la vertu diaphoretique que j'ay reconnuë fort sensiblement en luy, est un effet de son Sel volatile, lequel s'étant uni & concentré avec son Soufre, & l'obligeant à suivre & à seconder ses mouvemens, pouffe vers les pores de la peau les humeurs fuligineuses & rongeantes, qui interrompent le repos, & qui font la cause des douleurs".
96 See Salomon-Bayet 1972: 148: "on ne doit pas estre surpris de la grande diversité de sentimens de ceux qui en ont escrit, et que le défaut d'une jintime connaissance de l'Opium, et la crainte de s'y tromper, ait porté quelques uns a s'en abstenir et a en croire l'usage dangereux". This is the transcription of the "Nouvelles observations sur l'Opium par Charas", whose manuscript is in Paris, Archives de l'Académie des Sciences, *Procès-verbaux*, t.XV-*bis*, ff. 194v–195r. See Spary 2022.
97 Salomon-Bayet 1972: 148: "L'on eut pu', et l'on pouroit encore en estre plus jnstruit s'il se trouvoit quelque nombre de persones robustes, aussi curieuses et déterminées que ie le fus".

Because of the lack of accounts by others of experiments with the drug, Charas reported on a new experiment he had conducted on himself, this one having been carried out when he was sick.[98] He related how, during the three months of his sickness, he had taken a daily dose of one grain of opium with great benefit, which had convinced him to continue to take the medicament every day, even once he had been restored to health.[99] Charas hoped that his experience would help others to appreciate the "beautiful effect" of the drug, stimulating further "curiosity to observe and make new experiences".[100]

One of those who enthusiastically followed in Charas's footsteps was the British physician John Jones (1644–1709), who praised opium to such an extent in his treatise *The Mysteries of Opium Reveal'd* (1700) that he described it as the most useful and effective of all medicaments. He argued that all "magnanimous" effects of the drug could be explained by the fact that it caused a "pleasant Sensation", which led to a relaxation of all sensitive parts of the body. Based on this psychosomatic explanation, he believed that in order to learn about the true effects of opium, Western physicians should look to the peoples of the Middle East, who used it often, and not the Ancient authors. He also underlined that physicians used to administer opium only to "Sick people . . . when [they] are going to Bed", which had precluded any possibility of developing or observing "those brisk Effects".[101]

This revolutionary approach to the study of the virtues of opium originated from the enthusiastic auto-experimentation that Jones had undertaken several times, which in all likelihood had given him such an appreciation for the stimulating and euphoric effect of opium that resulted in dependency. His enthusiasm for the drug emerges clearly in this passage, where he argues that a moderate dose of opium can cause a 'pleasure' that is almost impossible to describe:

---

[98] Salomon-Bayet 1972: 148: "mais n'ayant pu' savoir qu'aucun m'ait imité dans de pareilles expériences, ie crois qu'on sera bien aise, que ie publie icy de bonne foy ce que i'ay depuis peu soigneusement observé sur ma persone pensant tout le cours d'une maladie de trois mois . . .".

[99] Salomon-Bayet 1972: 150: ". . . ie crus a propos de prendre, comme je le fis reglément tous les jours sans aucune heure limitée, un grain d'Extrait d'Opium . . . Le sincere récit des éfets extraordinaires, que l'Extrait d'Opium, vien de produire sur moy, merite dien ce me semble, qu'on y reflechisse, et qu'on admire de plus en plus avec moy, qu'un remede si simple, et donné en si petite quantité, puisse produire des éfets si notables, si diferens, et si eloignez de toute aparence; i'ay aussi lieu d'esperer que les agréables éfets que i'en ay ressenti, en en prenant un grain chaque jour pendant trois mois et ma resolution d'en user encore longtemps en même dose, en santé comme en maladie, rendront son usage beaucoup moins suspect . . .".

[100] Salomon-Bayet 1972: 150: ". . . et que ceux qui y auront recours se troucant charmez de ses beaux éfetz, seront curieux de les observer et d'en faire de nouvelles expériences".

[101] Jones, *The mysteries of opium*: 44: "I observed, That those brisk Effects of Opium were not taken notice by our Physicians . . . First, Opium is seldom (if ever) given in these Western Nations, but to Sick People, (as the Ancients did) who are utterly incapable of those brisk Effects . . . Secondly, We (as the Ancients did) generally give Opium when People are going to Bed, by which means all Opportunity of Observation is lost".

> If the person keeps himself in Action, Discourse or Business, it [opium] seems like a most delicious and extraordinary refreshment of the Spirits upon very good News, or any other great cause of Joy, as the sight of a dearly beloved Person thought to have been lost at Sea. . . . It is indeed so unexpressibly fine and sweet a Pleasure, that it is very difficult for me to describe or any to conceive it, but such as actually feel it.[102]

Because of his deep knowledge of the drug's effects, Jones was highly sceptical of physicians who did not experience opium personally, as he did, and challenged their theoretical explanations, claiming that the stupefying effect of the drug was caused by "Ovation and Pleasure of the sensitive Soul and Spirits":

> Dr. Willis and others, having no true Experience, or Knowledge of Opium, imagined that it caused Courage, Bravery, Equanimity, ecc by stupefying the Senses, brain ecc making People inadvertent, dull, and inapprehensive; which is a great Mistake, and groundless Conceit; for it is a most certain Truth (which millions can affirm) that it produces those Effects by an Ovation and Pleasure of the sensitive Soul and Spirits, as generous Wine does before Men are tuddled, or overcome with it.[103]

In his critiques he made no exception of Thomas Willis, who did not have *true Experience* of opium and whose theory of opium action was groundless. This suggests that Jones did not consider Willis's experiments with animals sufficiently valid to provide a basis for his theories and believed that human experimentation was crucial for that purpose.

The experiences of John Jones show that at the beginning of the seventeenth century, auto-experimentation with opium shaped the emergence of new theories that were intimately connected with this new practice. This new approach to the investigation of opium's properties had a profound impact on traditional knowledge, and provided evidence that in all likelihood influenced the spread of opium use in non-medical contexts.

Conclusion

In early modern Europe, experimentation with opium reflected different attitudes towards and understandings of this ambiguous drug. In the sixteenth century, the discrepancy between the quality and the taste of opium in the Galenic system of medicine provided an ideal impetus for experimentation. Within this new theoretical framework, Falloppio conducted an experiment on a condemned criminal to demonstrate the cold properties of the drug. The account of Lemnius, meanwhile, indicates that Italian

---

102 Jones, *The mysteries of opium*: 20.
103 Jones, *The mysteries of opium*: 22.

physicians believed that opium's coldness preserved humours and animal spirits in good condition for further dissection or vivisection.

In the seventeenth century, the practice of testing drugs on condemned criminals came to an end, and physicians and learned men started to rely heavily on animal experimentation. The emergence of the iatrochemical theory of opium action and the discovery of the circulation of blood provided the theoretical basis for experiments involving the administration of opiates to living animals both orally and by infusion. The collective interest in scientific research promoted by the newly formed Royal Society helped to cement the popularity of this practice. Experimenters such as Boyle, Elsholtz, and Courten administered opium to living animals in order to understand whether drugs worked properly if injected directly into the bloodstream. These experiments were not aimed specifically at the study of the pharmacology of opium, but nonetheless facilitated the emergence of new theories. Willis seems to have been the first to use experiments to prove a new theory of opium action, administering opiates to animals to test his assumption about the toxicology of the drug. Indeed, it was only after this animal experimentation that he was able to claim that opium was not "either an Hypnotick or deadly", thus contributing to the process of dismantling the classical Galenic belief in the unmitigated toxicity of the drug.

In the final decades of the seventeenth century, due to the proliferation of opium medicaments throughout Europe, physicians like Charas and Jones began testing the drug on themselves. This new way of experiencing the drug had an immense impact on the theory of drug action, as both these authors not only rejected the classical view of its narcotic and dangerous effects, but advocated its use even on healthy men. In this way, they contributed to a profound change in the medical status of opium, which shifted from being considered a highly dangerous drug for desperate situations to being viewed as a remedy that could be used on a daily basis by healthy men. This change had deep and long-lasting consequences, as this new reputation of the drug persisted until the beginning of the twentieth century.

In conclusion, while the methodologies and theories lying behind the early experiments reflected and confirmed contemporary knowledge relating to the pharmacological action of opium, experimentation gradually became more important for the elaboration of new theories, and for the investigation of the properties of drugs more broadly. The evolution of opium experimentation also reflects the shift between the classical pharmacology of Galen and Dioscorides and a new pharmacology shaped by chemical and mechanical influences. This latter approach continued throughout the eighteenth century, and ultimately led the opium poppy to become the first plant whose alkaloid was isolated.

Federica Rotelli

The Accommodation of New World Plants in Early Modern Pharmacology: The Case of Cinchona Bark and the Challenges to Seventeenth-Century Galenism

Between the fifteenth and the sixteenth century, the period of great overseas explorations began in Europe. As a result of the first expeditions promoted by the Castilian Crown towards the West Indies, several unknown plants began to arrive in Europe. Then, towards the end of the sixteenth century, the arrival of new naval powers – French, English and Dutch –, contributed to a further enrichment of the knowledge of exotic flora. With the expansion of trade in the early modern Atlantic world, not only raw vegetal materials but also previously unknown knowledge began to travel from one coast to the other. Viewed in this light, the production of the knowledge of American plants and their European assimilation was also the result of the encounter between two different systems of knowledge: the Pre-Columbian and the European. Information flows brought new bodies of knowledge that were eventually integrated in the cultural system of the Old World, influencing its transformation in a decisive way.

The introduction of American plants in Europe had important repercussions for nutrition with an increase and differentiation of edible plants; for agriculture with new types of cultivation that led to the introduction of more productive plants and eventually transformed the rural landscape; and for medicine, where the introduction of new medicinal plants enriched the possibility of therapeutic remedies. In this chapter, I focus on this latter field, especially dealing with the role played by American plants as a contribution to the transformation of medical knowledge from the half of the sixteenth century to the end of the seventeenth century. According to Harold Cook, contact with a new world promoted a new attention to nature, as well as different practices of knowledge. Indeed, the production and circulation of the knowledge of American natural products favoured the development of new cultural models and methods of studying nature, providing new strength to natural history, which ultimately contributed in the construction of a modern science.[1] In this context, and

---

[1] Cook 2007: 5.

Note: An earlier version of this work was presented at HSS Annual Conference 2019 in Utrecht, in a panel co-organized by Alain Touwaide and Fabrizio Baldassarri. I would like to thank them for convening this event and again them and the other participants for their comments and suggestions. Many thanks go also to Peter Mason for his English translation of this chapter.

within this line, the study of American plants and materia medica has made a decisive contribution to the transformation of medicine into empirical science.

The introduction of American plants in Europe was important not only because their assimilation gave European medicine access to a richer and more diverse therapeutic arsenal, but all the more because the methods adopted for discovering how to apply them medicinally contributed to the development of a new scientific methodology in the field of therapy. Research on the use of new medicinal products in the treatment of ailments, partly encouraged by the need to verify their medicinal efficacy, promoted the development of an empirical method of therapy. Experiments with the new remedies had important consequences for pharmacological theory.

In this chapter, I deal with a very specific case: the medical debates arising about the use of Cinchona bark, from a plant native to America, as a remedy for malaria, and its integration into the medical and philosophical system of that period succeeded in questioning the Galenic paradigm. The therapeutic success of Cinchona bark demonstrated that, where theoretical medicine had not led to satisfactory results, positive results in the treatment of several diseases depended on the practical experience of medicine. Its definitive inclusion in the seventeenth century European pharmacopoeia not only challenged the principles of traditional Galenic theory, but also gave rise to more doubts about seventeenth century Galenism.

Those who took part in the development of what was to become a new science in the seventeenth century were not only learned élites and humanist literati, but also pharmacists, travellers, merchants and missionaries who, within an ever increasing involvement, contributed to the production and circulation of knowledge about the American materia medica, to its assimilation within European medicine, and to the development of an empirical culture within the medico-botanical field of the time. As William Eamon has shown: "If we think of science in terms of people interacting with their natural environment instead of merely a body of intellectual concepts and ideas, then the sciences that defined the Scientific Revolution immediately become vastly expanded".[2]

The Gradual Assimilation of American Medicinal Products within the European Cultural System

> If a New World were discovered today, would we be able to see it?
> (Italo Calvino, *Com'era nuovo il Nuovo Mondo*, in *Collezione di sabbia*. Milan, 1984, 15.)

Thanks to the monopoly on trade with the Americas conferred on Seville by the Spanish Crown, from 1492 onwards this port was the terminal where plants from the New

---

2 Eamon 2018: 151.

World arrived, mainly in the form of seeds or dried herbs. The contemporary sources, such as those contained in the work of the Spanish physician Nicolás Monardes (ca. 1512–1588),[3] attest that, approximately in the first seventy years after the discovery of America, the following medicinal products were introduced into Spain: guaiacum (*Guaiacum officinale* L.), tobacco (*Nicotiana tabacum* L.), canafistula (*Cassia grandis* L.), several varieties of balsam (*Croton* sp. and *Myroxylon balsamum* (L.) Harms.), mechoacan root (*Convolvulus mechoacan* Vandelli), sarsaparilla (*Smilax officinalis* H.B.K. and related species), sassafras root (*Sassafras albidum* L.), tacamahac (*Elaphrium tecomaca* (D.C.) Standl.) and liquidambar (*Liquidambar styraciflua* L.). From this source we know that the main American medicinal products were already known in Europe in the first half of the sixteenth century. They continued to be included in the European pharmacopeia, with the inclusion in the following century of other new American drugs (cinchona bark, ipecacuanha root and contrayerva root). The research of Patrick Wallis on British trade in exotic medicinal drugs from the second half of the sixteenth century have shown how the introduction of some of these products to the port of London confirms that the circulation of some American drugs was already widespread in a large part of Europe by this time.[4]

Figure 1: *Planisphere, called "Mappa del Cantino"*, 1502, in Biblioteca Estense Universitaria of Modena, C.G.A.2.

These decades witnessed not only the arrival of the first American plants, but also the circulation and printing in Europe of reports by the chroniclers of the Indies and also those written by humanists and physicians who, from first-hand knowledge or drawing on other people's evidence, described and extolled the great richness of the

---

3 Monardes, *Historia Medicinal*.
4 Wallis 2012: 20–46.

American flora, the medicinal virtues of the American plants, and their economic potential.[5] Some 180 different species of American plants, including around fifty with medicinal properties, were described by the Spanish chroniclers in their works on the conquest of America published in Europe in the first half of the sixteenth century. Gonzalo Fernández de Oviedo (1478–1557) on the natural world of the Antilles and the Caribbean,[6] Francisco López de Gómara (1511-ca. 1566) on New Spain[7] and Pedro Cieza de Léon (1520–1554) on Peru and the Andes[8] were the main sources of information on which the medico-botanical culture of the time drew.[9] It was mainly due to the production and translation of these works into Latin, Italian, French, German and English that news of these medicinal plants used in foreign medical traditions that had been unknown until then began to spread.[10]

The physicians' eyes had suddenly been opened onto a scenario in which the range of medicinal plants for treating the main diseases had become considerably broader. However, the fact that only relatively few species of new plants were included in the European materia medica of the sixteenth century shows that medicine was very cautious about welcoming these novelties. An analysis of the various official pharmacopoeias of the cities of London, Augsburg and Amsterdam that were published between the end of the sixteenth century and the first decades of the seventeenth had led to the conclusion that, except for some universally-accepted American drugs, there is no evidence that their use was widespread at that time.[11] A limited presence of American plants seems also to characterise the Spanish pharmacopoeias of the same period.[12] That of Valencia from 1601 mentions only guaiacum, tacamahac, the new balsam and canafistula.[13] As regards to Italy, only guaiacum, sarsaparilla, tobacco, sassafras, Peruvian balsam, mechoacan and liquidambar were included in the main pharmacopoeias of the sixteenth century.[14] The presence of no more than six or seven species of American plants is also a feature of the treatises on natural history and materia medica published in the first half of the same century. The Italian translation with commentary by Pietro Andrea Mattioli (1501–1578) of Dioscorides (*fl.* 1st cent. A.D.), *De materia medica* published in 1544,[15] and the one in Spanish[16] by Andrés

---

5 Ubrizsy Savoia 1996: 165.
6 Oviedo, *La Historia general de las Indias*.
7 López de Gómara, *La Istoria de las Indias*.
8 Cieza de Léon, *Parte primera dela Chronica del Peru*.
9 López Piñero and López Terrada 1997: 21.
10 Pardo Tomás 1991; López Piñero and López Terrada 1997: 24–30.
11 Talbot 1976: 2.837.
12 Valverde 2010: 97–101.
13 *Officina Medicamentorum*: 13, 8, 14.
14 Rotelli 2018: 827–880.
15 Mattioli, *Di Pedacio Dioscoride Anazarbeo Libri Cinque*.
16 Laguna, *Acerca de la materia medicinal*.

Laguna (1499–1559) likewise contain scarce information on a few American medicinal products, always the same ones.[17]

As demonstrated by José Pardo Tomás, the sixteenth century was still the century of Asia. In spite of the Spanish Crown's attempt to create a commercial Atlantic network for American products, the Orient was still considered – not only by Venetians, Portuguese and Arabs but also by the Spaniards – the principal source of medicinal plants and spices. The principal aim of the Spanish-promoted expedition of Ferdinand Magellan (1480–1521) to the Moluccas in 1519 and that of Ruy López de Villalobos (1500–1544) to the Philippines in 1542 was to obtain the much-coveted drugs from the East. Besides, oriental medicinal products (balsams, purgatives, resins and spices) and their medicinal properties had been known for centuries and their study continued to play an important part in the European universities in the sixteenth century.[18]

The American drugs, on the other hand, were new remedies, not known to the ancient sources, and knowledge about them depended to a large extent on the writings of travellers and chroniclers of the New World. The humanist attention to classical sources also concerned medical and pharmacological knowledge in the Renaissance, as scholars and physicians generally referred to (although with several complications) the knowledge handed down in classical Greek tradition of Hippocrates (ca. 460 – between 375 and 351 B.C.) and Galen (A.D. 129-after [?] 216), and, in the field of therapy, by Dioscorides (*fl.* 1st cent. A.D.), whose works were available at last in the original.[19] In particular, Dioscorides' *De materia medica*[20] and Galen's *De simplicium medicamentorum temperamentis ac facultatibus*[21] were the main sources on which medical therapy drew for the uses and properties of medicinal plants. Both works by the two Greek authors were extensively circulated among the texts adopted by the medical faculties. In this sense, when American plants reached Europe, scholars found it difficult to classify these bodies. As Giuseppe Olmi has shown, to catalogue some plants that had no history because they had not been taken into consideration by the classical authors and to include them in the *systema naturae* meant, above all, to question the entire apparatus of knowledge.[22] For physicians, acceptance of the new specimen as a source of pharmacology would have called into question the accepted pharmacological system.

Many naturalists and physicians, fascinated by the exotic, had dedicated themselves to reading the works by the chroniclers that also dealt with American nature, but they had difficulty in using this material when they tried to insert the new species into their catalogues of materia medica. After all, the information on the new vegetal

---

17 Fresquet Febrer 1992: 53–82.
18 Pardo Tomás 2015: 196.
19 Reeds 1976: 519–542; Olmi 1992: 233.
20 Dioscorides, *De materia medica* (ed. Wellmann 1906–1914. Engl. transl.: Beck 2005).
21 Galen, *De simplicium medicamentorum temperamentis ac facultatibus* 1–11 (ed. Kühn 11,379–892, 12.1–377).
22 Olmi 1992: 234–235.

products did not have scientific authority. The description of the new medicinal products lacked specialised medical knowledge, and the information they contained was often vague and imprecise. This created confusion in the task of identifying and naming them, besides errors and uncertainties about their origin.[23] The situation improved in the course of the following century, but it was not until the eighteenth century, in the wake of the first scientific explorations of the American continent, that various species that had only been partially known until then were directly observed, studied and classified. As demonstrated by Mauricio Nieto Olarte, their assimilation in medicine depended to a large extent, though not solely, on their botanical systematisation within taxonomic criteria that became indispensable for the certification and legitimation of their medicinal use.[24]

In the sixteenth century, however, most of the American products were known only in the form of dried roots, seeds, fruits and leaves. To add to the confusion on the identity of the new remedies and the unreliability of the information about them was the fact that their introduction and distribution via the European ports was in the hands of merchants. They were often considered incapable of identifying or preserving specimens, or even suspected of falsifying the content by substituting other substances or products. One of the commonest results was that, once transferred and used in Europe, the medicinal product proved ineffective or failed to display the medicinal properties for which it was known and celebrated.[25] The problem of the lack of reliability of information on the new remedies persisted in the pharmacies too, where the relation of trust between apothecary and client was often compromised by the uncertain information about identity and provenance.[26]

Naturalists and physicians decided to limit the range of American species included in their own works to the same, very restricted group, consisting mainly of those vegetal substances that had been acclimatised in their own country for decades, or those plants introduced as substitutes for European or oriental medicinal plants, or those plants such as guaiacum and sarsaparilla associated with the cure of a new, determinate disease—syphilis—which was considered to be endemic in America. This delimitation was due to several reasons: impossibility of directly observing most of the plants described by the travellers, reticence on the reliability of the sources of information available, lack of classificatory instruments that would enable to create new genera and difficulty of determining the efficacy of a medicinal product whose botanical identity was still uncertain. Finally, a big problem concerned the absence of a clear medical knowledge about them, with made impossible to determine the

---

23 López Piñero and López Terrada 1997: 38–43.
24 Nieto Olarte 2000: 124, 166.
25 Risse 1984: 31.
26 Bleichmar 2007: 127; Pugliano 2017: 233–273.

properties of these American species and include them within Galenic system.[27] As a result, until the middle of the sixteenth century, most of the descriptions of new species were of a general kind, included in chapters on plants that were already known, often lacking any indication of their geographical origin. Even if their therapeutic use was indicated, it was a matter of secondary interest.

American Materia Medica and its Reconfiguration within Galenic System

Knowledge about these plants that began to circulate and to be assimilated in Europe was the result of the way in which travellers perceived the natural world of America and its products. They were the ones who observed the use of plants and roots in native cures, described their appearance and shape, tested their medicinal properties and recorded their results. And they were always the first to import the new medicinal remedies into Europe.[28] Information about the new plants that reached Europe was the result of a work of cognitive selection, of "requalification of knowledge and know-how",[29] of what the colonists observed and knew in the settled territories. On the American continent, the encounter with the reality of American nature caught them unprepared; they had difficulty in accepting a different natural dimension and ended up conceiving of the New World as an extension of the Mediterranean one. If the New World was a projection of the Old and of its nature conceived as a vast reserve of raw materials, then it would be possible to obtain medicinal plants too, equivalent or similar to those used in the European pharmacopeia.[30] These are the plants on which the travellers focused their attention: those they were looking for, which they expected to find. Pardo Tomás has called that attitude "an obsession to identify what they saw with what they knew from their training".[31] As a result, the process of the construction of knowledge about American nature and its products developed through a constant reception of the new as something familiar, categorised on the basis of previous knowledge.

The travellers' descriptions of American plants reflected this attitude, as they named and illustrated plants on their analogy and similarities with known vegetal species. When it was impossible to create an analogy between a New World and an Old World plant, the description of the new species was based on personal experience. In

---

[27] Munger 1949: 196–229; Grafton, Shelford, and Siraisi 1992: 177–193; López Piñero and López Terrada 1997: 38.
[28] Valverde 1982: 155.
[29] Boumediene 2016: 22.
[30] González Bueno 2007: 13.
[31] Pardo Tomás and López Terrada 1992: 203.

this sense, travellers applied the same descriptive method adopted by naturalists, as Lucie Čermáková and Jana Černá have argued, to describe the *herbae nudae* that they discovered in the Old World.[32]

To search for familiar plants and to appropriate their medical properties also meant reconfiguring knowledge of them on the basis of European medical models: the Spaniards interpreted the new medicinal plants with which they came into contact and that they tried to use, verifying their curative properties among the local populations, in terms of the Galenic doctrine of diseases and medicinal therapy. Assuming their criteria to be universal, they believed these could apply to plants that grew in non-European territories and climates.[33] Still, in order to be assimilated, every new remedy had to be subjected to a process of reconstruction that converted it into something familiar and in accordance with European medical doctrine. The consequence is evident. The knowledge of American materia medica involved a reduction to its significance according to the Galenic theory of humours.[34] Everything that was not suitable for being elaborated according to the "empirico-rational" theories was rejected.[35] The system of beliefs within which the use of the medicinal plant was interpreted according to the indigenous medical system was completely eliminated.

It is probable, however, that the European assimilation of the American materia medica was facilitated by the shared characteristics of the European and Pre-Columbian medical systems. In an article from 1987, "Why is humoral medicine so popular?", the anthropologist Eugene N. Anderson affirmed that "Humoral medicine is probably the most widely influential belief system in the world. No religion, no political ideology, no other scientific system had spread so widely".[36] Humoral notions like hot and cold that characterised the traditional European medical system could also be found in the classical scientific traditions of ancient China, India, and native American cultures before they came into contact with and were influenced by European medical culture. It is therefore highly likely that the hot/cold dichotomy used in the Pre-Hispanic medical and cosmological systems to classify bodily conditions, foodstuffs, medicinal plants, ailments and other aspects of the cosmos such as the stars or supernatural beings preceded the arrival of the Europeans and the humoral medicine that they spread in the New World, mainly through missionary medicine. Besides the classification of plants as hot or cold shared by both medical systems, other similar notions such as the concept of health and illness as states of equilibrium or imbalance of the human body and the importance attributed to physiological factors as the cause of the illness probably facilitated the transmission of some medicinal products to the European pharmacopeia.[37]

---

32 Čermáková and Černá 2018: 69–106.
33 Wear 1999: 151.
34 Schiebinger 2004: 87.
35 Huguet-Termes 2001: 361.
36 Anderson 1987: 331.
37 Del Pozo 1966: 9–18; Messer 1987: 339–340.

In any case, the tendency to understand American materia medica in terms of the Galenic pharmacological model led to the search for familiar medicinal products like the known ones or that could produce similar curative results. Most of the New World remedies accepted into the European materia medica in the sixteenth century were surrogates of European medicinal products, principally those deriving from oriental drugs that had become scare or unobtainable, with purgative, emetic or sudoriferous properties that could integrate or replace those plants described in the humoral pharmacopeia. For example, Peruvian balsam (*Myroxilon balsamum* (L.) Harms. var. *pereirae* Harms.) was initially considered a substitute for oriental balsam (*Commiphora opobalsamum* (L.) Engl.) once its resin had been shown to possess the same healing properties. In this sense, American plants were classified by European medicine following the Galenic method, which identified their qualities on the basis of their colour, taste, smell and characteristics to the touch.[38] Most of them were indicated as 'hot' medicines. Guaiacum, for instance, was recognised as a 'hot' and 'dry' medicine. It could purify the blood by removing, through sweating and urination, the corrupt flows generated by syphilis.[39]

Figure 2: *Hyacum et lues venerea*, in Jan Van Der Straet, *Nova reperta*. Antwerp: Philip Galle, c. 1600, Plate 7.

---

38 Estes 1995: 4–5.
39 Monardes, *Historia Medicinal*: ff. 15v-16r.

As we have seen, most of the American medicinal products were regarded as second best. On the other hand, as Samir Boumediene and Valentina Pugliano have shown, "comparison is already a process of hierarchisation. The old is worth more than the new because, particularly in the field of health care, the known is more reliable than the unknown".[40] Even in the eighteenth century, the search for medicinal plants promoted by the French and Spanish scientific expeditions to the American continent focused on those species that could replace medicinal products from the Orient, a market that had fallen into British and Dutch hands some time before.[41]

The Appropriation of New World Plants: A Contribution to the Development of New Methodologies of Studying Nature and its Medicinal Products

So the process of assimilation of American medicinal plants was slow. Interpreting American medicinal plants in terms of the Galenic paradigm considerably limited the field of inquiry into new medicinal products to potential substitutes, bypassing the enhanced therapeutic value that could have been generated from them.[42] Additionally, the European pharmacopeia was already well furnished with medicinal plants that had formed part of its solid medico-botanical tradition for centuries. However, the contact with American nature and its vegetal products was beginning to promote the development of a knowledge of that nature and a study of its materia medica in which experience was assuming a new role, not only downgrading the authority of the ancients but also calling it into question. This led to the emergence of new methods for studying nature and its medicinal products that contributed to the development of a new empirical culture that facilitated its diffusion and assimilation in medicine.

In his important work on Cinchona bark, an American remedy that entered European medicine in the middle of the seventeenth century, Matthew James Crawford has stressed how "early modern science was as much the product of the cross-cultural interactions provoked by European commercial and imperial expansion, as it was the product of a handful of enterprising minds in Europe".[43] In this context, the sixteenth century Spanish explorations of the New World played a fundamental role, contributing to the development of an empirical sensibility and to the emergence of new forms of knowledge. For the Spanish Crown, colonising the Americas politically and economically

---

40 Boumediene and Pugliano 2019: 48.
41 Nieto Olarte 2000: 124–125.
42 Risse 1984: 32.
43 Crawford 2016: 41.

meant, above all, colonising the knowledge of the native cultures of those lands. The appropriation of knowledge about the natural history and medicine of the New World was considered a fundamental instrument in order to control and exploit it, as well as a source of the medicinal remedies to preserve the health of colonists affected by unfamiliar tropical illnesses. European medicinal products were often not readily available in the colonised territories, and after a long journey by sea, those that did arrive might have lost their efficacy; anyway, they would have been ineffective against the new diseases.[44] So systematically collecting detailed information on the natural products, plants and local practices of the colonised lands came to play a role in the expansionist programme of the Iberian colonial empire and, in later centuries, in that of the other European colonial powers.[45]

Clearly, for the Europeans in the New World, finding out about the American materia medica also meant coming to terms with the knowledge systems of the indigenous populations. The collection of information on the new products and the production of knowledge about them were closely dependent on contact with the native populations and their knowledge. It was a question of looking for medicinal plants that had never been known or used before and were potentially profitable or could be effective in curing unfamiliar diseases. Drawing on the knowledge, experience and competence of the native populations was therefore considered a fundamental and necessary instrument: their wealth and survival depended on their capacity to appropriate native knowledge, what Samir Boumediene has called "the colonisation of knowledge" of the New World.[46] In recent years various historians of science have stressed the important role played by the local participation and knowledge of the indigenous people in the construction of European knowledge about the New World, recognising their influence on that knowledge.[47] The indigenous knowledge of the medicinal plants of their lands was for the Europeans an indispensable source of information about the medicinal products that they discovered. In a study of the American sassafras tree in the first global era, Clare Griffin has analysed the mechanisms at work in the production and circulation of European knowledge about this American medicinal root and attributed a decisive role in this respect to the Native American population. It was the Timucua, who lived in what are today South Georgia and Florida, who passed on to the first European colonists – first the French, then the Spaniards – the knowledge of the use of parts of the sassafras, among other uses, to cure fever and of its preparation as a medicinal product. European knowledge of its properties and use was thus the result of the appropriation of indigenous experience and skills by the colonists. Sassafras was one of the most important

---

44 Cook and Walker 2013: 343.
45 Cook 2005: 100–101.
46 Boumediene 2016: 132.
47 Safier 2010: 136.

medicinal specifics that were imported, along with other medicinal products, into Europe between the sixteenth and the seventeenth centuries.[48]

Within the Iberian colonial world, as later in the first modern European imperial colonies, contact zones were places of encounter and exchange, of collaboration and competition between geographically diverse and historically separate cultures, and between European and non-European medical systems.[49] These cultural exchanges were important for the production of scientific knowledge.[50] Like any other place of knowledge production, they were not hermetically closed, but their cultural borders were constantly recreated through dynamic processes of the construction of scientific knowledge. American knowledge was transmitted in these trading zones and became a part of the European scientific heritage. The colonists in these contact zones tried to acquire and collect from the locals mainly empirical information, specific facts about their medicinal products, their use and efficacy. Even though pre-Hispanic therapeutic practices were closely bound up with magic and religion, their cures were also, above all, empirical, aimed at empirical and rational treatments. The pre-Hispanic medical system boasted a profound knowledge and empirical research on medicinal plants and their many varieties thanks to the rich biodiversity of their territory and the variety of their climates.[51] The Spaniards drew on the indigenous empirical knowledge of medicinal plants, which became a point of reference for the construction of knowledge of the American materia medica. On the basis of that knowledge, the colonists themselves experienced the properties and efficacy of American medicine.

For the chroniclers and naturalists of the New World, it was no longer possible to refer to the classical cultural system as the new plants were absent from the ancient texts. Consequently, the production of knowledge about American natural history and its medicinal products led to an epistemological change once the knowledge handed down through the classical tradition had lost some of its authority. Faced with an unfamiliar nature whose vegetal products were absent from the classical tradition, personal experience became the instrument for the construction of their knowledge, and at the same time the means of confirming empirical native knowledge. As argued by Antonio Barrera-Osorio: "An avocado was nowhere to be found in Pliny or Aristotle, thus the empirical information about avocados [. . .] became an alternative and more reliable source of knowledge than the imperfect knowledge contained in classical sources".[52] The method to describe the plants of the New World was different from that used by medico-naturalists who studied the nature of the Old World. In fact, the research method adopted by the scholars in the description of the vegetal species

---

48 Griffin 2020: 1–18.
49 Pratt 1992: 6.
50 Slater, Pardo Tomás and López Terrada 2014: 2.
51 Estrella 1992: 13–33.
52 Barrera-Osorio 2002: 164.

used a descriptive procedure through which an attempt was made to attribute symbolic values and moral meanings to them. The content of the descriptions of the new species, on the other hand, lacked the symbolic references through which nature was customarily described. In the literature on the conquest of the New World, the absence of those symbolic, philosophical or religious associations from the description of plants transformed them into true objects of research, contributing to convert the study of natural history into an empirical science.[53]

For Gonzalo Fernández de Oviedo, the first chronicler of the Americas and the first to publish a natural history of the New World, and for the other chroniclers and colonists, knowing and describing American nature and its products meant describing what they had seen and experienced themselves. Personal experience was elevated to the role of an independent source of knowledge and a method for the study of American nature. Oviedo frequently referred to the empirical knowledge of the indigenous people in his work and based the validity of his knowledge on the description of his direct experience with the new American products. This method which used a descriptive language and tended to emphasise personal experience characterised all the natural historical literature on the New World.[54] The knowledge of new American species that began to circulate in Europe brought out the incomplete nature of the scientific knowledge of the classical tradition, its inability to describe the nature of a new world or to offer a complete understanding of the contemporary world. In short, it questioned the authority of that tradition. The information about American plants helped to speed up the process of the erosion of the classical system that was already under way with their discovery and appreciation.[55]

Oviedo's study of American nature and its products was for the Spanish Crown one of the first attempts to colonise the knowledge of the New World by collecting as much information as possible about its natural resources and botanical diversity. The search for American medicinal plants that would be profitable to trade and to incorporate in the European pharmacopeia, both in the New and in the Old World, led to a practice of systematic experimentation on them. It was above all the Spanish Crown that promoted such empirical practices based on personal observations and tests of new medicinal products. These empirical activities, organised in the form of reports compiled by colonists (mainly entrepreneurs and royal officials) were based on personal observations, empirical information and the description of individual products.[56]

In the production of knowledge about the American materia medica, personal experience became a source of knowledge and a method of study. The resulting empirical knowledge formed the basis for knowledge of nature itself. Empirical knowledge filtered on the basis of native knowledge was validated by means of two institutions

---

53 Eamon 2014: 237.
54 Pardo Tomás 2002: 60 61.
55 Barrera-Osorio 2006: 103.
56 Barrera-Osorio 2009: 222.

set up by the Spanish Crown: the House of Trade (*fl.* 1503) and the Council of the Indies (*fl.* 1524). As spaces for the production of knowledge of nature, they transformed it into formal knowledge. Information derived from the senses and experiments on natural products became the epistemological foundation of the study of American medicine; the empirical study of natural products became the main way to investigate nature. Physicians and surgeons operative in the hospitals of Toledo, Seville, Granada and Burgos were invited by the Spanish Crown to test the new remedies, to produce written reports of their results, in other words, to validate the empirical information of the colonists.[57] Personal experience with the American materia medica had become the centre of knowledge production. It was a method far removed from the traditional ways of studying nature, but it was the production of empirical knowledge and its circulation, the main source of knowledge of American natural products and the principal instrument for the study of American nature, that began to develop from the second half of the sixteenth century among a growing community of intellectuals interested in the study of American nature and its materia medica.

Slightly less than a century after the first Spaniards landed in America, the work that contributed most to the diffusion and commercialisation of American medicinal plants was printed in Seville. This was the *Primera y Segunda Y Tercera Partes de la Historia medicinal de las cosas que se traen de nuestras Indias Occidentales que sirven en medicina* (1574) by the Spanish physician Monardes. Although he never travelled to the New World, his work, the first to be entirely devoted to the American pharmacopeia, became the prime European authority in the field.[58] For more than two centuries it was the essential point of reference for the study and use of American medicinal products, and its success can be gauged from the numerous editions and translations that appeared during and after the author's life.[59]

These publications were a strong stimulus on the market of American medicinal remedies and probably encouraged the Spanish Crown to organise what was the first modern scientific expedition to America to collect information about the medicinal plants of the territory under its control.

In the years in which Monardes was engaged on the publication of the first two parts of his work, another Spanish doctor, Francisco Hernández (ca. 1515–1587), personal physician to Philip II (b. 1527, king 1556, d. 1598), was on a royal commission to investigate the American materia medica, but on the other side of the Atlantic.[60] The knowledge about American medicinal plants collected by the two physicians was therefore the result of different experiences, as Hernández was able to study the new species on the spot. While Monardes' focus was on the pharmacological aspect of the new plants, arranged in his work according to their medicinal properties, that of Hernández in the

---

57 Barrera-Osorio 2009: 223.
58 Bleichmar 2007: 131–132; Boumediene 2015: 137–140.
59 Bleichmar 2005: 85.
60 O'Malley Pierson 2000: 11–18.

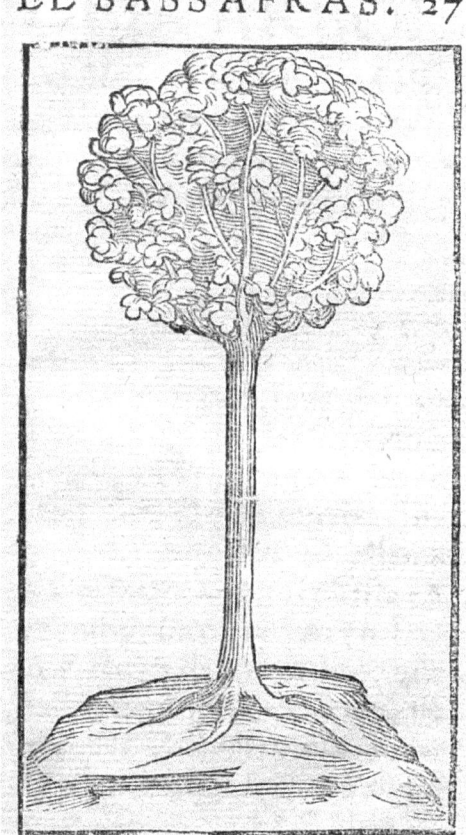

Figure 3: *Sassafras*, in Nicolás Monardes, *Segunda parte del libro, de las cosas que se traen de nuestras Indias Occidentales, que sirven al uso de medicina*. Sevilla: en casa Alonso Escrivano, 1571, f. 27 recto.

Historia de las plantas de Nueva España (written from 1571 to 1576) was on the morphological characteristics of the new species.[61] Despite these differences, however, they both preserved the knowledge acquired from the indigenous peoples as a source and personal experience as a study method and instrument required for the empirical confirmation of that knowledge.[62]

The systematic works of Hernández and Monardes heralded a new stage in the study of American botany and materia medica.[63] After the unification of Spain and Portugal in 1580, American plants began to attract the attention of other European countries.

---

61 López Piñero 1992b: 256–259; López Piñero and Pardo Tomás 2000: 124.
62 Risse 1987: 44–45; Marroquín Arredondo 2019: 59–63; Pardo Tomás 2007: 181–182.
63 López Piñero and Pardo Tomás 1994: 71–80, 145–153; López Piñero and Pardo Tomás 1996: 139–151, 189–203.

The new medicinal products, roughly those described by Monardes, began to circulate more and more widely in the cities of Seville, Valencia, Rome, Amsterdam and Exeter. Although they constituted only a minimal percentage of the remedies contained in the European pharmacopeia, they began to be indicated in very many composite medical preparations.[64] This wider circulation of American plants in the form of fresh samples, seeds, fruits, roots and leaves, was facilitated by the creative of ever growing circuits of accurate information among those who took a lively interest in exotic nature, naturalists, physicians, pharmacists, merchants and collectors.[65] Within these correspondence networks, the knowledge about American medicinal products that circulated was based much more on personal experience than on natural philosophy. Empirical research was beginning to establish itself as a research tool for the new flora. Personal experimentation had become the basis of the production of scientific knowledge, and empirical methods began to replace and exclude theoretical explanations in the production of knowledge about nature and medicine.

The Appropriation of Knowledge about Cinchona Bark

From the end of the sixteenth century onwards, the increase in the trade of exotic drugs, in which the English, Dutch and French appeared as new protagonists side by side with the Spanish and Portuguese, gave rise to an expansion in the diffusion and knowledge of these plants. Throughout the subsequent century, many new treatises on the exotic medicinal flora were published in Europe by physicians, naturalists and agents who had taken part in the journeys to the Americas organised by the European trading companies.[66] Moreover, works by explorers and missionaries who had personally tested the new medicinal plants in the overseas lands began to circulate and became quite widespread.[67] One of the circuits, and probably one of the most important, through which this information began to circulate in Europe was that of the missionaries in America. For them, as for the other colonists, the search for local medicinal remedies was primarily a necessity for their survival there, but the appropriation of that knowledge also became part of their missionary programme. Among the various orders, the Jesuits in particular played a fundamental role in the production of knowledge of the American materia medica and in the importation and distribution in Europe of

---

64 Boumediene 2016: 132.
65 Gómez López 2005: 13–40; Egmond 2007b: 109–113, 129–137, and 2017: 21–45; Pugliano 2018: 44–60.
66 Cook 2007: 210–225; Piso, Marggraf and Laet, *Historia naturalis Brasiliae*.
67 Pelayo 2003: 15–38.

various of its products. One of these, which came to assume vital global importance, was Cinchona bark.[68]

From the "discovery" of South American by Jesuit missionaries in the 1630s to the end of the eighteenth century, Cinchona bark played an important role in international trade, and the wish to obtain it was one of the main motives of the European scientific expeditions to the Americas.[69] Cinchona bark was important in the history of European medicine, not only because in the 1820s quinine was extracted from it (it was used for treating malaria down to the second half of the twentieth century), but also because "it was the prototype of a 'specific' remedy (against intermittent fevers, i.e. malaria) which challenged the principles of traditional, Galenic therapy".[70]

Known at the time as *quina, quinaquina,* or *cascarilla,* this bark from various species of trees of the *Cinchona* genus grew spontaneously in the Andean forests. From the 1640s, when it began to be introduced on the European medical market, it became the most widely used medicinal cure for intermittent fevers.[71] Its preventive and curative use for malaria is due mainly to the presence of quinine, but also to that of three other alkaloids: cinchonine, cinchonidine and quinidine. Although they were not identified until 200 years ago, that did not prevent physicians, missionaries and *indios* from using it for centuries as the main effective remedy against malaria.[72] It has been claimed that the Spaniards in America began to use it officially in the 1630s, but it is likely that they were already aware of its medicinal properties in the previous century. In fact, both Monardes (in the second part of his *Historia medicinal*) and the Spanish physician Juan Fragoso (ca. 1530–1597), whose treatise on American and oriental medicinal plants was published a year after Monardes' volume in 1572,[73] referred to a medicinal bark, though without naming it, from the Americas (Monardes gave its specific geographical place of origin as the New Kingdom of Granada). Its botanical characteristics, taste and colour, method of preparation and administration, and its use in medicine closely resemble the descriptions of Cinchona bark in the sources of seventeenth century chroniclers.[74] It is therefore possible that both physicians in Spain had already heard people talk about Cinchona bark some seventy years before its official arrival in Europe.

At any rate, reports by Spanish colonists, such as that by the Augustinian missionary Antonio de La Calancha (1584–1654) *Coronica moralizada del Orden de San Augustin en el Peru* (1638),[75] are evidence that the Jesuits had already begun to use Cinchona

---

68 Anagnostou 2005: 3–5; Boumediene 2020: 233–235.
69 Nieto Olarte 2000: 171, 181–185; Pelayo 2003: 38–48; Crawford 2007: 203–206, and Crawford 2016: 53–57.
70 Maehle 1999: 223.
71 Boumediene 2020: 242–243.
72 Aymard 2019: 239.
73 Monardes, *Las cosas*: 117; Fragoso, *Discursos de las cosas Aromaticas*.
74 Ortiz Crespo 1995: 169–181.
75 Calancha, *Coronica moralizada*: 59.

bark in America in the 1630s,[76] probably after observing its use by the indigenous populations. This hypothesis finds support in the writings of two other physicians, the Portuguese physician from Seville Gaspar Caldera de Heredia (ca. 1591–1669) and the Spaniard Gaspar Bravo de Sobremonte (ca. 1610–1683). The former in his *Tribunalis Medici Illustrationes* (1663)[77] and the latter in his *Disputatio Apologetica pro dogmatica Medicina praestantia* (1669)[78] reported that the Jesuits had learnt the properties of the American bark in calming cold tremors (fever) from the Indians who used it for this purpose after it had been ground, reduced to a powder, and drunk in an infusion with hot water.

Like the use of most of the other American medicinal products that had been included in the European pharmacopeia in the sixteenth century and which depended on indigenous knowledge, the knowledge of the efficacy of Cinchona bark in combating fever was derived from the local populations with whom the Europeans came into contact.[79] It is highly unlikely that the natives of Peru used Cinchona bark as a cure for intermittent malarial fevers before the arrival of the Europeans and, with them, African slaves since jungle fever was not widespread in South America before then.[80] On the other hand, as the Spanish physician Pedro Miguel de Heredia (ca. 1579–1655) argued,[81] it is probable that, following a classical line of reasoning based on analogy, in other words the use of similar treatments to cure similar diseases, the Jesuits considered that the same properties of the remedy as used in its place of origin could be effective in curing the tremors that preceded the outbreak of fevers, particularly intermittent fevers.[82] This is how the missionaries in America who began to test Cinchona bark on some patients suffering from intermittent fevers (tertian and quartan fever) obtained satisfactory results and promoted its diffusion, first in Peru, where the pharmacy of the Collegio di San Paolo in Lima became the principal distributor, and afterwards in Seville and Rome and elsewhere in Europe.[83]

Rome was the first European city to conduct pharmacological experiments with the new American remedy.[84] The Jesuit apothecary of the Collegium Romanum had become the main centre for the distribution of Cinchona bark and near its headquarters physicians and pharmacists began to prescribe Cinchona bark for their patients. The Roman Ospedale di Santo Spirito became one of the main laboratories for the production of this materia medica and for experimentation with it, a function that it

---

76 Guerra 1977a: 115.
77 Caldera de Heredia, *Tribunalis Medici*: 155–161; López Piñero and Calero 1992: 34–35.
78 Bravo de Sobremonte, *Disputatio apologetica*: 130.
79 Crawford 2016: 30–39.
80 Guerra 1977a: 112; Bruce-Chwatt and Zulueta 1980: 9–16.
81 Heredia, *Opera Medica*: 1.554.
82 Guerra 1977b: 135–136; Boumediene 2020: 244.
83 Anagnostou 2005: 13–14; Crawford 2014: 218.
84 Colapinto 2002: 173–184.

had already performed a century earlier for guaiacum and again, at the end of the century, for ipecacuanha root.[85] The pharmacist Domenico Auda (*fl.* 17th cent.) and the Jesuit cardinal Juan de Lugo (1583–1660), professor of the Collegium Romanum, conducted experiments there to test the efficacy of Cinchona bark on their patients and contributed to promoting its use and diffusion.[86] In Genoa, the Genoese physician Sebastiano Bado (*fl.* 1650–1676) introduced it in the Ospedale Maggiore di Pammatone.[87] During those years its preparation and methods of administration had been ratified by the Jesuits in a pamphlet known as *Schedula Romana* (1649–1651).[88] So both in America and in Europe, the Jesuits played a fundamental role in the production of knowledge about Cinchona bark and in experimenting with it.

The Incompatibility of Cinchona Bark with Humoral Pathology and its Challenge as a 'Specific' Remedy

From the early 1650s, when the use of Cinchona bark was already widespread in Europe, the acceptance of its curative efficacy in traditional therapy had turned out to be a goal that was not easy to achieve.[89] It was a question of the use of a new, exotic remedy. Knowledge about it and its diffusion were in the hands of the Jesuits, whom the European physicians regarded as lacking medical authority. Moreover, like most of the American products that reached Europe, it was a remedy that lacked a botanical identity. It was only partially known in the form of reddish and bitter bark until the first botanical description of the plant appeared in 1738. It was written by the French astronomer and mathematician Charles Marie de La Condamine (1701–1774) who had been able to observe the Cinchona plant on the spot near the city of Loja, Ecuador, during a scientific expedition in America.[90]

It was on the basis of this report that Linnaeus first described and classified the genus *Cinchona* in his *Genera Plantarum* (1742), followed by the species *Cinchona officinalis* in his *Species Plantarum* (1753).[91] Until then, the reports containing a description of it that were available to European pharmacists and physicians were few and not always reliable.

---

85 Jarcho 1993: 16–17; Boumediene 2016: 205–206; Boumediene 2020: 246, 249.
86 Jarcho 1993: 14–15.
87 Jarcho 1993: 1; Colapinto 2002: 178; Boumediene 2016: 205.
88 Jarcho 1993: 17–18, and 262–269.
89 Jarcho 1993: 27–34.
90 La Condamine, "Sur l'arbre du Quinquina": 226–243; Cuvi 2018: 4–5.
91 Linnaeus, *Species Plantarum*: 1.172.

Figure 4: *Cinchona tree*, in Charles-Marie de La Condamine, "Sur l'arbre du quinquina", in *Histoire de l'Académie Royale des Sciences avec les Mémoires de mathématique et de physiques*, Année 1738. Amsterdam: P. de Coup, 1745 [?]: plate 6, p. 346.

The earliest accounts of the American bark were that of the Jesuit Bernabé Cobo (1580–1657) in his *Historia del Nuevo mundo* (written in 1653)[92] and of the Genoese merchant Antonio Bollo (*fl.* 16th/17th cent.) in his *Anastasis corticis peruviae*, published

---

92 Cobo, *Historia del Nuevo Mundo* (ed. Mateos 1956: 2.274).

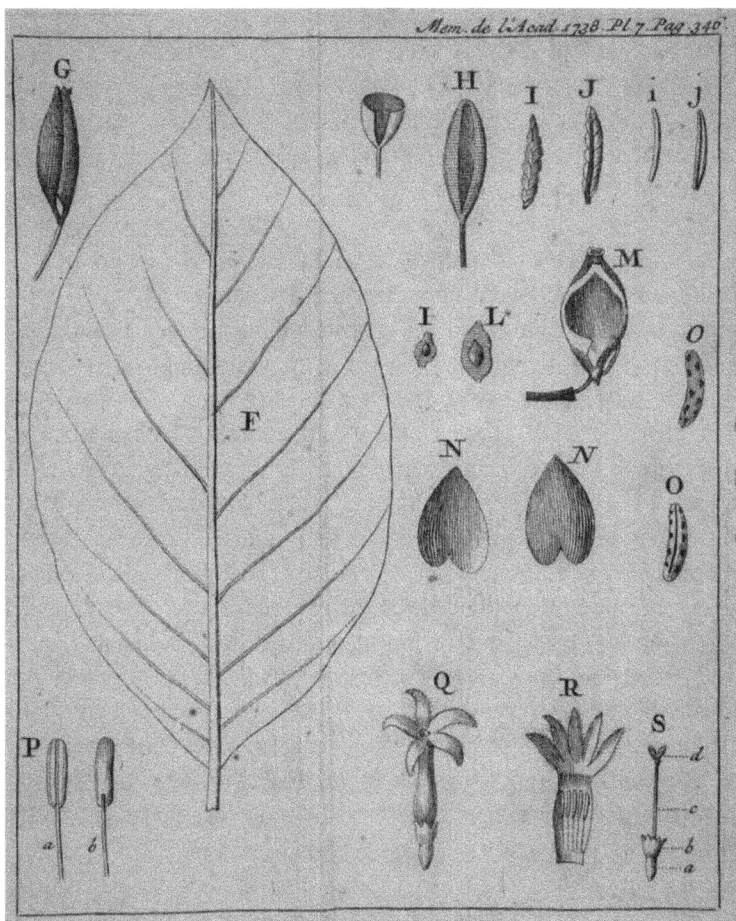

Figure 5: *Cinchona tree*, in Charles-Marie de La Condamine, "Sur l'arbre du quinquina", in *Histoire de l'Académie Royale des Sciences avec les Mémoires de mathématique et de physiques*, Année 1738. Amsterdam: P. de Coup, 1745 [?]: plate 7, p. 346.

by the physician Sebastiano Bado in 1663.[93] These works created confusion regarding the botanical identity of the new remedy. Cinchona bark was often confused with Peruvian balsam, a medicinal product with balsamic properties prepared using the resin of a tree of the *Myroxylon* genus that had been introduced into Europe by the Jesuits some decades earlier. Known for the ability of its reddish, bitter bark to combat fever, it was called *quina-quina,* a Quechua term with the same designation as Cinchona bark.[94] To add to the confusion, not only was Cinchona bark known under a

---

93 Bado, *Anastasis corticis peruviae*.
94 Haggis 1941a and 1941b.

multitude of names (Peruvian bark, Cascarilla, the Countess'powder, Jesuits'powder, Cardinal Lugo's powder) but there were various species of the genus Cinchona from which medicinal bark could be extracted, and various barks of different trees had the same colour and taste.[95] It often happened that, during its transfer from the Andean forests to the pharmacies of Europe, it was adulterated, for example by mixing different qualities of quina or the bitter barks of other trees. The bitter taste of the bark was its principal characteristic; in fact, the quinine present principally in the bark of *Cinchona calisaya* Wedd. and *Cinchona succirubra* Pav. Ex Klotzsch is very bitter.

Further difficulties arose from vagueness regarding the specific illness to which the American remedy could be applied. In the seventeenth century, as in antiquity, fevers, including malarial fevers caused by the natural surroundings, were not regarded as symptoms but as the illness itself, an excess of bodily heat whose symptoms resembled those of other illnesses.[96] In particular, traditional medicine had inherited from Hippocrates the theory that certain types of fever (what are now known to be symptoms of malaria) were caused by unhealthy air, within which there were *miasmas,* i.e. poisonous particles, that came into contact with human beings and started corruption or an accumulation of humours in their bodies. According to Galen's classification of fevers, they could be subdivided into three types: simple; putrid, which could be further subdivided into continuous and intermittent (tertian, quartan, quintan) and were traditionally associated with malaria; and malignant, connected with the plague. It was not until the nineteenth century that the term 'malaria' came to be used for a specific illness.[97] While tertian fever was considered a hot illness caused by the corruption of yellow bile, quartan fever was a cold illness due to the black bile of the spleen. It is evident that the two fevers were regarded as different types of disease, because they were attributed to two different types of humours. Consequently, different remedies were prescribed for them.[98]

In the middle of the seventeenth century, the remedies mainly used by the followers of Galen to treat malarial fever depended on the intensity of the body heat and the duration of the fever; the interval between the paroxysms was the only temporal criterion to distinguish the different types of fevers. To eliminate the faulty humours from the putrefied blood of the patients suffering from intermittent fevers, the main therapeutic procedure was phlebotomy, as Galen had prescribed, combined with some febrifuges. Until then, malarial fever had been treated principally with gentian, one of the most frequently used febrifuges.[99] Unlike guaiacum that was prescribed for an unknown illness at the time, Cinchona bark had to make its way amid

---

95 Guerra 1977b: 136.
96 Boumediene 2016: 180; Klein and Pieters 2016: 401.
97 Klein and Pieters 2016: 405; Boumediene 2016: 180.
98 Jarcho 1993: 218.
99 Jarcho 1993: 29.

therapeutic practices and medicinal plants that were already in widespread use for the treatment of a common ailment, fever.[100]

For several decades the Galenic doctrine and his theory of fever constituted the main obstacle to the acceptance of Cinchona bark.[101] First of all, the difficulty in fitting it into the Galenic system was due to the fact that European physicians and apothecaries who practised traditional medicine had classified the American plant, on the basis of its bitter taste, as a hot and dry medicine (while the Andean medical system considered Cinchona bark a cold medicine).[102] According to the theory of opposites, this quality made it unsuitable for curing a tertian fever and for removing the heat of the yellow bile humour. In other words, it was not easy to justify accepting that a hot remedy might cure a fever. In addition to this difficulty, there was another one: it was a matter of adopting a plant whose effects did not give rise to evacuations, i.e. those conditions that were needed to remove the faulty humours and cure the patient. Since Cinchona bark was not able to remove humours, the Galenists believed that the American plant, on the contrary, fixed the feverish material in the patient's body.[103] But Cinchona bark seemed to be effective against both tertian and quartan fever, even though it was neither purgative, nor sudorific, nor emetic. From the middle of the seventeenth century, the idea began to take hold that Cinchona bark was a "specific" remedy, i.e. a drug capable of curing a particular ailment, but the idea of the existence of "specific" remedies for "specific" ailments was not new. In the sixteenth century, the Swiss physician Theophrastus von Hohenheim known as Paracelsus (ca. 1493–1541) and his followers had advanced this hypothesis.[104]

Unlike traditional medicine, they supposed that the ailment was not caused by a humoral imbalance but by a determinate external cause that only a specific remedy could cure.[105] Thanks to experiments with Cinchona bark by the English pharmacist Robert Talbor (1642–1681) and the efficacy of his new treatment, the concept of a "specific" remedy was confirmed empirically in Europe.[106] Talbor had been apprenticed for a while to the English pharmacist Peter Dent (1629–1689) in Cambridge, who was experimenting with new medicinal preparations involving Cinchona bark in the 1660s. Talbor then moved to Essex, where he prepared a new treatment consisting of adding to the infusion of the Jesuits a tincture of Cinchona bark (made through its maceration in distilled alcohol and in which the concentration of quinine was higher) and a tincture of opium.[107] The efficacy of Cinchona bark began to be recognised all

---

100 Boumediene 2016: 207.
101 Maehle 1999: 227.
102 Crawford 2016: 34.
103 Jarcho 1993: 29.
104 Maehle 1999: 231–232.
105 Debus 1977; Smith 2004; Boumediene 2018: 70–75.
106 Boumediene 2016: 232.
107 Blégny, *Le Remede anglois*.

over Europe after Talbor had successfully cured Charles II (b. 1630, king 1660, d. 1685) with his new formula in 1679,[108] followed by the Dauphin and other nobles at the French court in Versailles.[109] The success of the result of Talbor's empirical experimentation with Cinchona bark came to prevail over theoretical considerations.

Cinchona bark entered the pharmacopeia in London in 1677.[110] It was beginning to be clear that the success in the treatment of several diseases, since theoretical medicine did not lead to satisfactory results, depended on the practical experience of medicine. Recognition of Cinchona bark as a specific remedy against intermittent fevers caused a crisis among the followers of Galen and its effectiveness no longer depended on the Galenic properties that acted to restore humoral equilibrium, but on its specific properties. It could successfully cure anyone irrespective of the individual temperament of the patient. This new medical vision entailed not only the transformation of the concept of the body from unique to common, but also transformed the concept of an ailment from a physiological to an ontological phenomenon.[111] In fact, the acceptance of the efficacy of Cinchona bark on the basis of experimental tests contributed to dent the humoral medical system. The debates on Cinchona bark, and the acceptance of its efficacy as a specific remedy for the treatment of certain diseases, obviously including malarial fevers, began to call into question the Galenic system.

The Contribution of Iatrochemistry and Iatromechanics in Pharmacological Experimentation with Cinchona Bark

During the seventeenth century, when Galenism still had an influence on most of medicine,[112] two new medical doctrines developed in Europe that partly opposed Galenism and supported the therapeutic use of Cinchona bark: iatrochemistry and iatromechanics. They gradually became a part of official medicine. Their ideas on living organisms, on the composition of plants and on diseases diverged from the humoral theory, albeit not always completely.[113] Although the iatrochemical theories and the iatromechanical ones were rivals from a conceptual point of view, they did supply the basis for the development of pharmacological experiments and clinical observations on Cinchona bark.[114]

---

108 Siegel and Poynter 1962: 82–85; Keeble 1997: 285–290.
109 Perez 2003: 25–30.
110 *Pharmacopoeia Collegii*: 4.
111 Cook 2011: 123–145.
112 Klerk 2015; Favaretti Camposampiero 2022.
113 Jarcho 1993: 221–231.
114 Maehle 1999: 230.

From the second half of the century, the followers of the iatrochemical doctrine carried out many pharmacological *in vitro* experiments with Cinchona bark, particularly on blood and animals, in order to analyse the action of the medicine in bodily fluids. Jacques Minot (*fl.* 17th cent.), for instance, through his *in vitro* experiments on animal and human blood, concluded that the alkalinity of Cinchona managed to dissolve the coagulated clots produced by excess acids and to destroy the acid ferments that were the cause of malarial fever.[115] This experimental demonstration, and others as well, enabled him to refute the criticism advanced by the Galenists of Cinchona and its 'negative' action in fixing the morbid material in the patient's body and condensing humours within his blood.[116] The physician Friedrich Hoffmann (1660–1742), who followed the iatromechanical doctrine, and his student Johann Balthasar Schondorff (b. 1672) carried out experiments on Cinchona bark, and in some cases used a microscope to analyse its action with chemical tests. They refuted the iatrochemical theories on the effectiveness of the alkalinity of Cinchona in destroying the acid ferments in fever, and proposed an iatromechanical interpretation of the action of Cinchona bark. According to this theory, the efficacy of the American plant on fevers was largely due to its astringent, tonic properties that gave the blood a more solid consistency.[117]

Thanks to the *in vitro* chemical experimentation and to the experiments on animals that the followers of the iatrochemical doctrine and those of the iatromechanical one had carried out with Cinchona bark, the descriptions of the causes of malarial fevers, of the components of the American plant, and of the way in which its particles acted on human blood, had been explained with arguments that, though different and contrasting, were able to support and justify its use in medicine, in spite of the fact that they were not compatible with the Galenic pathology. The therapeutic success of Cinchona bark and its definitive inclusion in the eighteenth-century European pharmacopoeia gave rise to more doubts about Galenism. A great number of chemical and pharmaceutical experiments were carried out on various species of Cinchona: they promoted the knowledge of the components of Cinchona and of the different qualities attributed to it, and furthered the improvement of the relevant medicinal preparations. Several of these results were published in the *Philosophical Transactions of the Royal Society* of London.

After the foundation, in 1660, of the Royal Society in Great Britain and, in 1665, of the Académie Royale des Sciences in France, both of which encouraged the study of nature, their journals published travel reports, information about exotic plants, and articles on the testing of medicinal plants and on the clinical cases in which these remedies had been used.[118] These contributions often consisted of descriptions based on the personal experiments of the author on a single plant, among which Cinchona

---

115 Minot, *De la nature, et des causes de la fièvre*: 153–340.
116 Minot, *De la nature, et des causes de la fièvre*: 167–169.
117 Schondorff, *Disputatio inauguralis*; Maehle 1999: 239–240.
118 Reveal 1992: 22; Pelayo 2003: 15–27.

bark occupied a special position.[119] The *Philosophical Transactions of the Royal Society* became the most important scientific journal in Europe in the eighteenth century. The Académie des Sciences was one of the European institutions that was most active in promoting the study of medicinal exotic plants, and ordered and organised scien-

Figure 6: Title page, in Francis Bacon, *Instauratio Magna*. London: Ioannem Billium, 1620.

---

119 Crawford 2016: 63–65.

tific expeditions to the overseas territories. These scientific institutions, like others created in the same century in Europe, were inspired by the model of the Iberian institutions – (House of Trade and Casa de Contratación) – and their method of scientific production.[120]

In fact, as we have seen, an empirical approach to the study of the new plants had already begun to emerge in the sixteenth century, thanks to the attempt of the Spanish Crown to colonize the knowledge about natural history and medicine of the native American cultures. However, as I pointed out in the first part, the appropriation and assimilation of American medicinal plants into traditional medicine was neither easy nor immediate, and this was chiefly due to the problem of checking whether the properties of the new plants were compatible with the Galenic theory of humours. The reconfiguration of the knowledge of the American materia medica on the basis of the models of the Galenic doctrine considerably limited the therapeutic arsenal on which to draw and its curative potential. Nevertheless, as I show, contact with a new nature had begun to promote the development of a knowledge of nature in which experience took on a new role, questioning the classical sources and sometimes depriving them of their authority. From the attempt to appropriate the materia medica of the New World, new methods of studying its medicinal plants emerged in which, no longer the knowledge handed down by the classics, but personal experience became a source of knowledge and a method of study. In the second part I deal with the specific case of one of the most important American medicinal plants, the Cinchona bark, because the history of its therapeutic success and its assimilation into European medicine was the result of pharmacological experiments that not only replaced theoretical explanations but also challenged the seventeenth century Galenism. As seen, although the qualities and properties of Cinchona bark were incompatible with humoral pathology, it was efficacy against what are known the symptoms of malaria and its therapeutic success was the result of empirical experiments in which the principles of the Galenic doctrine did not find space. Ultimately, this chapter attempts to emphasize the important phenomenon of the introduction of American medicinal plants into European pharmacopoeia of the early modern age. Their study contributed to the construction of a modern science.

---

120 Barrera-Osorio 2006: 2.

Bettina Dietz
Knots in a Web: Botany, Materia Medica, and South Asian Languages in the Publication of Paul Hermann's *Ceylon-Herbaria* (ca. 1690–1770)

Introduction

For botanists in the early modern period, publishing their works was often a protracted process to which many—sometimes countless—people contributed, and that could take several years, even decades. Reflecting the accumulative nature of botanical practice, authors and editors published their local, regional, and global flora in repeated cycles of updating and correcting[1]. As soon as relevant new information and plants became available, they were incorporated into an existing work, and an amended edition appeared on the market. As a result, most botanical publications were both provisional and iterative in nature. Often this process was continued by other botanists after the death of an author. Sometimes it only began after an author's death, when others took on the task of publishing the botanical material that he left behind. A case of this sort will provide the focus here, in order to allow discussion of another aspect of the correlation between the *modus operandi* of early modern botany and the characteristic strategies of its community-driven publication system. Material collected by botanists—plants, seeds, and plant-based products, as well as botanical, ethno-botanical (including ethno-medical) and linguistic information—was often published in a sequence of iterative books in a process that could take decades. It was successively linked with relevant information from older and more recent literature to form a dense web of information. Sometimes such a publication process proceeded on different tracks – in the case presented here, in publications on Ceylon's flora and on materia medica, which became increasingly intertwined over time.

The German botanist Paul Hermann (1646–1695), who had been in Ceylon (today Sri Lanka) from 1672 to 1680 in the service of the Dutch East India Company (VOC), returned to Leiden with an extensive herbarium. There he published a single work, a catalogue of the plants in Leiden's botanical garden (*Horti academici Lugduno-Batavi catalogus*, Leiden 1687), which also contained some of the plants he had brought from Ceylon. But by far the majority of the plants and material he had gathered during his eight years in Ceylon was only published later, by others[2]. Although I have discussed the stage-by-stage, posthumous publication of Hermann's material elsewhere, in a

---
1 See Dietz 2012, 2017, 2019 and 2022.
2 Dietz 2022.

piece looking at the practice of identifying so-called synonyms – the different names which various botanists gave the same plant over time – Paul Hermann will again be at the centre of this essay. The starting point will be two of his herbaria and the broad spectrum of botanical, ethno-botanical and linguistic information that they contained[3]. In addition, Hermann had brought back extensive notes from Ceylon, on which the *Musaeum Zeylanicum* (1717), published by William Sherard (1659–1728) after Hermann's death, was based. They are thus available in printed form, at least in part[4].

Paul Hermann was born in Halle an der Saale (Saxony-Anhalt) in 1646. He studied medicine in Leipzig and Padua, where he graduated in 1670[5]. In 1671 he continued his studies in Leiden, where he attended the botany lectures given by Arnold Syen (1640–1678), director of the botanical gardens. Through Syen, he met Hieronymus van Beverningh (1614–1690), a Dutch diplomat and plant lover, with whose help he was sent to Ceylon by the VOC as a medical doctor. From 1672 to 1680, Hermann headed the Dutch hospital in Colombo. When Syen died in 1678, Hermann was appointed his successor and returned to Leiden in 1680, where he concentrated especially on expanding the botanical gardens. Hermann died in 1695 at the age of 48. At that time, his Ceylon material was still unpublished.

We know of four herbaria that Hermann created in Ceylon. The biggest comprised five volumes—four with dried plants, and one with drawings—and is today held by the Natural History Museum in London (hereafter London Herbarium, LH). For a long time, its whereabouts was a mystery until, by chance, it fell into Linnaeus's hands. He identified it as Hermann's personal herbarium and published his *Flora Zeylanica* (1747) based on it[6]. The second—a single volume containing only dried plants and no illustrations—is today held by the library of the Institut de France in Paris (hereafter Paris Herbarium, PH). It had been in the possession of the Dutch botanist Johannes Burman (1707–1780), and had provided the basis of his *Thesaurus Zeylanicus* (1737)[7]. The third herbarium, in two volumes, is today held by the Naturalis Biodiversity Center in Leiden (Netherlands)[8]. Hermann had given it as a present to Beverningh, who had facilitated his stay in Ceylon. A hand-written dedication by Hermann in a fourth herbarium shows that he had sent it to Johannes Breyne, a less well-known member of the Danzig family of botanists. It is today held by the Gotha Research Library (Thuringia)[9].

---

[3] The question of how, or with whose help, Hermann accumulated this material will not be discussed here. For Georg Eberhard Rumpf's (Rumphius) information network see, Yoo 2018.
[4] Hermann, *Musaeum Zeylanicum*. Some of Hermann's papers can be found today with William Sherard's papers, in the library of the Department of Plant Sciences, University of Oxford (Sherard collection).
[5] For Hermann's biography, given the few sources that exist, see Rauschert 1970: 304; Veldman 2012.
[6] On this see below, p.##.
[7] On this see below, p.##.
[8] On this see Trimen 1887; Van Oostroom 1937; Van Andel and Barth 2018; Van Andel et al. 2018.
[9] According to van Andel, Johannes Breyne was a cousin of the Danzig botanist Jakob Breyne (1637–1697); see Van Andel and Barth 2018: 978. A list of the plants in the Gotha Herbarium can be found in Rauschert 1970.

This essay will investigate how this heterogeneous material found its way from Hermann's herbaria and notes into print via a series of iterative publications on the flora of Ceylon and the wider region on the one hand, and on materia medica on the other. Each consecutive editor who worked on the publication of Hermann's material tied new knots in an information web and undid some of those which had been tied by his predecessors. As both the authors and the readers of these publications constantly compared living and dried plants with descriptions and illustrations in the relevant published literature, each one came to slightly different conclusions in the process. Publishing and comparing these divergences were an important resource for the botanical community.

The First Knots in the Web

Two of Hermann's herbaria—that in Paris and that in London—contain numerous handwritten notes, both by Hermann himself and by two later owners, the Dutch botanist Johannes Burman (1706–1779) and the Swede Carolus Linnaeus (1707–1778). In what follows the early stages of the publication of Hermann's Ceylon material will be traced, using these notes as a starting point. In his personal herbarium (LH), for example, Hermann had noted next to a plant:

> "Pithawakka Cingala
> Urinaria indica erecta vulgaris.
> This plant is highly effective for expelling urine"[10].

The entry consists of the name of the plant in Sinhalese (plus the Latin adjective for 'Ceylonese'), a Latin descriptive name indicating the plant's diuretic qualities, and a note on its particular potency. Later, the Dutch botanist Johannes Burman supplemented this by adding in his hand "Thes. Zeyl. pag. 230 tab. 93. fig. 2", a reference to the *Thesaurus Zeylanicus* (1737), the first large work based on Hermann's plants from Ceylon that Burman had himself published[11]. On the page referred to in the *Thesaurus Zeylanicus*, we find that the entry under *Urinaria Indica erecta vulgaris* has grown from the original three lines to around fifteen. It begins with a list of so-called synonyms followed by a detailed description which ends with a comment that Hermann's name *Urinaria* has been retained because of the strong diuretic effect of the plant.

Anyone who looked up *Urinaria* in the *Thesaurus Zeylanicus* found much – although not all – that was available about this plant in the botanical literature at that time. This included references to the first two mentions that go back to Hermann, but

---

10 *Collectio plantarum* Ms 3912: 66r (available in digitized form at https://minerva.bibliotheque-institutdefrance.fr; accessed 14 May 2021). If not otherwise indicated, all quotations from the original Latin here and in the following were translated by me.
11 Burman, *Thesaurus Zeylanicus*: 230.

Figure 1: *Herbarium sheet of the Paris Herbarium*, in Paul Hermann, *Collectio plantarum Ceylanensium quas olim peritissimus Botanicus Paulus Hermannus in ipsa Ceylona observavit atque colligit [. . .] alias Thesaurum*

had not been published by him: the plant appears in the *Paradisi Batavi Prodromus*, an advance notice of Hermann's catalogue of Leiden's botanical gardens which the British botanist William Sherard had brought out, with Hermann's permission, while Hermann was still alive[12]. The plant is listed there as *Urinaria Zeylanica erecta*, with the additional information 'Jathaaembula Zeylanens., Orleana seu Orellana seu Urucu Marck. & Pis.' During his stay in Ceylon, Hermann had apparently noted two local names for this plant-*Pithawakka* (see above, on the herbarium page) and *Jathaaembula*, as it is also called in the *Musaeum Zeylanicum* (1717)[13]. The earliest information about *Urinaria* that Hermann presented to the botanical public therefore included its name in Sinhalese, as well as a reference to its medicinal properties conveyed by its Latin name. Beyond this, the entry on the plant in the *Musaeum Zeylanicum*, like almost all entries there, contained a brief linguistic commentary on the morphology of the name: *Jatha* meant below, *Aebula* meant solvent; the *Jatha-Aebula* thus had a purgative effect[14].

Back to the *Prodromus*. The Latin and Sinhalese name of the plant is supplemented there by a first synonym which would later disappear from the literature, but which Hermann considered accurate at this time. He had compared his plants from Ceylon with the contents of the *Naturalis Historia Brasiliae* (Leiden, 1648) by Georg Markgraf (or Marggraf, 1610–1644) and Willem Piso (1611–1678), and believed that their *Orleana* or *Urucu* was identical with his *Urinaria*[15]. Anyone who consulted the *Naturalis Historia Brasiliae* found an entry there, containing a precise description of the plant and a discussion of ways of using it[16]. This publication claimed that the plant's ground seeds, boiled up in water, produced a strong, deep red dye. Drunk, the decoction was said to be an antidote to various poisons. In conclusion, there was a long commentary on Francisco Ximenes' (ca. 1560–1620) description of the plant in his *Quatro libros de la naturaleza y virtudes de las plantas y animales [. . .] en la Nueva España* (Mexico, 1615), where it was listed under its local name of *Axiotl*. Ximenes specified a number of its medical applications, including as a diuretic[17].

Figure 1 (continued)
meum Zeylanicum conscripsi & edidi anno MDCCXXXVII. Paris, Bibliothèque de l'Institut de France, Collection Benjamin Delessert, Ms 3912 Réserve: f. 66. Bibliothèque de l'Instiut de France, Collection Benjamin Delessert, Ms 3912 Réserve.

12 Sherard, *Schola Botanica*: 385.
13 Hermann, *Musaeum Zeylanicum*: 11.
14 Hermann, *Musaeum Zeylanicum*: 11.
15 See *Paradisi Batavi Prodromus*, part of Sherard, *Schola Botanica*: 385.
16 See Piso, Marggraf and Laet, *Historia naturalis Brasiliae*: II.61–62 (*Historia plantarum*); IV.65 (*De facultatibus simplicium*).
17 See See Piso, Marggraf and Laet, *Historia naturalis Brasiliae*: II.61–62 (*Historia plantarum*); IV.65 (*De facultatibus simplicium*): II.62.

The first, brief mention of *Urinaria Indica* in the *Paradisi Batavi prodromus* thus directed the reader to places where information on this plant could be found in the emerging literature on non-European flora. Anyone who followed some or all of the references from book to book would find what was known about the plant at this time: more or less detailed descriptions, illustrations of varying quality, indigenous plant names and linguistic information, Latin synonyms and information on medicinal and other uses (here as a pigment) were scattered between various publications and linked by cross-references. 'Reading' the *Prodromus* meant following up these references, comparing the various authors with each other, or even with a dried or perhaps living specimen of the *Urinaria Indica*, and drawing one's own conclusions about the quality of the illustrations and descriptions, the reliability of the information about pharmaceutical efficacy, and the appropriateness of the synonyms suggested by various authors. Over the eighteenth century, this process generated an increasingly dense web of information.

After these two brief mentions in the *Paradisi Batavi prodromus* and the *Musaeum Zeylanicum*, it was Johannes Burman who, in 1737, published the first description of *Urinaria Indica* based on Hermann's herbarium in his *Thesaurus Zeylanicus*.

(a)

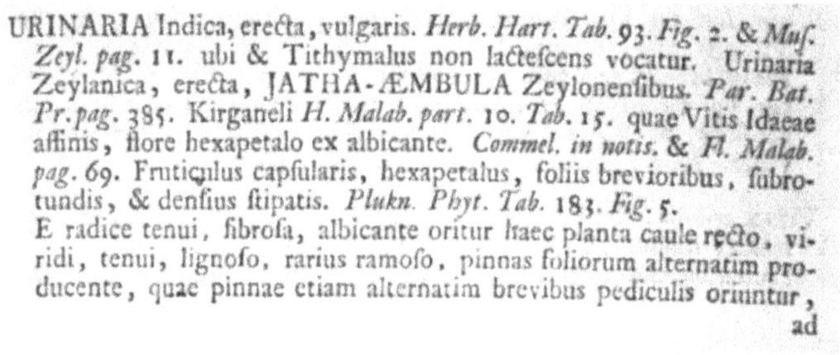

URINARIA Indica, erecta, vulgaris. *Herb. Hart. Tab.* 93. *Fig.* 2. & *Muf. Zeyl. pag.* 11. ubi & Tithymalus non lactescens vocatur. Urinaria Zeylanica, erecta, JATHA-ÆMBULA Zeylonenfibus. *Par. Bat. Pr. pag.* 385. Kirganeli *H. Malab. part.* 10. *Tab.* 15. quae Vitis Idaeae affinis, flore hexapetalo ex albicante. *Commel. in notis.* & *Fl. Malab. pag.* 69. Fruticulus capfularis, hexapetalus, foliis brevioribus, fubrotundis, & denfius ftipatis. *Plukn. Phyt. Tab.* 183. *Fig.* 5.
E radice tenui, fibrofa, albicante oritur haec planta caule recto, viridi, tenui, lignofo, rarius ramofo, pinnas foliorum alternatim producente, quae pinnae etiam alternatim brevibus pediculis oriuntur, ad

(b)

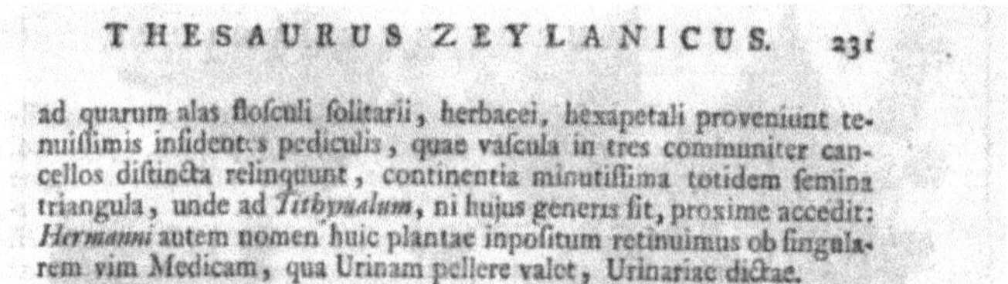

THESAURUS ZEYLANICUS. 231

ad quarum alas flofculi folitarii, herbacei, hexapetali proveniunt tenuiffimis infidentes pediculis, quae vafcula in tres communiter cancellos diftincta relinquunt, continentia minutiffima totidem femina triangula, unde ad *Tithymalum*, ni hujus generis fit, proxime accedit: *Hermanni* autem nomen huic plantae inpofitum retinuimus ob fingularem vim Medicam, qua Urinam pellere valet, Urinariae dictae.

Figure 2: *Urinaria Indica*, in Johannes Burman, *Thesaurus Zeylanicus*. Amsterdam: Janssonio-Waesbergios & Salomonem Schouten, 1737: pp. 231–232. Digitized by Bayerische Staatsbibliothek München.

Burman was also one of the first botanists to systematically compare their plants (in this case collected by Hermann) with the newly published botanical literature on the flora of south India and the islands of south-east Asia, thus establishing first insights into the plant geography of the region. The main reference work was initially the *Hortus Malabaricus* (Amsterdam, 1678–1693), a spectacular twelve-volume illustrated work on the flora of the west coast of India, which local scholars and plant experts had compiled on the initiative of the governor of Dutch Malabar, Hendrik Adriaan van Rheede tot Drakenstein (1636–1691)[18]. This was followed in 1741 by the appearance of Georg E. Rumpf's (1627–1702) *Herbarium Amboinense* on the plants of Ambon and other Maluku Islands, which Burman had published posthumously. In addition, there were various other smaller publications, which Burman and others also consulted and cited, as the following list of synonyms in the entry on *Urinaria Indica* in the *Thesaurus Zeylanicus* shows:

> *Herb Hart.* [Herbarium Hartog] *tab.* 93. *Fig.* 2 & *Mus. Zeyl.* [Musaeum Zeylanicum] *pag.* 11 [. . .]. Urinaria Zeylanica, erecta, JATHA-AEMBULA Zeylonensibus. *Par. Bat. Pr.* [Paradisi Batavi Prodromus] *pag.* 385. Kirganeli *H. Malab.* [Hortus Malabaricus] *part.* 10, tab. 15, quae Vitis Idaeae affinis, flore hexapetalo ex albicante. *Commel. in notis* [Commelin's comments in the Hortus Malabaricus] & *Fl. Malab.* [Flora Malabarica] *pag.* 69. Fruticulus capsularis, hexapetalus [. . .] *Plukn. Phyt.* [Plukenet, Phytographia] *Tab.* 183. *Fig.* 5[19].

Here Burman inserted new information about *Urinaria Indica* and potential synonyms which had become available in the twenty years since the publication of the *Musaeum Zeylanicum* into the web of references begun by Hermann and Sherard. In the first place he mentions a herbarium compiled by Jan Hartog, which was in Burman's possession, followed by the two publications already discussed, the *Paradisi Batavi prodromus* and the *Musaeum Zeylanicum*. *Kirganeli* is the first new synonym identified by Burman – the name of a plant in the *Hortus Malabaricus* which Burman thought was identical with *Urinaria Indica*. If we consult the passage cited in the tenth volume of the *Hortus Malabaricus*, we find not only a one-page description of the plant which gives detailed information about its medicinal uses, but also whole-page illustrations of two species of *Kinganeli*[20]. Burman's final reference refers to the *Phytographia* (London 1691–1696, 4 vols.) by the English botanist Leonard Plukenet (1642–1706), which contained another illustration.

Material on Ceylon in the Editions of Hermann's Materia Medica

The publication and reception of the plants from Hermann's Ceylon herbaria and the accompanying botanical, ethno-botanical and linguistic information started with the first references in Hermann's *Horti academici Lugduni-Batavo catalogus* (1687) and his notes,

---

18 See Henniger 1986: 97–176; Grove 1996; see also Fontes da Costa 2012.
19 Burman, *Thesaurus Zeylanicus*: 230–231, entry on *Urinaria Indica*. Here and in the following I have given the full titles–which are abbreviated in the original–in square brackets.
20 See Van Reede, *Hortus Malabaricus*: X.29, illustrations 15 and 16.

published posthumously as the *Musaeum Zeylanicum* (1717). From there, a dynamic of iterative, interlinked publications containing new or corrected information unfolded. It gained a significant boost from the publication of Burman's *Thesaurus Zeylanicus* in 1737, and later from Linnaeus' *Flora Zeylanica* (1747) and *Species plantarum* (1753), and will be traced here to Niklaas Laurens Burman's (1734–1793) *Flora Indica* (1768). First, however, a second publication track will be addressed, one which initially ran alongside the one outlined above before the two began to overlap – the publication of medicinal plants from Hermann's collection and of information on their medicinal use in a similarly iterative sequence of works on materia medica[21]. This sequence started from Hermann's notes for the lectures he delivered on materia medica at Leiden University, which were first published after his death by one of his students under the title *Lapis materiae medicae Lydius* (Touchstone of materia medica; 1703). *Lapis materiae medicae* is a small compendium, about 60 pages long, which was intended for the use of Dutch medical students and listed common medicinal plants and their effects. Although the main emphasis was on European plants, a few plants from Ceylon also made an appearance here. On the basis of this booklet, numerous expanded and corrected editions appeared over the next few decades under the editorship of various authors. Here I will trace the path of one plant – the Ceylonese cinnamon tree (which Hermann called *Cinamomum sive Canella Zeylanica*), and the remedy which, according to Hermann, could be derived from it, namely camphor – from his herbarium, through the iteration of the *Lapis materiae medicae*, and into the mesh of botanical and medical information pertaining to the flora of Sri Lanka.

My starting point is a number of sheets with dried twigs and individual leaves of the *Canella Zeylanica* which can be found in Hermann's personal herbarium (LH). One of these sheets contains the following note in Hermann's hand:

> Cinamomum sive Canella Zeylanica, B. Pin. [Caspar Bauhin, Pinax] pag. 408, Kurundu, [transcribed into Latin, and in Singhalese script][22].

The use of parts of the tree to produce a spice (cinnamon) and a medicine (camphor) is not explicitly addressed here but can be deduced from references to Caspar Bauhin's (1560–1624) *Pinax* (1623). If we look up the page cited, we find there, under the name *Cassia sive Canella & Cinamomum*, a long entry in which Bauhin initially tries to explain which plants Dioscorides and other authors from Antiquity called *Cassia*

---

21 On European materia medica, see Adlung and Urdang 1935; Kremers and Urdang 1951; Reeds 1976; Palmer 1985b; Siraisi 1990; Aliotta et al. 2003; Minuzzi 2016.

22 Hermann, *Herbarium*, [LH]: ID 408. The PH herbarium contains a single page on the *Canella Zeylanica* with the same handwritten information (see Paulus Hermannus, *Collectio plantarum Zeijlanesium*, [PH]: ID 21, *Canella Zeylanica*). The London Herbarium was Hermann's personal herbarium, which he hoped to use as a basis for his publications on the flora of Ceylon. This could explain why it contains several examples of a number of plants. The other Ceylon herbaria, including the Paris Herbarium, were gifts from Hermann to other botanists.

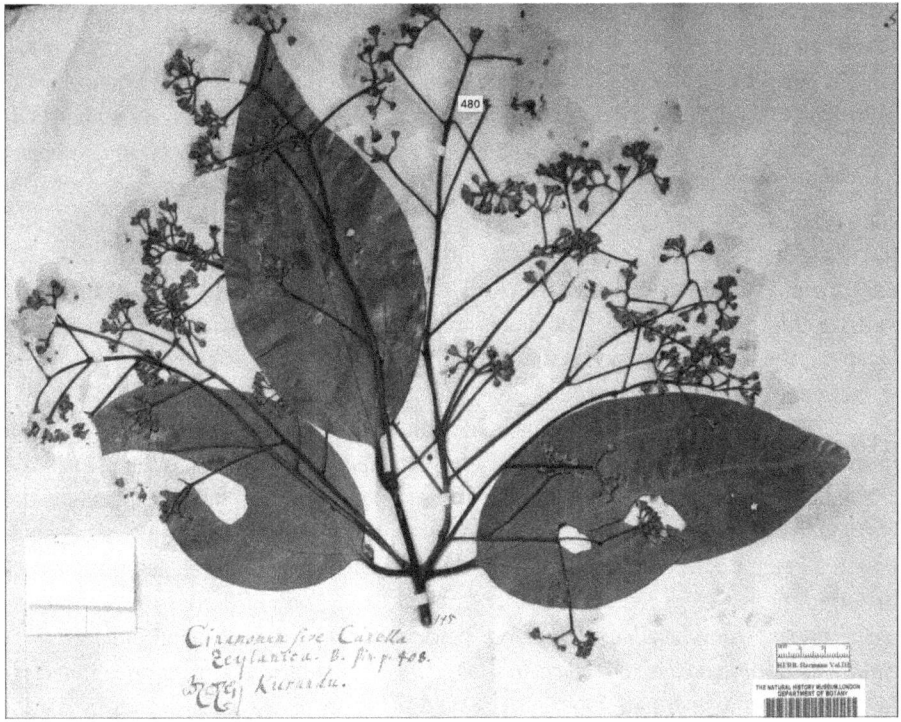

Figure 3: *Cinnamon sive Canella Zeylanica*, in Paul Hermann, *Herbarium*. Ceylon 1670s, now London, Natural History Museum: ID 480.

and *Cinamomum*[23]. He also lists about ten early modern authors who mention the cinnamon tree in their works. While the plant was already known to European botanists, Hermann contributed new information about its occurrence and use in Ceylon, on its local name and additional synonyms. In 1717, interested botanists found much of this in the entry on *Kurudu* in the *Musaeum Zeylanicaum*, which was based on Hermann's notes:

> Cinamomum seu Canella Ceylanica C.B. [Caspar Bauhin]. Cuurdo Pisonis [Piso, Marggraf: *Historia naturalis Brasiliae*], Laurus Ceylanica glandifera, folio trinervio, which provides the best, true cinnamon. It is pronounced: Kurundu. Camphor, and camphor oil are distilled from the skin of the roots; cinnamon oil from the bark of the trunk. An oil similar to clove oil is made from the leaves, and from the fruit, an oil that resembles juniper oil[24].

---

23 Bauhin, Pinax: 408.
24 Hermann, *Musaeum Zeylanicum*: 12.

The medical applications of camphor derived from the cinnamon tree in general, and of the camphor derived from the root of the Ceylonese cinnamon tree in particular are left out here. But readers who were interested knew where to look up pharmacological information about Hermann's Ceylonese plants, namely, in the successively updated editions of notes taken at Hermann's lectures on materia medica. The entry on *Camphora* (camphor) in the first edition (*Lapis materiae medicae*, 1703) did not comment on *Canella Zeylanica*, but discussed the method of producing camphor, its sweat-inducing and pain-relieving properties, and its indication in cases of malignant fever, delirium, melancholy and twitching[25]. In the next edition of 1710 (under the new title of *Cynosura materiae medicae*), the entry had grown to fill four pages. It discusses four species of camphor, the second of which is that derived from the Ceylonese cinnamon tree, from the skin of the root, as it states[26].

In 1726 a third, considerably expanded edition of the *Cynosura* came out. It was produced by a different editor, and the entry for *Camphora* now filled six pages[27]. As in the 1710 edition, four species of camphor are distinguished, and the editor supplements the entry with new observations by a number of authors. Another new feature is the addition of a lengthy commentary pointing to relevant linguistic, botanical, and pharmaceutical information from the literature. A detailed description of the Japanese camphor tree is quoted from Engelbert Kaempfer's account of his journey to Japan (*Amoenitatum exoticarum politico-physico-medicarum fasciculi*; 1712), which is said to be better than Jakob Breyne's, which had apparently served botanists as a standard reference until then[28]. The entry mentions in this context that Hermann had sent Beverningh and Syen, both Dutch, samples of camphor prepared from different parts of the cinnamon tree[29]. These samples found their way into print via the posthumous editions of Hermann's *Materia medica*, taking a different route from the botanical material. But the two types of publication were so closely interlinked by cross-references that readers themselves could assemble the full range of information available about Hermann's plants.

Inscribing Hermann's Plants from Ceylon into the Linnaean System

The next stage in the publication of Hermann's Ceylonese plants was marked by the appearance of two works by Linnaeus, the *Flora Zeylanica* (1747) and *Materia medica* (1749), in which he merged the two strands which had hitherto run separately – the

---

25 [Hermann], *Lapis materiae medicae*: 46.
26 [Hermann], *Cynosura materiae medicae*: 255.
27 [Hermann], *Cynosura materiae medicae*: 617–623.
28 See [Hermann], *Cynosura materiae medicae*: 620.
29 See [Hermann], *Cynosura materiae medicae*: 620.

materia medica and the Ceylon flora. An apothecary in Copenhagen, August Günther, had acquired a five-volume herbarium under unexplained circumstances, and sent it to Linnaeus with a request to identify the plants it contained. Linnaeus recognized that the order in which the plants were listed was the same as that in the *Musaeum Zeylanicum* and concluded that this must be Hermann's personal herbarium[30]. Linnaeus's contribution was to inscribe Hermann's plants, along with the information web in which they had become enmeshed, into his taxonomic system, and thus to gather all the available botanical and medical-pharmaceutical information in one place. At the same time, Linnaeus – like all the preceding authors – added what he considered relevant. In the case of the *Flora Zeylanica* this was mainly references to his own earlier publications, such as the *Hortus Cliffortianus* (1737). So, what did the *Flora Zeylanica* make of the cinnamon tree and camphor?

As *Laurus foliis ovato-oblongis trinerviis, basi nervos unientibus*, a new descriptive name, it was here identified as one of two species of the genus *Laurus*[31].

The list of synonyms used by Linnaeus points to Bauhin's *Pinax*, Burman's *Thesaurus Zeylanicus*, his own *Hortus Cliffortianus*, Hermann's catalogue of the botanical gardens in Leiden and the *Musaeum Zeylanicum*, where the plant was called by its local name of *Kurudu*, which Linnaeus also gives. This was followed by the name commonly used in early modern pharmacopoeias (*Cinnamomum*) and a list of the various remedies that could be obtained from the bark of the tree. While the production process and medical indications are left out here, they are added in great detail in Linnaeus' *Materia medica* (1749) two years later.

There Linnaeus lists seven species of *Laurus*, with the Ceylonese cinnamon tree in first place.

The number of synonyms is reduced to a minimum in order to keep the compendium to a practical size, and the only references that are made are to the *Flora Zeylanica*, Hermann's catalogue of the botanical gardens and Burman's *Thesaurus*[32]. But anyone who checks the *Flora Zeylanica*, will find the extensive list of synonyms mentioned above. It also gives the pharmaceutical name, *Cinnamomi cortex* (bark of the cinnamon tree), followed by the location (Ceylon), smell and taste, the range of effects (stimulating, strengthening, warming, fever-reducing, anti-flatulence), indications (against weakness, vomiting, flatulence, heavy blood loss during menstruation, and other gynaecological complaints), and a list of *composita*, that is, of medicines that can be produced by adding further ingredients.

A plant from Rheede's *Hortus Malabaricus, Katou-Carva*, is listed here as a variant of this species of *Laurus* which had not yet appeared in the *Flora Zeylanica* published

---

30 On this see Linnaeus's Foreword to Linnaeus, *Flora Zeylanica*: (unpaginated).
31 Linnaeus, *Flora Zeylanica*: 61 (no. 145).
32 Linnaeus, *Materia medica*: 64.

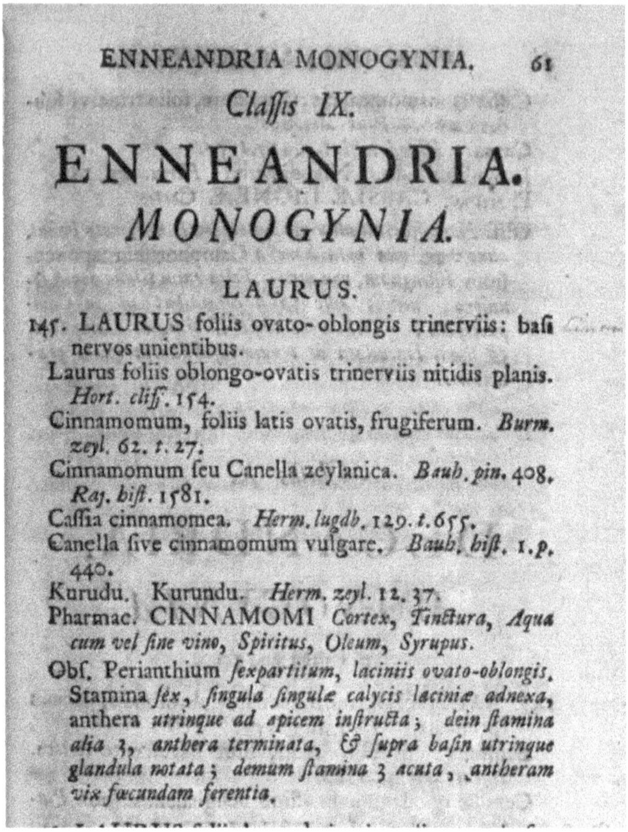

> ENNEANDRIA MONOGYNIA. 61
>
> *Classis IX.*
>
> # ENNEANDRIA.
> ## MONOGYNIA.
>
> ### LAURUS.
>
> 145. LAURUS foliis ovato-oblongis trinerviis: bafi nervos unientibus.
> Laurus foliis oblongo-ovatis trinerviis nitidis planis. *Hort. cliff.* 154.
> Cinnamomum, foliis latis ovatis, frugiferum. *Burm. zeyl.* 62. *t.* 27.
> Cinnamomum feu Canella zeylanica. *Bauh. pin.* 408. *Raj. hift.* 1581.
> Caffia cinnamomea. *Herm. lugdb.* 129. *t.* 655.
> Canella five cinnamomum vulgare. *Bauh. hift.* 1. *p.* 440.
> Kurudu. Kurundu. *Herm. zeyl.* 12. 37.
> Pharmac. CINNAMOMI *Cortex, Tinctura, Aqua cum vel fine vino, Spiritus, Oleum, Syrupus.*
> Obf. Perianthium *fexpartitum, laciniis ovato-oblongis. Stamina fex, fingula fingulæ calycis laciniæ adnexa, anthera utrinque ad apicem inftructa; dein ftamina alia 3, anthera terminata, & fupra bafin utrinque glandula notata; demum ftamina 3 acuta, antheram vix fœcundam ferentia.*

Figure 4: *Laurus*, in Carolus Linnaeus, *Flora Zeylanica sistens plantas indicas Zeylonae insulae* Stockholm: Salvius, 1747: p. 61. Digitized by the Bayerische Staatsbibliothek München.

just two years earlier. While working on the *Materia medica*, Linnaeus had apparently once again compared Hermann's plants with those in the *Hortus Malabaricus*, identified a further correspondence, one which he had not noticed while working on the *Flora Zeylanica*. All these intertextual knots were only provisionally tied, and could be untied again at any time by the authors themselves, or by others in the endless process of comparing plants, herbaria, and publications.

Unlike the editions of Hermann's *Cynosura*, Linnaeus in his *Materia medica* listed only a single *Laurus* as suitable for the extraction of camphor – the Japanese one, citing Engelbert Kaempfer's (1651–1716) *Amoenitatum exoticarum fasciculi* (1712). But the perpetuum mobile of iterating, supplementing and correcting did not come to a halt. In the fifth – now posthumous – edition of Linnaeus' *Materia medica*, the editor, Johann Christian Daniel Schreber (1739–1810), commented that another camphor-containing tree

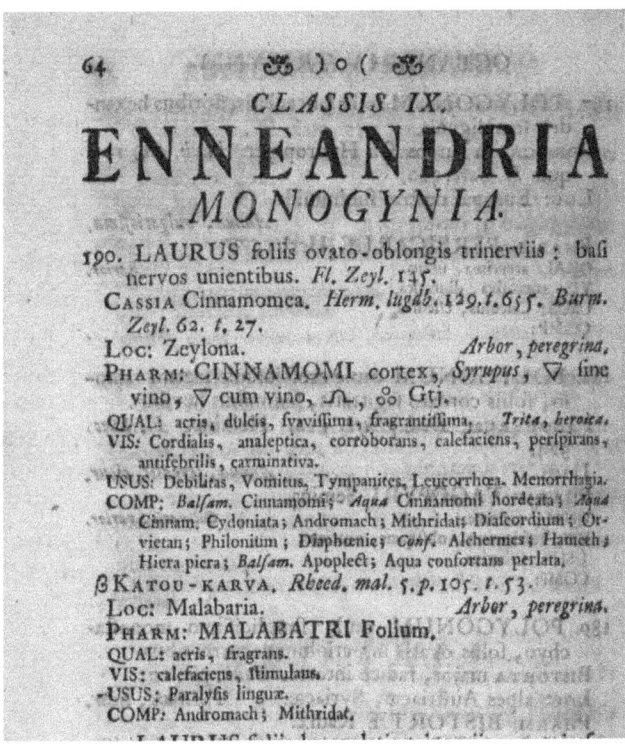

Figure 5: *Laurus*, in Carolus Linnaeus, *Materia medica, liber 1 de plantis*. Stockholm: Laurentius Salvius, 1749: p. 64. Digitized by the Bayerische Staatsbibliothek München.

unrelated to the *Laurus* had meanwhile been reported in an essay published in the *Philosophical transactions* of 1778[33].

Conclusion

The publication of the *Flora Indica* in 1768 marked the provisional end of this process of iterative publishing, which tied Hermann's Ceylonese plants into a mesh of descriptions, illustrations, living and dried plants. In the introduction, Niklaas Laurens Burman characterized the work as a synthesis of the plants of India and the south Asian and southeast Asian islands collected by various botanists over the last hundred years along with the relevant published information. The material basis for this was provided by the extensive collection which his father, J. Burman, had created and enriched with the herbaria of other well-known botanists, including Willem Piso, Paul Hermann, Laurent Garcin

---

33 Linnaeus, *Materia medica* [Editio quinta, Christ. Dan Schreber, 1787]: 124.

(1683–1752), Breyne (which of the Breynes is unclear) and others[34]. N. L. Burman based his *Flora Indica* on Linnaeus' taxonomy and nomenclature, and added descriptions of the rarest plants and illustrations of those of which none – or only inadequate ones – were available. Like all other posthumous editors of Hermann before him, he put together a selection of already published and unpublished material.

The characteristic qualities of botanical practice were reflected in the strategies of botanical publishing. While the accumulative nature of botany demanded iterative publication which did justice to the consecutive growth and constant correction of its material, this essay has concentrated on the publication tracks by which the material collected by Hermann – botanical, ethno-botanical and linguistic information, from his herbarium and notes – found its way into print over the course of time. Plants and information from the herbarium were published in successively iterated works on the flora of Ceylon (then India) on the one hand, and on materia medica on the other, and in the process became increasingly interlinked with relevant material from other publications to form an ever denser web. Each consecutive author repeatedly compared the material available to him with the existing literature, coming to conclusions that differed from those of others to a greater or lesser extent. Not all botanical observations were considered of equal value, but many of them contained information that could become relevant for others at a later date. Divergent observations on the same plant in different botanical publications reflected the long-term, community-driven process of taxonomic identification and were an essential resource for the botanical community.

---

34 Burman, *Flora Indica*: Proemium (unpaginated).

Bibliography

Archival Documents

Bologna, Biblioteca universitaria, Aldrovandiano 136:
- tomo V, ff. 208–220.
- tomo XIX.

Florence, Archivio di Stato, Mediceo del Principato, vol. 241.
Leiden, Erfgoed Leiden en Omstreken, Boerhaave, Herman, NL-LdnRAL-1658/I Herman Boerhaave/I.B.5.
Paris, Archives de l'Académie des Sciences, *Procès-verbaux*, t. XV-*bis*, 1696, ff. 194v–195r.
Venice, Archivio delle Istituzioni di Ricovero e di Educazione, DER E 189.1–6.
Venice, Biblioteca Nazionale Marciana:
    Marc. It. VI, 505 (coll. 12299), fasc. 20.
    Marc. It. VII, 2361 (coll. 9717).

Manuscripts

Florence, Biblioteca Medicea Laurenziana, Ashburnham 1448:
    ff. 57r–87r: [Galenus], *De dinamidiis*, tr. Nicolò Roccabonella.
    f. 155v: Nicolò Roccabonella, nota ad Arnaldi de Vilanova, *De vinis*.
Leiden, Universiteit Bibliotheek, BPL 3654: Herman Boerhaave, *Index Seminum Satorum*.
London, British Library:
    Ms 22332, ff. 3r–184v: Pietro Andrea Matthioli, *Discorsi Dioscoridei*.
    Egerton MS 747, ff. 1r–106v: Bartholomeus Mini de Senis, *Tractatus de herbis*.
    Egerton MS 2020: Serapion (Ibn Sarabiyun), *Libro agregà, Herbolario volgare; De medicamentis*.
London, Wellcome Library, MS. 136, f. 138v: Pierre Boaistuau, *Histoires prodigieuses*.
Montpellier, Bibliothèque universitaire, MS 277, f. 161r: Nicolò Roccabonella, *Nota de puero nato 1411 die 24 iulii*.
Paris, Bibliothèque de l'Arsenal, Ms 2808, f. 7r: *Recueil de figures de plantes colorées*.
Paris, Bibliothèque nationale de France:
    Latin 6823: *Liber de herbis et plantis*.
    Latin 9333: Elbocasim de Baldach (Ibn Butlân), *Tacuinum sanitatis*.
    Latin 17848: *Liber de plantis*.
Paris, Bibliothèque de l'Institut de France, Collection Benjamin Delessert, Ms 3912 Réserve: Paulus Hermannus, *Collectio plantarum Zeijlanesium*.
Oxford, Bodleian Library, MS. Ashmole 1431: [Apuleius], *Herbarium*. Sevilla, Biblioteca de la Universidad de Sevilla, A 331/102: Bartolomé Cobo y Peralta, *Historia del Nuevo mundo*, 1653.
Venice, Biblioteca Nazionale Marciana, Marc. Lat. VI, 59 (coll. 2548): Nicolò Roccabonella, and Andrea Amadio, *Liber de simplicibus*.
Wolfenbüttel, Herzog August Bibliothek, Cod. Guelf. 254/1 Helmst.:
    (ff. 9–12) *Fragmentum breviarii vel capitularis imperatoris Carli Magni*;
    (ff. 12v–16) *Capitulare Caroli Magni de villis vel curtis imperii*.

Sources

Aldrovandi, *Dendrologiae*: Ovidius Montalbanus (ed.), *Ulyssis Aldrovandi . . . Dendrologiae naturalis scilicet arborum historiae libri duo, sylva glandaria, acinosumque, pomarium ubi eruditiones omnium generum una cum botanicis doctrinis ingenia quaecumque non parum iuvant et ablectant.* Bononiae: ex typographia Ferroniana, 1668.

Alpini, *De balsamo*: Prosperi Alpini, *De balsamo dialogus. In quo verissima balsami plantae, opobalsami, & xilobalsami cognitio plerisque antiquorum atque iuniorum medicorum occulta nunc elucescit.* Venetiae: Sub Signum Leonis, 1591.

French translation: Antoine Colin, *Histoire du Baulme. Ou il est prouvé que nous avons vraye cognoissance de la plante qui produict le Baulme, & par consquent de son fruict, & de son bois . . . Version Françoise, tirée de Prosper Alpin*, in Antoine Colin, *Histoire des drogues episceries, et de certains medicamens simples, qui naissent és Indes & en l'Amerique. Ceste matière comprise en six livres: dont il y en a cinq tirés du latin de Charles de l'Escluse: & l'histoire du baulme adioustee de nouveau: où il est proué, que nous avons le vray baulme d'Arabie, contre l'opinion des ancies & modernes . . .* Seconde edition reveuë & augmentée, Lyon: Iean Pillehotte, 1619.

Alpini, *De medicina Aegyptiorum*: Prosperi Alpini, *De medicina Aegyptiorum: Libri qvatvor. In quibus multa cum de vario mittendi sanguinis usu per venas, arterias, cucurbitulas, ac scarificationis nostris inusitatas, deque inustionibus, & aliis chyrurgicis operationibus, tum de quamplurimis medicamentis apud Aegyptios frequentioribus, elucescunt. Quae cum priscis medicis doctissimis, olim notissima, ac pervulgatissima essent, nunc ingenti artis medicae iactura à nostris desiderantur . . .* Venetiae: apud Franciscum de Franciscis Senensem, 1591.

French translation: Raymond de Fenoyl, *Prosper Alpini, La médecine des Egyptiens, 1581–1584* (Voyageurs 21). Le Caire: Institut français d'archéologie orientale, 1980.

Alpini, *De plantis Aegypti*: Prospero Alpini, *De plantis Aegypti Liber. In quo non pauci, qui circa herbariam materiam irrepserunt, errores, deprehenduntur, quorum causa hactenus multa medicamenta ad usum medicinae admodum extepetenda, plerisque medicorum, non sine artis iactura, occulta, atque obsoleta iacuerunt . . .* Venetiae: apud Franciscum de Franciscis Senensem, 1592.

Italian translation: Elsa Mariella Cappelletti, Luciano Cremonini, and Giuseppe Ongaro (eds), *Prospero Alpini. Le piante dell'Egitto – Il balsamo (Venezia 1592)* (Centro studi Prospero Alpini-Marostica 1). Treviso: Antilia, 2009.

[Aristotle], *De plantis*: Hendrik Joan Drossart Lulofs, and Evert Lubbertus Jacobus Poortman (eds), *Nicolaus Damascenus:*
De Plantis. Five Translations (Aristoteles Semitico-Latinus 4). Amsterdam: North-Holland, 1989.

Astruc, *De morbis venereis*: Jean Astruc, *De morbis venereis libri sex. In quibus disseritur tum de Origine, Propagatione & Contagione horumce affectuum in genere; tùm de singulorum Naturà, Aetologiâ & Therapeiâ, cum brevi Analysi & Epicrisi Operum plerorumque quae de eodem argumento scripta sunt . . .* Lutetiae Parisiorum: Apud Guillelmum Cavelierum, 1736.

Bacon, *Instauratio Magna*: Francisci Bacon de Verulamio, *Instauratio Magna*. Londini: apud Ioannem Billium, 1620.

Bado, *Anastasis corticis Peruviae*: Sebastianus Badus, *Anastasis corticis Peruviae, seu Chinae Chinae defensio*. Genuae: Typis Petri Ioannis Calenzani, 1663.

Bauhin, *PINAX*: Caspari Bauhini, *ΠΙΝΑΞTheatri Botanici sive Index in Theophrasti Dioscoridis Plinii et Botanicorum qui à Seculo scripserunt Opera: Plantarum circiter sex millium ab ipsis exhibitarum nomina cum earundem Synonymiis & differentiis Methodicè secundum earum & genera & species proponens . . .* Basileae: Sumptibus & typis Ludovici Regis, 1623.

Belon, *Les Observations de plusieurs singularitez*: Pierre Belon, *Les Observations de plusieurs singularitez & choses memorables, trouvées en Grece, Asie, Iudée, Egypte, Arabie & autres pays estranges, redigées en trois livres. Reveuz de nouveau & augmentez de figures.* Paris: Guillaume Cavellat, 1554.

Latin translation: Carolus Clusius, *Petri Bellonii Cenomani, Plurimarum singularium & memorabilium rerum in Graecia, Asia, Aegypto, Iudaea, Arabia, aliisque exteris prouinciis ab ipso conspectarum obseruationes, tribus libris expressae*. Antwerpiae: ex officina Christophori Plantini, 1589.

Bidloo, *Anatomia humani corporis*: Godefridi Bidloo, *Anatomia humani corporis, Centum & quinque Tabulis, Per artificiosissimum G. de Lairesse ad vivrum delineatis, Demonstrata, Veterum Recentiorumque Inventis explicata plurimisque, hactenus non detectis, illustrata*. Amstelodami: Sumptibus viduae Joannis à Someren, Haeredum Joannis à Dyk, Henrici & Viduae Theodori Boom, 1685.

Blégny, *Le Remede anglois*: Nicolas de Blegny, *Le Remede anglois pour la Guerison des fievres; publié par Ordre du Roy. Avec les Observations de Monsieur le Premier Medecin de sa la Majesté, sur composition, les vertus, & l'usage de ce Remede*. Paris: chez l'auteur. Et La Veusve d'Antoine Padeloup, 1682.

Boerhaave, *Index plantarum*: Hermannus Boerhave, *Index plantarum, quae in Horto Academico Lugduno Batavo reperiuntur*. Lugduni Batavorum: Apud Cornelium Boutestein, 1710.

Boerhaave, *Institutiones et experimenta chemiae*: Hermanni Boerhaave, *Institutiones et experimenta chemiae*, 2 vols. Parisiis [i.e. Leiden]: [s.n.], 1724 (re-published: Venetiis: Apud Sebastianum Coleti, 1726).

Boerhaave, *Index alter plantarum*: Hermannus Boerhaave, *Index alter plantarum quae in Horto Academico Lugduno-Batavo aluntur*, 2 vols. Lugduni Batavorum: Apud Janssonios Vander Aa, 1727.

Boerhaave, *Elementa Chemiae*: Hermannus Boerhaave, *Elementa Chemiae, quae anniversario labore docuit, in publicis, privatisque, scholis*, 2 vols. Lugduni Batavorum: Sumptibus Joannis Rudolphi Im-hoff, 1732. English translation: Peter Shaw, *A New Method of Chemistry; including the History, Theory, and Practice of the Art: Translated from the Original Latin of Dr. Boerhaave's Elementa Chemiae*, 2 vols. London: Printed for T. Longman, 1741.

Boyle, *Of the Usefulness of Experimental Naturall Philosophy*: Robert Boyle, *Some Considerations touching the usefulnesse of experimental naturall philosophy propos'd in familiar discourses to a friend, by way of invitation to the study of it*. Oxford: Hen. Hall, 1663.
(Michael Hunter, Edward B. Davis, Antonio Clericuzio, and Lawrence M. Principe (eds), *Robert Boyle, The Usefulness of Natural Philosophy and sequels to Spring of Air, 1662–1663*, in Michael Hunter, Edward B. Davis, Antonio Clericuzio, and Lawrence M. Principe (eds), *The Works of Robert Boyle*, 14 vols. London: Pickering & Chatto, 1999–2001, vol. 3, 1999, pp. 189–548).

Bravo de Sobremonte, *Disputatio apologetica*: Gaspar Bravo de Sobremonte Ramires, *Disputatio apologetica, Pro Dogmaticae Medicinae praestantia; & omnium Scientiarum, & Artium dignitate, ex omnigenae litteraturae decretis. Huic accesserunt tractatus duo, Quorum primus continet X. Consultationes Medicas nusquam hactenus editas. Alter verò Tyrocinium Practicum Artis Curatricis exhibet. Cum Indice triplici: Primo Sectionum, & Consultationum; Altro Autorum in hoc Opere laudatorum; Tertio denique Rerum praecipuarum*. Lugduni: Sumptibus Petri Chevalier, 1669.

Brunfels, *Herbarum vivae Eicones*: Otho Brunfelsius, *Herbarum vivae Eicones ad naturae Imitationem, summa cum diligentia & artificio effigiatae, unà cum effectibus earundem, in gratiam veteris illius, & iamiam renascentis Herbariae Medicinae. Tomus I. Quibus adiecta ad calcem, Appendix isagogica de usu & administratione simplicium. Item Index Contentorum singulorum*. Argentorati: apud Ioannem Schottum, 1530.

Burman, *Thesaurus Zeylanicus*: Johannes Burmannus, *Thesaurus Zeylanicus, exhibens plantas in insula Zeylana nascentes; Inter quas Plurimae novae species, & genera inveniuntur. Omnia Iconibus illustrata, ac descripta*. Amstelaedami: Apud Janssonio-Waesbergios & Salomonem Schouten, 1737.

Burman, *Flora Indica*: Nicolaus Laurentius Burmannus, *Flora Indica: cui accedit series zoophytorum Indicorum, nec non Prodromus Florae Capensis*. Lugduni Batavorum: Apud Cornelium Haek. Amstelaedami: Apud Johannem Schreuderum, 1768.

Calancha, *Coronica moralizada*: Antonio de la Calancha, *Coronica moralizada del Orden de San Augustín en el Peru, con sucesos egenplares en esta Monarquia*. Barcelona: Pedro Lacavalleria, 1638.

Caldera de Heredia, *Tribunalis Medici*: Gasparis Calderae de Heredia, *Tribunalis Medici illustrationes et observationes practicae. Accessit liber aureus de facile parabilibus, E Veterum & Recentiorum observatione comprobatis, & ex arcanis Naturae, Chymico artificio & Artis magisterio eductis*. Antverpiae: Apud Iacobum Meursium, 1663.

Cassini, "Philosophical Particulars": Giovanni Cassini, "An Intimation of divers Philosophical particulars, now undertaken and consider'd by several Ingenious and Learned men; here inserted to excite others to joyn with them in the same or the like Attempts and Observations", *Philosophical Transactions* 6 (1671), pp. 2216–2219.

Catelan, *Rare et curieux discours*: Laurens Catelan, *Rare et curieux discours de la Plante appellée Mandragore; de ses especes, vertus et usage. Et particulierement de celle qui produict une Racine, representant de figure, le corps d'un homme; qu'aucuns croyent celle que Iosephe appelle Baaras; & d'autres, les Teraphins de Laban, en l'Escriture Sainte*. Paris: Aux despens de l'Autheur, 1638.

Celsus, *De Medicina*: Walter George Spencer (ed.), *Aulus Cornelius Celsus, De Medicina*, 3 vols. (Loeb Classical Library 292, 304, 336). London: William Heinemann, and Cambridge, MA: Harvard University Press, 1935–1938 (reprint: Cambridge, MA: Harvard University Press, 1971–1979).

Charas, *Thériaque d'Andromachus*: Moyse Charas, *Histoire Naturelle des Animaux, des Plantes, & des Mineraux qui entrent dans la Composition de la Theriaque d'Andromachus . . . Avec les reformations & les Observations de l'Autuer, tant sur l'Election, & sur la Préparation, que sur le dernier Melange de tous les Ingrediens de cette grande Composition*. Paris: Chez Olivier de Varennes, 1668.

Charas, *Pharmacopée royale*: Moyse Charas, *Pharmacopée royale galénique et chymique*. Paris: Chez l'Auteur, 1676.

Cieza de León, *Parte primera dela Chronica del Peru*: Pedro Cieça de León, *Parte primera. De la chronica del Peru. Que tracta la demarcacion de sus provincias: la descripcion dellas. Las fundaciones de las nuevas ciudades. Los ritos y costumbres de los indios. Y otras cosas estrañas dignas de ser sabidas*. Sevilla: [Martin de Montesdoca], 1553.

Clifford, *Herbarium*: George Clifford, *Herbarium*. Hartelamp, 1720s (now preserved at London, Natural History Museum).

Clusius, *Exoticorum*: Carolus Clusius, *Exoticorum libri decem: Quibus Animalium, Plantarum, Aromatum, aliorumque peregrinorum Fructuum historiae describuntur: Item Petri Bellonii Observationes*. Lugduni Batavorum: ex officina Plantiniana Raphelengii, 1605.

Collegio degli Speziali di Napoli, *Parere*: Collegio degli Speziali di Napoli, *Parere dell'almo Collegio de' Spetiali di Napoli sopra l'opobalsamo Mandatoli dalli Signori Consoli del Collegio de' Spetiali di Roma. Con un picciolo Trattato dell'Opobalsamo Orientale di Gioseppe Donzelli*. Napoli: Francesco Savio, 1640.

Commelin, *Plantarum usualium*: Casparus Commelin, *Plantarum Usualium Horti Medici Amstelodamensis catalogus*. Amstelodami: in Horto Medico, [1698].

Commelin, *Rariores et exoticae*: Casparus Commelin, *Horti Medici Amstelaedamensis plantae rariores et exoticae ad vivum aeri incisae*. Lugduni Batavorum: Apud Fredericum Haringh, 1706.

Commelin, *Horti Medici Amstelodamensis*: Frederico Ruyschio & Francisco Kiggerlario (eds), *Horti Medici Amstelodamensis rariorum Tam Orientalis, quàm Occidentalis Indiae, aliarumque Peregrinarum Plantarum Magno studio ac labore, sumptibus Civitatis Amstelodamensis, longâ annorum Serie collectarum, Descriptio et icones Ad vivum œri incisae. Auctore Joanne Commelino . . . opus posthumum, Latinitate donaturm, Notisque & Observationius Illustratum*, 2 vols. Amstelodami: Apud P. & J. Blaeu, et Abramus à Someren, 1697–1701.

Courten and Sloane, "Experiments and Observations": William Courten, and Hans Sloane, "Experiments and Observations of the Effects of several sorts of Poisons upon Animals, &c. Made at Montpellier in the Years 1678 and 1679, by the late William Courten Esq; Communicated by Dr. Hans Sloane, R. S. Secr. Translated from the Latin MS", *Philosophical Transactions* 27 (1710–1712), pp. 485–500.

[Cuba], *Herbarius*: [Johannes von Cuba], *herbarius*. Mencz [Mainz]: [Peter Schöffer], 1485.

da Orta, *Colóquios*: Garcia d'Orta, *Coloquios dos simples, e drogas he cousas mediçinais da India, e assi dalgunas frutas achadas nella onde se tratam algunas cousas tocantes amediçina, pratica, e outras cosas boas, pera saber*. Goa: Ioannes de endem 1563.

Della Porta, *Magiae naturalis*: Io. Baptista Porta Neapolitanus, *Magiae naturalis libri viginti*. Francofurti: Apud Andreae Wecheli heredes, Claudium Marnium, & Joann. Aubrium, 1591.

Dioscorides, *De materia medica*: Max Wellmann (ed.), *Pedanii Dioscuridis Anazarbei de materia medica libri quinque*, 3 vols. Berlin: Weidmann, 1906–1914.

English translation: Lily Y. Beck, *Dioscorides. De Materia Medica* (Altertumswissenschaftliche Texte und Studien 38). Hildesheim: Olms, 2005.

English translation of the preface: John Scarborough, and Vivian Nutton, "The Preface to Dioscorides' *Materia Medica*: Introduction, Translation, and Commentary", *Transactions and Studies of the College of Physicians of Philadelphia*, ser. 5, 4 (1982), pp. 195–197.

Dodonaeus, *Stirpium historiae*: Rembertus Dodonaeus, *Stirpium historiae pemptades sex sive Libri XXX*. Antverpiae: Ex officina Christophori Plantini, 1583.

Doni, *Les Mondes Celestes*: Gabriel Chappuis (ed.), *Doni, Les Mondes, Celestes, Terrestres Et Infernaux. Le Monde petit, Grand, Imaginé, Meslé, Risible, des Sages & Fols, & le Tresgrand. L'Eenfer des Escoliers, des mal Mariez, des Putains & Ruffians, des Soldats & Capitaines poltrons, des pietres Docteurs, des Usuriers, des Poëtes & Compositeurs ignorans: Tirez des œuvres de Doni Florentino, par Gabriel Chappuis Tourangeau*. Lyon: Estienne Michel, 1578.

Donzelli, *Petitorio*: Giuseppe Donzelli, *Petitorio Napolitano spiegato, et illustrato . . . nel quale si contiene quanto deve, per obbligo tenere ciascheduno Spetiale di questa Città, e Regno nella sua Spetiaria, E mostrare nelle Regie Visite, che si faranno dal Protomedico. Con due aggionte, in una delle quali si contiene quanto puo usarsi in una bene ordinata Spetiaria: e nell'altri si propone il modo di Balsamare i Cadaveri de' Corpi humani, Et Aggiuntoui in fine un Discorso della Dignità del Regio Reneral Protomedico*. Napoli: Novello de Bonis, 1663.

Dorn, *Congeries Paracelsicae chemiae*: Gerardo Dorneo, *Congeries Paracelsicae chemiae de transmutationibus metallorum, ex omnibus quae de his ab ipso scripta reperire licuit hactenus. Accesit genealogia mineralium, atque metallorum omnium*. Francofurti: Apud Andream Wechelum, 1581.

Ely, *Dissertatio*: Pinas Ely, *Dissertatio Inavgvralis Physico-Medica De Opobalsamo Et Oleo Balanino quam indultu gratiose facultatis medicae . . . . pro gradu doctoris summisque in medicina honoribus et privilegiis rite atque legitime obtinendis defendet . . .* Francofurti ad Viadrum: typis Ioannis Christiani Winteri, 1770.

Elsholtz, *Clysmatica nova*: Johann Sigismund Elsholzii, *Clysmatica nova, sive Ratio, qua in venam sectam medicamenta immitti possint, ut eodem modo, ac si per os assumta fuissent, operentur: in animantibus per drastica, in homine per leniora hactenus probata, & adserta*. Berolini: Ex typographia Rungiana. Sumptibus Danielis Reichelii, 1665.

English translation: Eric Gladstone, "Johann Sigismund Elsholtz: Clysmatica Nova (1665): Elsholtz' Neglected Work on Intravenous Injection: Part I", *California and Western Medicine*, 38/6 (1933), pp. 432–434.

Fabricius, *De formato foetu*: Hieronymi Fabricii ab Aquapendente, *De formato foetu*. Venetiis: per Francescum Bolzettam, 1600.

Fabricius, *Tractatus Quatuor*: Hieronymi Fabricii ab Aquapendente, *Tractatus Quatuor. Quorum I. De Formato Foetu. II. De Locutione & eius instrumentis. III. De Loquela Brutorum. IV. De venarum ostiolis, loquitur. Duplici Indice donati, Figurisque aeneis ornati*. Francofurti: Impensis Iacobi de Zetteri; Typis Hartm. Palthenij, 1624.

Falloppio, *de Tumoribus*: Gabrielis Falloppii, *Libelli duo: alter de ulceribus: alter de Ttmoribus praeter naturam*. Venetiis: Apud Donatum Bertellum, 1563.

Flavius Josephus, *Histoire*: Robert Arnauld D'Andilly (transl.), *Histoire de la Guerre des Juifs contre les Romains par Falvius Joseph. Et sa vie écrite par lui-meme. Ecrite par Flavius Joseph sous le titre de Antiquitez Uidaiques*, 5 vols. Paris: Le Petit, 1668.

Fludd, *Utriusque cosmi maioris*: Roberto Flud, *Utriusque Cosmi Maioris scilicet et minoris metaphysica, physica atque technica historia. In duo Volumina secundum Cosmi differentiam diuisa. Tomus Primus De Macrocosmi Historia in duos tractatus divisa* ... Oppenhemii: AEre Johan-Theodori de Bry, typis Hieronymi Galleri, 1617.

Fracastoro, *Syphilis*: Hieronymus Fracastorius, *Syphilis, sive Morbus gallicus*. Verona [i.e., Venice?]: [Stefano Nicolini da Sabbio], 1530.
English translation: Nahum Tate (transl.), *Syphilis: or, A Poetical History of the French Disease. Written in Latin by Fracastorius. And now Attempted in English*. London: Printed for Jacob Tonson, 1686.

Fragoso, *Discursos de las cosas Aromaticas*: Iuan Fragoso, *Discursos de las cosas Aromaticas, arboles y frutales, y de otras muchas medicinas simples que se traen de la India Oriental, y sirven al uso de medicina*. Madrid: en casa de Francisco Sanchez, 1572.

Fuchs, *De historia stirpium*: Leonhartus Fuchsius, *De Historia stirpium commentarii insignes, maximis impensis et vigiliis elaborati, adjectis earundem vivis plusquam quingentis imaginibus, nunquam antea ad naturae imitationem artificiosius efficitis & expressis*. Basileae: In Officina Insingriniana, 1542.
French translation: Eloi de Maignan (transl.), *Commentaires tres excellens de l'hystoire des plantes, Composez premierement en latin par Leonarth Fousch, medecin tres renommé: Et depuis, nouvellement traduictz en langue Françoise*. Paris: Iacques Gazeau, 1549.

Galen, *De antidotis*: Karl Gottlieb Kühn (ed.), *Galenus, De Antidotis*, in Karl Gottlieb Kühn (ed.), *Claudii Galeni opera omnia* (Medicorum Graecorum Opera quae exstant 1–20), vol. 14. Leipzig: K. Knobloch, 1827, pp. 1–209.

Galen, *De simplicium medicamentorum temperamentis ac facultatibus*: Karl Gottlieb Kühn (ed.), *Galenus, De simplicium medicamentorum temperamentis ac facultatibus lib. VI*, in Karl Gottlieb Kühn (ed.), *Claudii Galeni opera omnia* (Medicorum Graecorum Opera quae exstant 1–20), vol. 12. Leipzig: K. Knobloch, 1826, pp. 379–892.

Galen, *De Theriaca*: Karl Gottlieb Kühn (ed.), *Galenus, De Theriaca ad Pisonem*, in Karl Gottlieb Kühn (ed.), *Claudii Galeni opera omnia* (Medicorum Graecorum Opera quae exstant 1–20), vol 14. Leipzig: K. Knobloch, 1827, pp. 210–294.

Gellius, *Noctes Atticae*: Thomas E. Page, Edward Capps, and William H. D. Rouse (eds), and John C. Rolfe (tr.), *The Attic Nights of Aulus Gellius*, 3 vols (Loeb Classical Library 195, 200, 212). Cambridge MA: Harvard University Press, and London: W. Heinemann, 1927.

Gerard, *The Herball*: John Gerarde, *The Herball or Generall Historie of Plantes*. London: John Norton, 1597.

Gleditsch, "Sur la mandragore": Johannes Gottlieb Gleditsch, "Sur la mandragore, dont l'histoire a été fort altérée dans l'Antiquité", *Nouveaux Mémoires de l'Académie royale des sciences et belles-lettres*, année 1778 (1780), pp. 36–61.

Glisson, *Anatomia hepatis*: Francisci Glissonii, *Anatomia hepatis. Cui praemittuntur Quaedam ad rem Anatomicam universe spectantia. Et Ad calcem Operis subjiciuntur nonnulla de Lymphae-ductibibus nuper repertis*. Londini: Typis Du-Guardianis, Impensis Octaviani Pullein, 1654.

[Grasshoff], *Dyas chymica*: [Johann Grasshoff (ed.)], *Dyas Chymica Tripartita, Das ist: Sechs herzliche Teutsche Philosophische Tractatlein. Deren II. Von an jtzo noch im Leben: II. Von mitlern Alters: und II. Von ältern Philosophis beschrieben worden. Nunmehr aber Allen Filiis Doctrinae zu Nuss an Taggeben und mit shoenen Figuren gezieret*. Franckfurt am Mayn: Luca Jennis, 1625.

Grew, *Musaeum regalis societatis*: Nehemiah Grew, *Musaeum regalis societatis. Or A Catalogue & Description Of the Natural and Artificial Rarities Belonging to the Royal Society And preserved at Gresham College . . . Whereunto is Subjoyned the Comparative Anatomy of Stomachs and Guts*. London: W. Rawlins, 1681.

Grew, *Anatomy of Plants*: Nehemiah Grew, *The Anatomy of Plants. With an Idea of a Philosophical History of Plants. And several other Lectures, Read before the Royal Society*. London: W. Rawlins, 1682.

Haak, *Catalogus*: Cornelius Haak, *Catalogus praestantissimorum & insignium Librorum. praecipue Theologicorum, Juridicorum, & Miscellaneorum, &c. Quibus usus est. N.B.D. Accedit Appendix I. quae continet libros Jacobi Ligtvoet, Dum in vivis versaretur, Hortulani vigilantissimi Horti Academici Leidensis. Inter quos praecipue eminet insignis Herbarius Vivus, opera atque labore defuncti in 13 fascibus praegrandibus collectus. Nec Non Appendix II. Librorum Theologicorum, Juridicorum, Medicorum, Historicorum nec non Miscellaneorum. Waar onder veel beste Nederduitse Boeken zyn. Quorum omnium Publica fiet Auctio in AEdibus Cornelii Haak, Die 23. & 24. Octobris 1752*. Lugduni Batavorum: Apud Cornelium Haak, 1752.

Hales, *Vegetable Staticks*: Stephen Hales, *Vegetable Staticks: Or, An Account of some Statical Experiments on the Sap in Vegetables: Being an Essay towards a Natural History of Vegetation. Also, a Specimen of An Attempt to Analyse the Air, By a great Variety of Chymico-Statical Experiments; Which were read at several Meetings before the Royale Society*. London: Printed for W. and J. Innys, at the West End of St. Paul's; and T. Woodward, over-against St. Dunstan's Church in Fleetstreet, 1727.

Harvey, *De motu cordis*: Guilielmus Harveus, *Exercitatio anatomica de motu cordis et sanguinis in animalibus*. Francofurti: Sumptibus Guilhelmi Fitzeri, 1628.
English translation: Robert Willis (transl.) and Arthur C. Guyton (introduction), *The Works of William Harvey*. London: Sydenham Society, 1847 (reprints: 1867, 1949, 1965, 1989), pp. 1–86.

Harvey, *Exercitationes de Generatione Animalium*: Guilielmus Harveus, *Exercitationes de Generatione Animalium. Quibus accedunt quaedam De Partu: de Membranis ac humoribus Uteri: & de Conceptione*. Londini: Typis Du-Guardianis; impensis Octaviani Pulleyn in Coemeterio Paulino, 1651.
English translation:
Martin Lluelyn, *Anatomical Exercitations, Concerning the Generation of Living Creatures by William Harvey*. London: Printed by James Young, for Octavian Pulleyn, 1653.
Robert Willis (transl.) and Arthur C. Guyton (introduction), *The Works of William Harvey*. London: Sydenham Society, 1847 (reprints: 1867, 1949, 1965, 1981, 1989), pp. 169–592.

Heinsius, *Dissertatio medica De opobalsamo*: Ioannes Samuel Heinsius, *Dissertatio Medica Inauguralis De Opobalsamo quam . . . Pro licentia summos in arte medica honores et privilegia more maiorum rite consequendi solenni Philiatrorum examini submittit . . . Ad die XXX Septembris MDCCXVII in Auditorio Medico horis ante et post meridianis*. Ienae: Literis Wertherianis, [1717].

Heredia, *Opera Medica*: Petrus Michaelis de Heredia, *Opera medica: in quatuor Tomos divisa: I. In duas partes divisus universalem continet doctrinam de Febribus. II. Historias Epidemicas Hippocratis elucidat. III. De acutis tractatus Morbis. IV. et Ultimus Affectuum particularium aliquot Tractationes perlustrat, et de Morbis Mulierum, et Utero gerentium disserit*, 4 vols. Lugduni: Sumptibus Philippi Borde, Laurentii Arnaud, Petri Borde et Guill. Barbier, 1655.

Hermann, *Herbarium*: Paul Hermann, *Herbarium*, Ceylon 1670s, now London, Natural History Museum, 5 volumes; cited here: plant *Cinamonum sive Canella Zeylanica*, ID 408, specimen nr 621999.

Hermann, *Horti academici*: Paul Hermann, *Horti academici Lugduno-Batavi catalogus exhibens Plantarum omnium Nomina, quibus ab anno MDCLXXXI ad annum MDCLXXXVI Hortus fuit instructus ut & Plurimarum in eodem cultarum & à nemine hucusque editarum descriptiones & icones*. Lugduni Batavorum: Apud Cornelium Boutesteyn, 1687.

[Hermann], *Florae Lugduno-Batavae*: Lotharius Zumbach (ed.), *Florae Lugduno-Batavae Flores sive Enumeratio Stirpium horti Lugduno-Batavi Methodo, Naturae vestigiis insistente, dispositarum, & Anno 1689. in lectionibus tam publicis quam privatis expositarum a Paulo Hermanno . . . nunc vero In gratiam Botanophilorum primum in lucem editarum opera*. Lugduni Batavorum: Apud Fredericum Haaring, 1690.

[Hermann], *Cynosura materiae medicae*: Sigismund Henninger (ed.), *Pauli Hermanni . . . Cynosura materiae medicae, seu brevis ac succincta methodus notitiam simplicium medicamentorum comparandi, desumta ab interna partium constitutione, ubi simplicia iuxta partes suas constitutivas in certas classes* *distribuuntur, post denominationem succincta descriptio traditur, vires atque virtutes enarrantur, et modus dextre usurpandi monstratur*. Argentorati: Typis et Sumptibus Joh. Frid. Spoor, 1710.

[Hermann], *Lapis materiae medicae*: Christian Ludwig Welsch (ed.), *Lapis materiae medicae Lydius seu accuratum medicamentorum simplicium examen secundum ductum partium eadem constituentium eorumque virtutes formantium institutum & Brevissima Methodo dispositum à Paulo Hermanno* . . . Lipsiae: typis et impensis Jo. Chr. Brandenburgeri, 1703.

Hermann, *Musaeum Zeylanicum*: Paul Hermann, *Musaeum Zeylanicum, sive catalogus plantarum, In Zeylana sponte nascentium, observatarum & descriptarum*. Lugduni Batavorum: Apud Isaacum Severinum, 1717.

[Hermes Trismegistus], *Tabula smaragdina*: [Hermes Trismegistus], *Tabula smaragdina Hermetis Trismegistu πεερὶ χημείας. Incerto interprete*, in Chrysogonous Polydorus (ed.), *In hoc volumine de Alchemia continentur haec, Gebri Arabis, Philosophi solertissimi, rerumque naturalium, praecipue mtallicarum peritissimi. De inuestigatio[n]e pfectionis metallorum. Liber I. Summæ perfectionis metallorum, siue perfecti magisterij. Libri II. . . . Eiusdem de inuentione ueritatis seu perfectionis metallorum. Liber I. De fornacibus construendis. Liber I. Item. Speculum alchemiæ doctissimi viri Rogerij Bachonis. Correctorium alchemiæ doctissimi viri Richardi Anglici. Rosarius minor, de alchemia, incerti authoris. Liber secretorum Alchemiæ Calidis filij Iazichi Iudæi. Tabula smaragdina de Alchemia, Hermetis Trismeg. Hortulani philosophi, super tabulam Smaragdinam Hermetis commentarius. Omnia collatis exemplaribus, emendatissima, nouoque; modo ad menten authorum distincta, & argumentis atque; picturis necessarijs illustrata, ita ut merito iam renata uideri que[m]at*. Norimbergae: apud Ioh. Petreium, 1541, pp. 363–373.

Hooke, *Micrographia*: Robert Hooke, *Micrographia,or some Physiological Descriptions of Minute Bodies made by Magnifying glasses with Observations and Inquiries thereupon*. London: John Martyn and James Allestry, 1665.

"**Inquiries for Turky**": "Inquiries for Turky", *Philosophical Transactions* 1 (1665–1666), pp. 360–362.

Jones, *The mysteries of opium*: John Jones, *The mysteries of opium reveal'd by Dr. John Jones . . . ; who, I. Gives an account of the name, make, choice, effects, &c. of opium, II. Proves all former opinions of its operation to be meer chimera's, III. Demonstrates what its true cause is, by which he easily, and mechanically explains all (even its most mysterious) effects, IV. Shews its noxious principle, and how to separate it, thereby rendering it a safe, and noble panacea, whereof, V. He shews the palliative, and curative use*. London: Printed for Richard Smith, 1700.

Josephus, *Bellum Judaicum*: Henry St. John Thackeray, *Josephus III, The Jewish War*, Books IV–VII (Loeb Classical Library 487). London: William Heinemann, and Cambridge: Harvard University Press, 1961.

Kircher, *Mundus subterraneus*: Athanasius Kircherius, *Mundus subterraneus, In XII Libros digestus; quo Divinum Subterrestris Mundi Opificium, mira Ergasteriorum Naturae in eo distributio, verbo παντάμορφον Protei Regnum, Universæ denique Naturæ Majestas & divitiæ summa rerum varietate exponuntur. Abditorum effectuum causae acri indagine inquisitae demonstrantur; cognitae per Artis & Naturae conjugium ad humanae nitae necessarium usum vario experimentorum apparatu, nec non novo modo, & ratione applicantur*, 2 vols. Amstelodami: Apud Joannem Janssonium & Elizeum Weyerstraten, 1665.

Kircher, *Ars magna sciendi*: Athanasius Kircherius, *Ars magna sciendi, in XII Libros Digesta, qua nova & universali methodo Per Artificiosum Combinationum contextum de omni re proposita plurimis & prope infinitis rationibus disputari, omniunque summaria quaedam cognitio comparari potest*. Amstelodami: Apud Joannem Janssonium à Waesberge, 1669.

Khunrath, *Amphitheatrum*: Heinrich Khunrath, *Amphitheatrum sapientiae aeternae solius verae, christiano-kabalisticum, divino-magicum, nec non. Physico-Chymicum, Tertriunum, Catholicon*. s.l., 1595 (second edition: Hanoviae, Antonius Wilhelm, 1602).

La Bruyère, *Caractères*: Robert Garapon (ed.), *La Bruyère, Les Caractères de Théophraste traduits du grec avec Les Caractères ou les Moeurs de ce siècle. Texte établi, avec introduction, notes, relevé de variantes, glossaire et index* (Classiques Garnier). Paris: Garnier Frres, 1962.

La Condamine, "Sur l'arbre du Quinquina": Charles-Marie de La Condamine, "Sur l'arbre du Quinquina", *Mémoires de l'Académie Royale des Sciences* (1738), pp. 226–243.

Laguna, *Acerca de la materia medicinal*: Andrés de Laguna, *Pedacio Dioscorides Anazarbeo, Acerca de la materia medicinal, y de los venenos mortiferos, Traduzido de lengua Griega, en la vulgar Castellana, & illustrado con claras y substantiales Annotationes, y con las figuras de innumeras plantas exquisitas y raras*. Anvers: En casa de Iuan Latio, 1555.

Le Maistre de Sacy, *La Sainte Bible*: Isaac-Louis Le Maistre de Sacy, *La Sainte Bible traduite en françois sur la Vulgate; Avec de Notes courtes tirées des SS. Peres & des meilleurs Interprétes, pour l'intelligence des endroits les plus difficiles*, 3 vols. Brusselles: Chez François Foppens, 1700.

[Lemnius], *Les Occultes merveilles*: Levinus Lemnius, *Occulta naturae miracula, ac varia rerum documenta, probabili ratione atque artifici coniectura explicata. Quibus praeter priores fusissimè recognitos ac locupletatos, accesserunt Libri duo novi, mira rerum ac sententiarum varietate exornati, qui studioso avidoque Lectori usui sunt futuri & oblectamento. Elenchus operis, & capitum enumeratio, omnium gustum exhibebunt*. Antverpiae: Apud Guilielmum Simonem, 1564.

French translation: Jacques Gohory (transl.): *Les Occultes merveilles et secretz de Nature, avec plusieurs enseignements des choses diverses tant par raison probable que par coniecture artificielle: exposées en deux livres de non moindre plaisir que proufit au lecteur studieux. Par Levin Lemne . . . & nouvellement traduit de Latin en François, par I.G.P. Avec deux tables, l'une des argumens des chapitres, l'autre des singulieres matieres d'iceux*. Paris: Pierre du Pré, 1567.

Levens, *Manipulus Vocabulorum*: Peter Levens, *Manipulus Vocabulorum*, Londini: Henrie, 1570, in Henry B. Wheatley (ed.), *Manipulus Vocabulorum. A Dictionary of English and Latin Words, arranged in the Alphabetical Order of the Last Syllables, by Peter Levins First printed A. D. 1570, now re-edited, with a preface and alphabetical index*. Westminster: Camden Society, 1867, pp. 1–230.

Liger, *Dictionnaire*: Louis Liger, *Dictionnaire pratique du bon menager de campagne et de ville, qui apprend generalement la manière de nourrir, élever & gouverner, tant en santé que malades toutes sortes de Bestiaux, Chevaux & Volailles, de savois mener à son profit tout ce qui provient de l'Agriculture; de faire valoir toutes sortes de Terres, Prez, Vignes & Bois; de cultiver les Jardins, tant Fruitiers, Potagers, que Jardin Fleuristes; de conduire les Eaux, & faire generalement tout ce qui convient aux Jardins d'ornements.Avec un traité de tout ce qui concerne la Cuisine, les Confitures, la Pâtisserie, les Liqueurs de toutes formes, les Chasses differentes, la Pêche, & autres divertissements de la Campagne; les mots Latins de tout ce qu'on traite dans ce Livre, & quelques Remarques curieuses sur la plupârt; le tout en faveur des Etrangers, & de tous ceux qui se plaisent à ces fortes lectures. Ouvrage tres-utile dans les Familles*. 2 vols. Paris: Chez Pierre Ribou, 1715.

Linnaeus, *Hortus Cliffortianus*: Carolus Linnaeus, *Hortus Cliffortianus: Plantas exhibens quas in Hortis tam Vivis quam Siccis, Hartecampi in Hollandia, coluit vir nobilissimus & generosissimus Georgius Clifford Juris utriusque Doctor, Reductis Varietatibus ad Species, Speciebus ad Genera, Generibus ad Classes, Adjectis Locis Plantarum natalibus Differentiisque Specierum*. Amstelaedami: [s.n.], 1737.

Linnaeus, *Flora Zeylanica*: Carolus Linnaeus, *Flora Zeylanica Sistens plantas indicas Zeylonae insulae; quae Olim 1670-1677. lectae fuere. A Paulo Hermanno . . .]; Demum, Post 70 annos Ab Augusto Günthero, . . . Orbi redditae; hoc vero opere Revisae, Examinatae, Determinatae & Illustratae Generibus certis, Differentiis specificis, Synonymis propriis, Descriptionibus, Compendiosis, Iconibus paucis*. Holmiae: Laurentius Salvius, 1747.

Linnaeus, *Materia medica*: Carolus Linnaeus, *Materia medica, liber I. de Plantis, Secundum Genera, Differentias, Synonyma, Loca, Durationes, Culturas, Nomina, Simplicia, Praeparata, Qualitates, Modos, Potentias, Vires, Usus, Composita, Digestus*. Holmiae: Laurentius Salvius, 1749.

Fifth edition: Christian Daniel Schrebero (ed.), *Carolus Linnaeus, Materia medica*. Leipzig and Erlangen: apud Wolfgagnum Waltherum, 1787.

Linnaeus, *Species plantarum*: Carolus Linnaeus, *Species plantarum. Exhibentes Plantas rite cognitas. Ad genera relatas. Cum Differentiis Specificis, Nominibus Trivialibus, Synonymis Selectis, Locis Natalibus, Secundum Systema sexuale digestas*, 2 vols. Holmiae: Impensis Laurentii Salvii, 1753.

Lister, "An Ingenious account of Veins": Martin Lister, "A letter . . . written to the Publisher from York, Januar. 10. 1671, containing an Ingenious account of Veins, by him observ'd in Plants, analogous to Human Veins", *Philosophical Transactions* 6 (1671), pp. 3052–3055.

Lister, "Circulation of Sap": Martin Lister,"Extracts of divers Letters, . . . ; Touching some Inquiries and Experiments of the Motion of Sap in Trees, and relating to the Question of the Circulation of the same", *Philosophical Transactions* 6 (1671), pp. 2119–2128.

Lister, "Juice in Vegetables": Martin Lister, "A further Account concerning the Existence of Veins in all kind of Plants; together with a Discovery of the Membranous substance of those Veins, and of some Acts in Plants resembling those of Sense; as also of the Agreement of the Venal Juice in Vegetables with Blood of Animals, &c. Communicated . . . in a Letter of Januar 8. 1672/73, and exhibited to the R. Society", *Philosophical Transactions* 7 (1672), pp. 5132–5137.

López de Gómara, *La Istoria de las Indias*: Francisco López de Gómara, *La istoria de las Indias y conquista de Mexico*. Zaragoza: en casa de Augustín Millán, 1552.

Luchtmans, *Bibliotheca Boerhaaviana*: Samuel Luchtmans, *Bibliotheca Boerhaaviana, sive Catalogus Librorum Instructissimae Bibliothecae viri summi D. Hermanni Boerhaave, Dum in vivis esset, . . . Nullis aliorum Libris intermixtis. Quorum publica fiet Auctio In Officina Luchtmanniana Die Lunae 8. Junii & seqq. Diebus 1739*. Lugduni in Batavis: Apud Samuelem Luchtmans, 1739.

Machiavelli, *La Mandragola*: Niccolo Machiavelli, *La Mandragola*. Florenza: [no name], 1524.
English translation: Peter Constantine (ed.), *Niccolo Machiavelli. The Mandrake*. New York, NY: The Modern Library, 2007.

Madathanus, *Aureum Seculum Redivivum*: Hinricus Madathanus (Adrian von Mynsicht), *Aureum Seculum Redivivum Das ist: Die uhralte entwichene Güldene Zeit So nunmehr Wieder auffgangen / lieblich geblühet / und wollrichenden güldenen Samen gesetzet. Welchen tewren und edlen Samen Allen wahren Sapientiae & doctrinae filiis zeig und offenbahret*. [in Monte Abiegno]: no name, 1621.
Reproduced in: [Johann Grasshoff (ed.)], *Dyas Chymica Tripartita, Das ist: Sechs herzliche Teutsche Philosophische Tractatlein. Deren II. Von an jtzo noch im Lebe; II. Von Mitlern Alterns; und II. Von ältern Philosophis beschireben worden.. Nunmehr aber Allen Filiis Doctrinae zu Nuss an Taggeben und mit shoenen Figuren gezieret*. Franckfurt am Mayn: Luca Jennis, 1625, pp. 67–87.
The German text is also reproduced as the first instalment of *Geheime Figuren der Rosenkreuze, aus dem 16ten und 17ten jahrhundert*, 1. Heft. Aus einem alten Mscpt zum erstenmal aus Licht gestellt. Altona: Eckhardt, 1785.
Latin version (first edition): Hinricus Madathanus, *Aureum Seculum Redivivum, Quod nunc iterum apparvit, suaviter floruit et odoriferum aureumque semen peperit. Carum pretiosumque illud semen omnibus verae Sapientiae et Doctrinae filiis monstrat et revelat*, in *Musaeum Hermeticum, Omnes Sopho-Spagyricae artis discipulos fidelissime erudiens, quo pacto summa illa veraque Medicina, qua res omnes, qualescumque defectum patientes, instaurati possunt (quae alias Benedictus Ladips Sapientium appelatur) inveniri ac haberi queat. Continens tractatus Chymicos novem praestantissimos, quorum nomina et seriem versa pagella indicabit . . .* Francoforti: Luca Jennisius, 1625, pp. 75–99
Second edition: Hinricus Madathanus, *Aureum speculum redivivum, quod nunc iterum apparvit, suaverit floruit, & odoriferum aureumque semen peperi*, in *Musaeum hermeticum, reformatum et amplificatum: omnes sopho-spagyricae artis discipulos fidelissime erudiens, quo pacto summa illa veraque lapidis philosophici medicina*. Francofurti: Hermann Sande, 1678, pp. 53–72.

Maeier, *Atalanta fugiens*: Michael Majerus, *Atalanta fugiens, hoc est, Emblemata nova de secretis naturae chymica, Accomodata partim oculis & intellectui, figuris cupro incisis, adjectisque sententiis, Epigrammatis & notis, partim auribus & recreationi animi plus minus 50 Fugis Musicalibus trium Vocum, quarum duae ad unam simplicem melodiam distichis canendis peraptam, correspondeant, non absque singulari jucunditate* *videnda, legenda, meditanda, intelligenda, dijudicanda, canenda & audienda*. Oppenheimi: Ex typographia Hieronymi Galleri, Sumpibus John. Theodori de Bry, 1618.

Malpighi, *Anatome plantarum*: Marcellus Malpighus, *Anatome plantarum. Cui subjungitur Appendix, Iteratas & auctas ejusdem Authoris de ovo incubato Observationes continens. Regiae societati, Londini ad scientan naturalem promovendam institutae, dicata*, 2 vols. Londini: Johannes Martyn, 1675–1679.

Malpighi, "Letter to Spon": Marcellus Malpighus, "Praeclarissimo et Eruditissimo Viro D. Jacobo Sponio Medicinae Doctori, & Lugdunensi Anatomico Accuratissimo. Marcellus Malpighius S.P.", *Philosophical Transactions* 14 (1684–1685), pp. 601–608, pp. 630–646.

Malpighi, *Opera posthuma*: Marcellus Malpighus, *Opera posthuma Figuris Aeneis Illustrata*. Londini: A.&J. Churchill, 1697.

Maranta, *Della theriaca*: Bartolomeo Maranta, *Della theriaca et del mithridato libri due . . . Ne quali s'insegna il vero modo di comporre i sudetti antidoti, et s'esaminano con diligenza tutti i medicamenti che v' entrano*. Vinegia: Marcantonio Olmo, 1572.

Mattioli, *Di Pedacio Dioscoride Anazarbeo Libri cinque*: Pietro Andrea Mattioli, *Di Pedacio Dioscoride Anazarbeo Libri cinque Della historia, & materia medicinale tradotti in lingua volgare italiana . . . . Con amplissimi discorsi, et comenti, et dottissime annotationi, et censure . . . .* Venetia: Per Nicolo de Bascarini, 1544.

Mattioli, *Commentarii*: Petri Andreae Matthioli, *Commentarii, in libros sex Pedacii Dioscoridis Anazarbei, De materia medica: adjectis quam plurimis plantarum & animalium imaginibus eodem authore*. Venetiis: apud Vincentium Valgrisium, 1554.

Revised edition: Petri Andreae Matthioli, *Commentarii in sex libros Pedacii Dioscoridis Anazarbei de Medica materia, iam denuo ab ipso autore recogniti, et locis plus mille aucti. Adiectis magnis, ac novis plantarum, ac animalium Iconibus, supra priores editiones longe pluribus, ad vivum delineatis. Accesserunt quoque ad margines Graeci contextus quam plurimi, ex antiquissimis codicibus desumpti, qui Dioscoridis ipsius depravatam lectionem restituunt. Cum locupletissimis indicibus, tum ad rem Herbariam, tum Medicamentariam pertinentubus*. Venetiis: ex officina Valgrisiana, 1565.

Mattioli, *Discorsi*: *I discorsi di M. Pietro And. Matthioli . . . ne i sei libri di Pedacio Dioscoride Anazarbeo della materia medicinale: i quali discorsi in diversi luoghi dall'auttore medesimo sono stati accresciuti di varie cose, con molte figure di piante, & d'animali nuovamente aggiunte*. Venetia: Vincenzo Valgrisi, 1559 [Fourth revised edition of Mattioli, *Di Pedacio Dioscoride Anazarbeo Libri cinque*].

Mexía, *Lección*: Pedro Mexía, *Silva de varia leccion*. Madrid: Por Luis Sanchez. A costa de Iuan Berrillo, 1602.

Minot, *De la nature, et des causes de la fièvre*: Jacques Minot, *De la nature et des causes de la fievre, avec quelques Experiences sur le Quinquina, & des Reflexions sur l'action de ce Remede*. Paris: Chez Robert Pepie, 1684.

Monardes, *Dos libros*: Nicolás Monardes, *Dos libros: el uno trata de todas las cosas q[ue] trae[n] de n[uest]ras Indias Occidentales, que sirven al uso de Medicina, y como se ha de usar de la rayz del Mechoacan, purga excelentissima. El otro libro, trata de dos medicinas maravillosas que son contra todo Veneno, la piedra Bezaar, y la yerva Escuerçonera. Con la cura de los Venenados. Do veran muchos secretos de naturaleza y de medicina, con grandes experiencias. Agora nuevamente compuestos*. Sevilla: Sebastian Trugillo, 1565.

Italian translation: Giordano Ziletti, *Delle cose che vengono portate dall'Indie Occidentali pertinenti all'uso della medicina. Raccolte, & trattate dal Dottor Nicolò Monardes, Medico in Siviglia. Parte Prima. Novamente recata dalla Spagnola nella nostra lingua Italiana. Dove ancho si tratta de Veneni, & della lor cura. Aggiunti dei Indici; uno de' Capi principali; l'altro delle cose piu rilevanti, che si ritrovano in tutta l'opera*. In Venetia: Presso di Giordano Ziletti, 1575.

Monardes, *Las cosas*: Nicolás Monardes, *Segunda parte del libro, de las cosas que se traen de nuestras Indias Occidentales, que sirven al uso de medicina. Do ſe trata del Tabaco, y de la Saffafras: y del CarloSancto, y de otras muchas yervas y Plantas, Simientes, y Licores: que agora nuevamente han venido de* *a quellas partes, de grandes virtudes, y maravillo los effectos . . . Va anedido un libro de la Nieve. Do veran los que beven frio conella, cosas dignas de faber, y de grande admiracion, cerca del uso del enfriar con ella*. Sevilla: En casa Alonso Escrivano, 1571.

Monardes, *Historia Medicinal*: Nicolás Monardes, *Primera y segunda y tercera partes de la Historia Medicinal, de las cosas que se traen de nuestras Indias Occidentales, que sirven en Medicina; Tratado de la Piedra Bezaar y de la yerva escuerçonera; Dialogo de las grandezas del hierro y de sus virtudes medicinales; Tratado de la nieve y del bever frio*. Sevilla: En casa de Alonso Escrivano, 1574.

Morison, *Plantarum Historiae*: Roberto Morison, *Plantarum Historiae Universalis Oxoniensis pars tertia seu Herbarum Distributio Nova, per tabulas Cognationis & Affinitatis Ex Libro Naturae Observata & Detecta*. Oxonii: E Theatro Sheldoniano, 1699.

***Officina Medicamentorum*:** *Officina Medicamentorum, et methodus recte eadem componendi, cum variis scholiis, et aliis quamplurimis, ipsi operi necessariis; ex sententia Valentinorum Pharmacopolarum*. Valentiae: Apud Iohannem Chrysóstomum Garriz, 1601.

Oldenburg, *Letter to Boyle*: Henry Oldenburg, *[Letter] to Boyle (3 December 1667)*, In Alfred Rupert Hall, and Marie Boas Hall (eds), *The Correspondence of Henry Oldenburg*, 13 vols. Madison, WI: University of Wisconsin Press; London: Mansel; London: Taylor & Francis, 1965–1986, vol. 4, 1965, pp. 6–7.

Ovidius, *Metamorphoses*, 1: Georges Lafaye (ed. and Fr. transl.), *Ovide, Métamorphoses*, Livre 1 (Collection des Universités de France). Paris: Les Belles lettres, 1928 (2nd ed.: Georges Lafaye and Olivier Sers [ed. and French transl.], *Ovide, Métamorphoses*, Livre 1 [Classiques en poche 93]. Paris: Belles Lettres, 1989).

French translation: Anne-Marie Boxus and Jacques Poucet (eds), *Ovide, Métamorphoses*, Livre 1. Traduction nouvelle annotée (Bibliotheca classica Selecta, Traductions [BCS-TRA] – Itinera electronica – Bibliothèque virtuelle). Louvain-la-Neuve: Université catholique de Louvain, 2005, e-publication.

Oviedo, *La Historia general de las Indias*: Gonzalo Fernández de Oviedo, *La Historia general de las Indias*. Sevilla: en la emprenta de Juan Cromberger, 1535.

Palsgrave, *Lesclarcissement*: John Palsgrave, *Lesclarcissement de la Langue francoyse*. London?: [Richard Pynson], 1530 (2nd ed.).

Susan Baddeley (ed.), *John Palsgrave, L'éclaircissement de la langue française* (Textes de la Renaissance 69). Paris: Classiques Garnier, 2003.

Paracelsus, *De gradibus*: Paracelsus, *De gradibus et Compositionibus Receptorum et Naturalium Libri VII*, in Karl Sudhoff and Kurt Goldammer (eds), *Sämtliche Werke*, 1. Abteilung, Medizinische naturwissenschaftliche und philosophische Schriften, Bd 4. München: R. Oldenbourg, 1931, pp. 9–12.

Pegolotti, *La pratica della mercatura*: Francesco Balducci Pegolotti, *Della decima e di varie altre gravezze imposte dal comune di Firenze, della moneta e della mercatura de' Fiorentini fino al secolo XVI*, vol. 3: *La pratica della mercatura . . . e copiata da un codice manoscritto esistente in Firenze nella Biblioteca Riccardiana*. Lisbona e Lucca: [no name], 1766.

Allan Evans (ed.), *Francesco Balducci Pegolotti, La pratica della mercatura*. Cambridge, MA: The Mediaeval Academy of America, 1936.

***Pharmacopoeia Collegii*:** *Pharmacopoeia Collegii Regalis Londini*. Londini: Impensis Thomas Newcomb, Thomas Basset, John Wright & Richard Chiswel, 1677.

***Pharmacopoea Hagana*:** *Pharmacopoea Hagana ex auctoritate magistratus poliatrorum opera instaurata et aucta*. Hagae-Comitum: apud Fredericum Boucquet, 1738.

Piso, Marggraf and Laet, *Historia naturalis Brasiliae*: Willem Piso, Georg Marggraf, and Johannes de Laet, *Historia naturalis Brasiliae . . . In qua non tantum Plantae et Animalia, sed et Indigenarum morbi, ingenia et mores describuntur et Iconibus supra quingentas illustrantur*. Lugduni Batavorum: Hackius, et Amstelodami: Elzevirius, 1648.

Plaat and Scheurleer, *Bibliotheca*: Christiann Plaat, and Bernardus Scheurleer, *Bibliotheca Schwenkiana: sive catalogus collectionis egregiae librorum*. The Hague: Christiaan Plaat & Bernardus Scheurleer, 1785.
Plato, *Protagoras*: Walter R.M. Lamb, *Plato in Twelve Volumes*, vol. 3 (Loeb Classical Library 165). Cambridge: MA, Harvard University Press; London: William Heinemann, 1967.
Pliny, *Naturalis Historia*: Harris Rackham, David E. Eichholz, William H. S. Jones (eds), *Pliny's Natural History*, 9 vols (Loeb Classical Libary 330, 352, 353, 370, 371, 392, 393, 394, 418, 419). London: Heinemann; Cambridge MA: Harvard University Press, 1938–1963.
Plukenet, *Almagestum botanicum*: Leonard Plukenet, *Almagestum botanicum sive Phytographiae Pluc'netianae onomasticon Methodo Syntheticâ digestum Exhibens Stirpium exoticarum, rariorum, novarumque Nomina, quae Descriptionis locum supplere possunt. Cui (ad ampliandum regnum Vegetabilium) accessere Plantae circitèr Quingentae suis Nominibus similitèr insignitae; Quae nullibi nisi in hoc opere (sex ferè Plantarum Chiliadas complectente) memorantur. Adjiciuntur & aliquot novarum Plantarum Icones In gratiam Phylophytosophorum in lucem nùnc editae*. Londini: Sumptibus auctoris, 1696.
Polydorus, *De alchemia*: Chrysogonus Polydorus (ed.), *In hoc volumine de Alchemia continentur haec, Gebri Arabis, Philosophi solertissimi, rerumque naturalium, praecipue metallicarum peritissimi. De ineftigatione perfectionis metallorum. Liber I. Summæ perfectionis metallorum, siue perfecti magifterii. Libri II. Quae sequuntur,omnia nunc primum excus sunt. Eiusdem De inventione veritatis seu perfectionis metallorum. Liber I. De Fornacibus construendis. Liber I. Item. Speculum Alchemiae, doctissimi viri Rogerij Bachonis. Correctorium Alchemiæ doctissimi viri Richardi Anglici. Rosaríus minor, de Alchemia, Incerti authoris Liber Secretorum Alchemie Calidis filij Iazichi Iudœi. Tabula Smaragdina de Alchemia, Hermetis Trismegisti, Hortulani philosophi, super Tabulam Smaragdinam Hermetis Commentarius. Omnia collatis exemplaribus, emendatissima, nouoq; muedo ad mentem authorum distincta, & argumentis atque picturis necessarijs illustrata, ita ut merito iam renata uideri queant*. Norimbergae: Apud Ioh. Petreium, 1541.
Ruscelli, *Les secrets*: [Girolamo Ruscelli], *Secreti del revendo donno Alessio Piemontese. Nuovamente posti in luce. Opera utile, et necessaria universalmente à ciascuno*. In Venetia: Per Sigismondo Bordogna, 1555. French translation: *Les secrets de reverend signeur Alexis Piemontois. Contenans excellens remedes contre plusieurs maladies, playes, & autres accidens, Auec la maniere dv faire distillations, parfuns, confitures, teintures, couleurs, & fusions: Oeuure bien approuué, tresutile & necessaire a un chacun. Traduit d'Italien en François*. En Anvers: Christophe Plantin, 1557.
Rycaut, *The Ottoman Empire*: Paul Rycaut, *The Present State of the Ottoman Empire. Containing the Maxims of the Turkish Politie, The most material Points of the Mahometan Religion, Their Sects and Heresies, their Convents and Religious Votaries. Their Military Discipline, With an Exact Computation of their Forces both by Land and Sea. Illustrated with divers Pieces of Sculpture, representing the variety of Habits amongst the Turks. In three Books*. London: Printed for John Starkey and Henry Brome, 1667.
Schondorff, *Disputatio inauguralis*: Johannes Balth. Schondorff, *Disputatio Inauguralis Medica De Chinae Chinae modo operandi, usu et abusu, Quam Serenissimi ac Potentissimi Electoris Brandenburgici Clementia, . . . Pro Gradu Doctorali Solennissimo inaugurationis actu rite obtinendo, Eruditorum diquisitioni subjicit, in auditorio majori. Hallae Saxonia, Die 24. Jun. Anno M. DC. XCIV, horis ante & pomeridi*. Hallae, Literis Viduae Salfeldianae.
Schwencke, *Officinalium Plantarum Catalogus*: Martinus Guill. Schwencke, *Officinalium Plantarum Catalogus,quae in Horto Medico, qui Hagae Comitum est, aluntur*. Hagae- Comitum: Apud Petrum de Hondt, 1752.
Scribonius, *Compositiones*: Sergio Sconocchia (ed.), *Scribonius Largus, Compositiones medicae* (Bibliotheca scriptorum Graecorum et Romanorum Teubneriana). Leipzig: B.G. Teubner, 1983.
Italian translation: Loredana Mantovanelli, *Scribonio Largo, Ricette Mediche. Traduzione e commento*. Padua: S.A.R.G.O.N., 2012.
Sherard, *Schola Botanica*: William Sherard (ed.), *Schola Botanica sive catalogus plantarum, quas ab aliquot annis in Horto Regio Parisiensi Studiosis indigitavit vir clarissimus Joseph Pitton Tournefort, D. M. ut et Pauli*

Hermanni P. P. Paradisi Batavi Prodromus. In quo Plantæ rariores omnes, in Batavorum Hortis hactenus cultae, & plurimam partem à nemine antea descriptae recensentur. Amstelaedami: Apud Henricum Wetstenium, 1689.

Spigelius, *Isagoges*: Adriani Spigelii, *Isagoges in rem herbariam Libri Duo. Ad Illustrissimam quae Patavii est Germanicam Nationem*. Patavii: Apud Paulum Meiettum, 1606.

Swinden, *An Enquiry*: Tobias Swinden, *An Enquiry into the Nature and Place of Hell*. London: Printed by W. Bowyer, for W. Taylor, and H. Clements, 1714.

Theophrastus, *Historia plantarum*: Arthut Hort (ed.), *Theophrastus, Enquiry into Plants*, 2 vols (Loeb Classical Library 70–79). Cambridge, MA: Harvard University Press, 1916.

Tournefort, *Institutiones rei herbariæ*: Josephi Pitton de Tournefort, *Institutiones rei herbariae. Editio altera, Gallicâ longè auctior*, 3 vols. Parisiis: E Typographia Regia, 1700–1703.

Turner, *A new herball*: William Turner, *A new herball, wherin are conteyned the names of herbes in Greke, Latin, Englysh, Duch, Frenche, and in the potecaries and herbaries Latin, with the properties degrees, and naturall places of the same*. London: Steven Mierdman, 1551 (reprint: George T. L. Chapman, Marilyn N. Tweddle, Frank McCombie, and Anne Wesencraft (eds), *William Turner. A New Herbal: Parts 1, 2 and 3*, 2 vols. New York, NY, and Cambridge: Cambridge University Press, 1995–1996).

Van Haecht Goidtsenhoven, *Parvus mundus*: Laurent Van Haecht Goidtsenhoven, Μικροκόσμοσ: *Parvus mundus*. [Antwerpiae: Apud Gerard de Iode, 1579] (second edition: Arnhem: Jansonius, & Amstelodami: Theodorus Petrus, ca. 1610).

French translation: Laurens van Haecht Goidtsenhoven, *Le Microcosme Contenant divers tableaux de la vie humaine. Representez En figures avec une brieve exposition en vers françois*. Amsterdam: Chez Théodore Pierre, [1613].

Van Helmont, *Ortus Medicinae*: Franciscus Mercurius Van Helmont (ed.), *Ioannes Baptista van Helmont, Ortus Medicinae. Id est, initia physicae inaudita. Progressus medicinae novus, in morborum ultionem, ad vitam longam*. Amsterodami: Apud Ludovicum Elzevirium, 1648.

English translation: John Chandler (ed.), *Van Helmont's Works containing his most excellent philosophy, physick, chirurgery, anatomy: wherein the hilosophy of the schools is examined, their errors refuted, and the whole body of physick reformed and rectified: being a new rise and progresse of philosophy and medicine, for the cure of diseases, and lengthening of life*. London: Printed for Lodowick Lloyd, 1664.

van Reede, *Hortus Malabaricus*: Vol. I–XII: Henricus van Rheede, van Draakenstein [sic] et Johannes Casearius; vols III–V: Johannes Munnicks; vol. VI: Theodor J. van Almeloveen et Arnoldus Syen; vols II–XII: Johannes Commelinus; vols VII–XII: Abrahamus a Poot, *Hortus Indicus Malabaricus, continens regni Malabarici apud Indos cereberrimi onmis generis plantas rariores, Latinis, Malabaricis, Arabicis, et Bramanum characteribus hominibusque expressas . . . addita . . . descriptione, qua . . . praecipuae in medicina vires ... demonstrantur*, pars I–XII. Vols I–II: Amsteladami: Sumptibus viduae Joannis van Someren, et Haeredum Joannis van Dyck; vols III–XII: Amsteladami: Sumptibus Viduae Joannis van Someren, Haeredum Joannis van Dyck, Henrici et Viduae Theodori Boom, 1673 [i.e., 1682]–1703.

English translation: Kattungal Subramaniam Manilal (ed.), *Van Rheede's Hortus Malabaricus. English Edition. With Annotations and Modern Botanical Nomenclature*, 12 vols. Thiruvanthapuram: University of Kerala, 2003.

van Royen, *Florae Leydensis Prodromus*: Adrianus van Royen, *Florae Leydensis Prodromus, exhibens Plantas quae in Horto Academico Lugduno-Batavo aluntur*. Lugduni Batavorum: Apud Samuelem Luchtmans, 1740.

Vesling, *De plantis aegyptiis*: Ioannes Veslingius, *De Plantis Aegyptiis observationes et notae ad Prosperum Alpinum, cum additamento aliarum eiusdem regionis*. Patavii: Apud Paulum Frambottum, 1638 [i.e., 1640].

Weissmann, *Balsamum*: Joannes Fridericus Weissmann, *Balsamum verum quod vulgo Opobalsamum dicitur . . . Pro Licentia Doctoris Gradum et Privilegia More Majorum legitime capessendi publice eruditorum* *disquisitioni exponet . . . In Auditorio Majori horis ante et pomeridianis C. IV. Feb, MDCCV*. Jenae: Literis Gollnerianis [s.d.].

Willis, *Pharmaceutice rationalis*: Thomas Willis, *Pharmaceutice rationalis: Or, an Exercitation of the Operations of Medicines in Humane Bodies. Shewing The Signs, Causes, and Cures of most Distempers incident thereunto. In Two Parts. As also a Treatise of the Scurvy, and the several sorts thereof, with their Symptoms, Causes, and Cure*. London: Printed for T. Dring, C. Harper, and J. Leigh, 1679.

Zalužanský, *Methodi Herbariae*: Adamus Zaluziansky à Zaluzian, *Methodi Herbariae, libri tres. Theophra[stus]: de Historia plantrum lib. 4. Cap II. Propria itaque privatim, communica publicè contemplari congruum est*. Pragae: in Officina Georgij Daczieeni, 1592.

Secondary Literature

Adelmann 1966: Howard B. Adelmann (ed.), *Marcello Malpighi and the Evolution of Embryology*, 5 vols. Ithaca and New York, NY: Cornell University Press, 1966.

Adlung and Urdang 1935: Alfred Adlung, and Georg Urdang, *Grundriß der Geschichte der deutschen Pharmazie*. Berlin: Springer, 1935.

Ago 2006: Renata Ago, *Il gusto delle cose. Una storia degli oggetti nella Roma del Seicento*. Rome: Donzelli, 2006.

Agrimi and Crisciani 1994: Jole Agrimi, and Chiara Crisciani, *Les Consilia médicaux* (Typologie des sources du Moyen Âge Occidental 69). Turnhout: Brepols, 1994.

Al-Sieni 2014: All Al-Sieni, "The Antibacterial Activity of Traditionally Used Salvadora Persica L. (Miswak) and Commiphora Gileadensis (Palsam) in Saudi Arabia", *African Journal of Traditional, Complementary and Alternative Medicine* 11/1 (2014), pp. 23–27.

Aliotta et al. 2003: Gianni Aliotta, Daniele Piomelli, Antonino Pollio, and Alain Touwaide, *Le piante medicinali del Corpus Hippocraticum* (Hippocratica civitas 5). Milan: Guerini, 2003.

Ambrosoli 1992: Mauro Ambrosoli, S*cienziati, contadini e proprietari: botanica e agricoltura nell'Europa occidentale 1350-1850* (Biblioteca di cultura storica 190). Turin: Giulio Einaudi, 1992 (English translation: Mary McCann Salvatorelli, *The Wild and the Sown: Botany and Agriculture in Western Europe, 1350-1850* [Past and present publications]. Cambridge and New York, NY: Cambridge University Press, 1997).

Anagnostou 2005: Sabine Anagnostou, "Jesuits in Spanish America: contributions to the exploration of the American materia medica", *Pharmacy in History* 47/1 (2005), pp. 3–17.

Anagnostou et al. 2011: Sabine Anagnostou, Florike Egmond, and Christoph Friedrich (eds), *A Passion for Plants: Materia Medica and Botany in Scientific Networks from the 16th to 18th Centuries* (Quellen und Studien zur Geschichte der Pharmazie 95). Stuttgart: Wissenschaftliche Verlagsgesellschaft, 2011.

Anderson 1987: Eugene N. Anderson, "Why is humoral medicine so popular?", *Social Science and Medicine* 25/4 (1987), pp. 331–337.

Arber 1912/1970: Agnes Arber, *Herbals: Their Origins and Evolution. A Chapter in the History of Botany, 1470-1670*. Cambridge: Cambridge University Press, 1912 (second edition: Cambridge: Cambridge University Press, 1938; reprint: Darien, CT: Hafner, 1970).

Arber 1942: Agnes Arber, "Nehemiah Grew (1641-1712) and Marcello Mapighi (1628-1694): An Essay in Comparison", *Isis* 34/1 (1942), pp. 7–16.

Arber 1950: Agnes Arber, *The Natural Philosophy of Plant Form*. Cambridge: Cambridge University Press, 1950.

Arnold 2001: David Arnold, *Science, Technology, and Medicine in Colonial India*. New York, NY: Cambridge University Press, 2001.

Aymard 2019: Gerardo A. Aymard, "Breve reseña de los aspectos taxonómicos y nomenclaturales actuales de género *Cinchona* (Rubiaceae-Cinchoneae)", *Revista de la Real Academia de Ciencias Exactas, Físicas y Naturales* 43 (2019), pp. 234–241.

Baldassarri 2020: Fabrizio Baldassarri, "Early Modern Philosophy of Plants and the Unwelcome Guest: Pseudo-Aristotle's *De plantis*", in Giglioni and Ferrini 2020, pp. 237–264.

Baldassarri 2021: Fabrizio Baldassarri, "Before *Vitalism*: Libertine Botany and the Non-Obscure Life of Plants", *Scienza & Filosofia* 25 (2021), pp. 218–231.

Baldassarri 2022a: Fabrizio Baldassarri, "In the Beginning was the Plant: The Plant-Animal Continuity in the Early Modern Medical Reception of Galen", in Matteo Favaretti Camposampiero, and Emanuela Scribano (eds), *Galen and the Early Moderns* (International Archives of the History of Ideas 236). Cham: Springer, 2022a, pp. 55–81.

Baldassarri 2022b: Fabrizio Baldassarri, "Botany and Medicine", in Dana Jalobeanu, and Charles T. Wolfe (eds), *Encyclopedia of Early Modern Philosophy and the Sciences*. Cham: Springer, 2022b, pp. 224–229.

Baldassarri and Blank 2021: Fabrizio Baldassarri, and Andreas Blank (eds), *Vegetative Powers: The Roots of Life in Ancient, Medieval, and Early Modern Natural Philosophy* (International Archives of the History of Ideas 234). Cham: Springer, 2021.

Baldassarri and Martin 2023: Fabrizio Baldassarri, and Craig Martin (eds), *Andrea Cesalpino and Renaissance Aristotelianism: Natural Philosophy and Medicine in the Sixteenth Century* (Bloomsbury Studies in Aristotelian Tradition). London and New York, NY: Bloomsbury, 2023, forthcoming.

Baldassarri and Matei 2018: Fabrizio Baldassarri, and Oana Matei (eds), "Manipulating Flora: Seventeenth-Century Botanical Practices and Natural Philosophy", *Early Science and Medicine* 23/5–6 (2018), pp. 413–583.

Bamji 2016: Alexandra Bamji, "Medical Care in Early Modern Venice", *Journal of Social History* 49/3 (2016), pp. 483–509.

Bandinelli 2013: Angela Bandinelli, *Le origini chimiche della vita. Legami tra la Rivoluzione di Lavoisier e la Biologia di Lamarck* (Biblioteca di Nuncius, Studi e testi 71). Florence: Olschki, 2013.

Barker 2002: Miles Barker, "Putting Thoughts in Accordance with Things: The Demise of Animal-Based Analogies for Plant Functions", *Science & Education* 11 (2002), pp. 293–304.

Barrera-Osorio 2002: Antonio Barrera-Osorio, "Local Herbs, Global Medicines: Commerce, Knowledge, and Commodities in Spanish America", in Smith and Findlen 2002, pp. 163–181.

Barrera-Osorio 2006: Antonio Barrera-Osorio, *Experiencing Nature: The Spanish American Empire and the Early Scientific Revolution*. Austin, TX: University of Texas Press, 2006.

Barrera-Osorio 2009: Antonio Barrera-Osorio, "Knowledge and Empiricism in the Sixteenth-Century Spanish Atlantic World", in Daniela Bleichmar, Paula De Vos, Kristin Huffine, and Kevin Sheehan (eds), *Science in the Spanish and Portuguese Empires, 1500–1800*. Stanford, CA: Stanford University Press, 2009, pp. 219–232.

Bartolommea and Cavagna 1834: Franco Bartolommea, and Antonio Cavagna Sangiuliani di Gualdana, *Degli illustri marosticensi: Discorso*. Bassano: Baseggio, 1834.

Barreiros and Fontes da Costa 2021: Bruno Barreiros, and Palmira Fontes da Costa, "Materia Medica and the History of Book in Seventeenth-Century Portugal", *Nuncius* 36/2 (2021), pp. 394–430.

Basse Eriksen 2018: Christoffer Basse Eriksen, *Beneath the Visible: Nature and the Sub-Visible World in Early Modern Microscopy*. Unpublished PhD thesis. Aarhus: University of Aarhus, 2018.

Baumann 1974: Felix Andreas Baumann, *Das Erbario Carrarese und die Bildtradition des Tractatus de Herbis. Ein Beitrag zur Geschichte der Pflanzendarstellung im Übergang von Spätmittelalter zur Frührenaissance* (Berner Schriften zur Kunst 12). Bern: Benteli, 1974.

Bayle and Gauvin 2019: Ariane Bayle, and Brigitte Gauvin (eds), *Le siècle des vérolés. La Renaissance européenne face à la syphilis. Une anthologie*. Grenoble: Jérôme Millon, 2019.

Begley 2022: Justin Begley, "Stephen Hales (1677–1761) and the Uses and Abuses of Plant-Animal Analogies", in Alessandra Dattero (ed.), *Il bosco: Biodiversità, diritti e culture dal medioevo al nostro tempo* (I libri di Viella 411). Rome: Viella, 2022, pp. 257–274.

Bellorini 2016: Cristina Bellorini, *The World of Plants in Renaissance Tuscany: Medicine and Botany* (The History of Medicine in Context). Farnham, and Burlington, VT: Ashgate, 2016.

Bély 2009: Lucien Bély, *La France au XVIIe siècle. Puissance de l'État, contrôle de la société*. Paris: Presses Universitaires de France, 2009.

Ben-Yehoshua et al. 2011: Shimshon Ben-Yehoshua, Carole Borowitz, and Lumír Ondřej Hanuš, "Frankincense, Myrrh, and Balm of Gilead: Ancient Spices of Southern Arabia and Judea", *Horticultural Reviews* 39 (2011), pp. 1–76.

Berkhout 2020: Lenneke Berkhout, *Hoveniers van Oranje*. Hilversum: Verloren, 2020.

Bertin 2016: Alice Bertin, *The historic herbarium of Simone D'Oignies (1780) and its current scientific importance*. Unpublished master thesis. Leiden: Leiden University, 2016.

Bertoloni Meli 2011: Domenico Bertoloni Meli, *Mechanism, Experiment, Disease: Marcello Malpighi and Seventeenth-Century Anatomy*. Baltimore: Johns Hopkins University Press, 2011.

Bertoloni Meli 2016: Domenico Bertoloni Meli, "Machines of the Body in the Seventeenth Century", in Peter Distelzweig, Benjamin Goldberg, and Evan R. Raglan (eds), *Early Modern Medicine and Natural Philosophy* (History, Philosophy and Theory of the Life Sciences 14). Dordrecht: Springer, 2016, pp. 91–116.

Bertoloni Meli 2019: Domenico Bertoloni Meli, *Mechanism: A Visual, Lexical, and Conceptual History*. Pittsburgh, PA: University of Pittsburgh Press, 2019.

Bertoloni Meli 2022: Domenico Bertoloni Meli, "Images & Color: The Strasbourg Printer Johann Schott and His Circle", *Early Science and Medicine* 27/6 (2022), pp. 527–571.

Beukers and Bierman 2013: Harm Beukers, and Annette Bierman, "Kloostergeneeskunde", *Bulletin van de Kring voor de Geschiedenis van de Pharmacie in Benelux-Bulletin du Cercle Benelux d'histoire de la pharmacie* 63 (2013), pp. 2–8.

Bianchi 1987: Massimo Bianchi, *Signatura rerum. Segni, magia e conoscenza da Paracelso a Leibniz*. Rome: Edizioni dell'Ateneo, 1987.

Bible Plants: "Bible Plants", *Plant S.i.t.e.* Old Dominion University, Norfolk [online resource].

Bifulco et al. 2020: Maurizio Bifulco, Giuseppe Marasco, Luca Colucci-D'Amato, and Simona Pisanti, "Headaches in the medieval Medical School of Salerno", *Cephalalgia* 40/8 (2020), pp. 871–877.

Bitter 1914: Hendrik Bitter, *De "Hortus Medicus" of stadstuin van het collegium medico-pharmaceuticum te Haarlem, een en ander uit de notulenboeken van het collegium-medicum te Haarlem / voor de Vereeniging "Haerlem" uitgeg.* Haarlem: Erven F. Bohn, 1914.

Blair 2004: Ann Blair, "Note taking as an art of transmission", *Critical Inquiry* 31/1 (2004), pp. 85–107.

Blair 2010a: Ann Blair, "The rise of note-taking in Early Modern Europe", *Intellectual History Review* 20/3 (2010a), pp. 303–316.

Blair 2010b: Ann Blair, *Too Much to Know: Managing Scholarly Information Before the Modern Age*. New Haven, CT, and London: Yale University Press, 2010b.

Blair 2013: Ann Blair, "Revisiting Renaissance encyclopaedism", in Jason König, and Greg Woolf (eds), *Encyclopaedism from Antiquity to the Renaissance*. Cambridge: Cambridge University Press, 2013, pp. 379–397.

Blamey and Grey-Wilson 2008: Marjorie Blamey, and Christopher A. Grey-Wilson, *Wild Flowers of the Mediterranean*. London: A & C Black, 2008.

Blank 2018: Andreas Blank, "Sixteenth-Century Pharmacology and the Controversy between Reductionism and Emergentism", *Perspectives on Science* 26/2 (2018), pp. 157–184.

Bleichmar 2005: Daniela Bleichmar, "Books, Bodies, and Fields: Sixteenth-Century Transatlantic Encounters with New World Materia Medica", in Schiebinger and Swan 2005, pp. 83–99.

Bleichmar 2007: Daniela Bleichmar, "The Trajectories of Natural Knowledge in the Spanish Empire (*ca.* 1550–1650)", in Victor Navarro Brotóns, and William Eamon (eds), *Más allá de la Leyenda Negra: España y la Revolución Científica/Beyond the Black Legend: Spain and the Scientific Revolution*. Valencia: Instituto de historia de la ciencia y documentación López Piñero, 2007, pp. 127–134.

Bolam 1973: Jeanne Bolam, "The Botanical Works of Nehemiah Grew, F.R.S. (1641–1712)", *Notes and Records of the Royal Society of London* 27/2 (1973), pp. 219–231.

Boudon-Millot and Cobolet 2004: Véronique Boudon-Millot, and Guy Cobolet (eds), *Lire les médecins grecs à la Renaissance, aux origines de l'édition médicale. Aux origines de l'édition médicale. Actes du Colloque international de Paris (19–20 septembre 2003)* (Collection Medic@). Paris: De Boccard, 2004.

Boumediene 2015: Samir Boumediene, "L'acclimatation portuaire des savoirs sur le lointain: les drogues exotiques à Séville, Cadix et Livourne (XVIe–XVIIe siècles)", in Pilar González Bernaldo, and Liliane Hilaire-Pérez (eds), *Les savoirs-mondes. Mobilité et circulation des savoirs depuis le Moyen Âge*. Rennes: Presses Universitaires de Rennes, 2015, pp. 133–145.

Boumediene 2016: Samir Boumediene, *La colonisation du savoir. Une histoire des plantes médicinales du "Nouveau Monde" (1492–1750)*. Vaulx-en-Velin: Éditions des Mondes à faire, 2016.

Boumediene 2018: Samir Boumediene, "La conquête du quinquina", *L'Histoire* 448 (2018), pp. 70–75.

Boumediene 2020: Samir Boumediene, "Jesuits recipes, Jesuits receipts: the Society of Jesus and the introduction of exotic materia medica into Europe", in Linda A. Newson (ed.), *Cultural Worlds of the Jesuits in Colonial Latin America*. London: University of London Press, Institute of Latin American Studies, 2020, pp. 227–252.

Boumediene and Pugliano 2019: Samir Boumediene, and Valentina Pugliano, "La route des succédanés. Les remèdes exotiques, l'innovation médicale et le marché des substituts au XVIe siècle", *Revue d'histoire moderne et contemporaine* 66/3 (2019), pp. 24–54.

Bouras-Vallianatos 2013: Petros Bouras-Vallianatos, "Simon of Genoa's *Clavis sanationis*: a Study of Thirteenth-Century Latin Pharmacological Lexicography", in Zipser 2013, pp. 31–48.

Boutroue 2002: Marie-Élisabeth Boutroue, "'Ne dites plus qu'elle est amarante': les problèmes de l'identification des plantes et de leurs noms dans la botanique de la Renaissance", *Nouvelle Revue du XVIe Siècle* 20/1 (2002), pp. 47–64.

Brancher 2015: Dominique Brancher, *Quand l'esprit vient aux plantes. Botanique sensible et subversion libertine (XVIe-XVIIe siècles)* (Les seuils de la modernité 19). Geneva: Droz, 2015.

Bruce-Chwatt and Zulueta 1980: Leonard Jan Bruce-Chwatt, and Julian de Zulueta, *The Rise and Fall of Malaria in Europe: A Historico-Epidemiological Study*. Oxford: Oxford University Press, 1980.

Burke 1989: Peter Burke, "The Renaissance Dialogue", *Renaissance Studies* 3/1 (1989), pp. 1–12.

Burnett 2016: Charles Burnett, "The *Synonyma* Literature in the Twelfth and Thirteenth Centuries", in Sonja Brentjes, and Jürgen Renn (eds), *Globalization of Knowledge in the Post-Antique Mediterranean, 700–1500*. London and New York, NY: Routledge, 2016, pp. 131–140.

Campbell 2004: Stephen J. Campbell, *The Cabinet of Eros: Renaissance Mythological Painting and the Studiolo of Isabella d'Este*. New Haven, CT, and London: Yale University Press, 2004.

Camporesi 1985: Piero Camporesi, *Le officine dei sensi*. Milan: Garzanti, 1985 (second edition: Turin: Garzanti, 1991).

Cañizares-Esguerra 2001: Jorge Cañizares-Esguerra, *How to Write the History of the New World: Histories, Epistemologies, Identities in the Eighteenth-Century Atlantic World*. Palo Alto, CA: Stanford University Press, 2001.

Cappelletti 1989: Elsa M. Cappelletti, "I semplici a Venezia nel secolo XV: sofisticazioni, succedanei ed errori di interpretazione", in Giuseppe Maggioni (ed.), *Atti del congresso internazionale di storia della farmacia, Piacenza, Italia, Accademia italiana di storia della farmacia*. Padua: Accademia italiana di storia della farmacia, 1988, pp. 89–93.

Carlino and Jeanneret 2009: Andrea Carlino, and Michel Jeanneret, *Vulgariser la médecine, du style médical en France et en Italie*. Geneva: Droz, 2009.

Carruthers 1990: Mary Carruthers, *The Book of Memory: A Study of Memory in Medieval Culture* (Cambridge Studies in Medieval Literature 10). Cambridge and New York, NY: Cambridge University Press, 1990.

Casini 2018: Emanuele Casini, "Rethinking the multifaceted aspects of mandrake in Ancient Egypt", *Egitto e Vicino Oriente* 41 (2018), pp. 101–116.

Castiglioni 1953: Arturo Castiglioni, "The School of Ferrara and the Controversy on Pliny", in Edgar A. Underwood (ed.), *Science, Medicine and History: Essays on the Evolution of Scientific Thought and*

Medical Practice written in Honour of Charles Singer, 2 vols. London and Oxford: Oxford University Press, 1953, vol. 1, pp. 269–279.

Cavallo and Gentilcore 2007: Sandra Cavallo, and David Gentilcore, "Introduction: Spaces, Objects and Identities in Early Modern Italian Medicine", *Renaissance Studies* 21/4 (2007), pp. 473–479.

Céard 2013: Jean Céard, "De la racine de Baara et de quelques autres plantes merveilleuses à la Renaissance", *Curiositas, Les cabinets de curiosités en Europe [Online]*, University of Poitiers and Espace Mendès France, 27 janvier 2013 (online resource).

Celenza 2004: Christopher S. Celenza, "Creating Canons in Fifteenth-Century Ferrara: Angelo Decembrio's *De politia litteraria* 1.10", *Renaissance Quarterly* 57/1 (2004), pp. 43–98.

Celenza 2016: Christopher S. Celenza, "An Imagined Library in the Italian Renaissance: the Presence of Greek in Angelo Decembrio's *De politia literaria*", in Ann Blair, and Anja-Silvia Goeing (eds), *For the Sake of Learning: Essays in Honor of Anthony Grafton* (Scientific and Learned Cultures and Their Institutions 18). Leiden and Boston, MA: Brill, 2016, pp. 393–403.

Čermáková and Černá 2018: Lucie Čermáková, and Jana Černá, "Naked in the Old and the New World: Differences and Analogies in Descriptions of European and American *herbae nudae* in the sixteenth-century", *Journal of the History of Biology* 51/1 (2018), pp. 69–106.

Cerrito 2021: Amalia Cerrito, "Disclosing the Hidden Life of Plants. Theories of the Vegetative Soul in Albert the Great's *De vegetabilibus et plantis*", in Baldassarri and Blank 2021, pp. 105–122.

Cevolini 2016: Alberto Cevolini (ed.), *Forgetting Machines: Knowledge Management Evolution in Early Modern Europe* (Library of the Written Word 53). Leiden and Boston, MA: Brill, 2016.

Chavannes-Mazel 2016: Claudine A. Chavannes-Mazel, "De Lange Schaduw van de Oudheid: Geneeskunst en Planten", in IJpelaar and Chavannes-Mazel 2016, pp. 126–141.

Chavannes-Mazel and Van Uffelen 2016: Claudine A. Chavannes-Mazel, and Gerda van Uffelen, "Namen Noemen: Planten in de Tijd van Karel de Grote", in IJpelaar and Chavannes-Mazel 2016, pp. 112–125.

Ciancio 2015: Luca Ciancio, "*Per questa via s'ascende a maggior seggio*. Pietro Andrea Mattioli e le scienze mediche e naturali alla corte di Bernardo Cles", *Studi Trentini. Storia* 94 /1 (2015), pp. 159–184.

Cicogna 1824–1853: Emanuele Antonio Cicogna, *Delle inscrizioni veneziane. Raccolte ed illustrate da Emmanuele Antonio Cigogna*, 6 vols. Venice: G. Orlandelli, 1824–1853.

Clebsch 2003: Betsy Clebsch, *The New Book of Salvias: Sages for Every Garden*. Portland, OR: Timber Press, 2003.

Clericuzio 2018: Antonio Clericuzio, "Plant and Soil Chemistry in Seventeenth-Century England: Worsley, Boyle and Coxe", *Early Science and Medicine* 23/5–6 (2018), pp. 550–583.

Cohen 1927: Hendrik Cohen, *Bijdrage tot de Geschiedenis der Geneeskruidcultuur in Nederland*. Rotterdam: W.L. & J. Brusse's Uitgevermaatschappij, 1927.

Cohen 2008: Simona Cohen, *Animals as Disguised Symbols in Renaissance Art*. Leiden and Boston, MA: Brill, 2008.

Colapinto 2002: Leonardo Colapinto, "La spezieria dell'Ospedale di Santo Spirito", *Il Veltro* 46/1–4 (2002), pp. 173–184.

Collard and Samama 2006: Franck Collard, and Evelyne Samama (eds), *Pharmacopoles et apothicaires, les pharmaciens de l'Antiquité au Grand Siècle*. Paris: L'Harmattan, 2006.

Collins 2000: Minta Collins, *Medieval herbals: The Illustrative Traditions* (British Library Studies in Medieval Culture). London: British Library, 2000.

Cook 1990: Harold J. Cook, "The New Philosophy and Medicine in Seventeenth-Century England", in David C. Lindberg, and Robert S. Westman (eds), *Reappraisals of the Scientific Revolution*. Cambridge: Cambridge University Press, 1990, pp. 397–436.

Cook 2005: Harold J. Cook, "Global Economies and Local Knowledge in the East Indies. Jacobus Bontius Learns the Facts of Nature", in Schiebinger and Swan 2005, pp. 100–118.

Cook 2007: Harold J. Cook, *Matters of Exchange: Commerce, Medicine, and Science in the Dutch Golden Age*. New Haven, CT: Yale University Press, 2007.

Cook 2011: Harold J. Cook, "Markets and Cultures: Medical Specifics and the Reconfiguration of the Body in early Modern Europe", *Transactions of the Royal Historical Society* 21 (2011), pp. 123–145.

Cook and Walker 2013: Harold J. Cook, and Timothy D. Walker, "Circulation of Medicine in the Early Modern Atlantic World", *Social History of Medicine* 26/3 (2013), pp. 337–351.

Cooper 2007: Alix Cooper, *Inventing the Indigenous: Local Knowledge and Natural History in Early Modern Europe*. Cambridge and New York, NY: Cambridge University Press, 2007.

Corradi 1886: Alfonso Corradi, "Degli esperimenti tossicologici in *anima nobili* nel cinquecento" *Annali Universali di Medicina e Chirurgia* 277/830 (1886), pp. 73–100.

Crawford 2007: Matthew James Crawford, "'Para desterrar las dudas y adulteraciones': Scientific Expertise and the Attempts to make a Better Bark for the Royal Monopoly of *Quina* (1751–1790)", *Journal of Spanish Cultural Studies* 8/2 (2007), pp. 193–212.

Crawford 2014: Matthew James Crawford, "An Empire's Extract: Chemical Manipulations of Cinchona Bark in the Eighteenth-Century Spanish Atlantic World", *Osiris* 29/1 (2014), pp. 215–229.

Crawford 2016: Matthew James Crawford, *The Andean Wonder Drug: Cinchona Bark and Imperial Science in the Spanish Atlantic, 1630–1800*. Pittsburgh, PA.: University of Pittsburgh Press, 2016.

Crisciani 1999: Chiara Crisciani, "'Il giardino di carta' di Pietro Andrea Mattioli", in Chiara Crisciani et al. (eds), *Scienziati a Siena* (Memorie dell'Accademia dei Fisiocritici 7). Siena: Accademia delle Scienze, 1999, pp. 7–18.

Crisciani 2005: Chiara Crisciani, "Aspetti del dibattito sull'umido radicale nella cultura del tardo medioevo (secoli XVIII–XV)", in Josep Perarnau (ed.), *Actes de la "II Trobada Internacional d'Estudis sobre Arnau de Vilanova" (Barcelona, 30 de setembre al 30 d'octubre de 2004)*. Barcelona: IEC, 2005, pp. 333–380.

Cule 1997: John Cule, "The Devil's Apples", *Vesalius* 3/2 (1997), pp. 95–105.

Cunningham 2010: Andrew Cunningham, "The Bartholins, the Platters, and Laurentius Gryllus: the *peregrinatio medica* in the Sixteenth and Seventeenth Centuries", in Ole Peter Grell, Andrew Cunningham, and Jon Arrizabalaga (eds), *Centres of Medical Excellence?: Medical Travel and Education in Europe, 1500–1789* (The History of Medicine in Context 21). Farnham, and Burlington, VT: Ashgate, 2010, pp. 3–16.

Curran 2007: Brian A. Curran, *The Egyptian Renaissance: The Afterlife of Ancient Egypt in Early Modern Italy*. Chicago, IL: University of Chicago Press, 2007.

Cuvi 2018: Nicolás Cuvi, "Tecnociencia y colonialismo en la historia de las *Cinchona*", *Asclepio* 70/1 (2018), pp. 1–13.

Daunay et al. 2007: Marie-Christine Daunay, Henri Laterrot, and Jules Janick, "Iconography of the Solanaceae from Antiquity to the XVIIth Century: A Rich Source of Information on Genetic Diversity and Uses", *Acta Horticulturae* 745/3 (2007), pp. 59–88.

Davidson 2012: Jane P. Davidson, *Early modern supernatural, The dark side of European culture, 1400–1700*. Santa Barbara, CA: Praeger, 2012.

Delaporte 1982: François Delaporte, *Nature's Second Kingdom*, (trans. A. Goldhammer). Cambridge, MA: MIT Press, 1982.

De Toni 1919–1925: Ettore De Toni, "Il libro dei semplici di Benedetto Rinio", *Memorie della Pontificia Accademia dei Nuovi Lincei* ser. II, 5 (1919), pp. 171–279; 7 (1924), pp. 275–398; and 8 (1925), pp. 123–264.

De Vivo 2007: Filippo de Vivo, "Pharmacies as Centres of Communication in Early Modern Venice", *Renaissance Studies* 21/4 (2007), pp. 505–521.

De Vos 2010: Paula De Vos, "European materia medica in historical texts: Longevity of a tradition and implications for future use", *Journal of Ethnopharmacology* 132 (2010), pp. 28–47.

Debus 1977: Allen G. Debus, *The Chemical Philosophy: Paracelsian Science and Medicine in the Sixteenth and Seventeenth Centuries*, 2 vols. New York, NY: Science History Publications, 1977.

DeLancey 2011: Julia A. DeLancey, "'In the Streets Where They Sell Colors': Placing 'vendecolori' in the urban fabric of Early Modern Venice", *Wallraf-Richartz-Jahrbuch* 72 (2011), pp. 193–232.

Del Pozo 1966: Efrén C. Del Pozo, "Aztec pharmacology", *Annual Review of Pharmacology* 6/1 (1966), pp. 9–18.

Di Gennaro Splendore 2018: Barbara Di Gennaro Splendore, "Prospero Alpini, il Balsamo, e le sue fonti Egiziane", in Giuseppe Ongaro (ed.), *Alpiniana Studi e Testi 3*. Treviso: Antilia, 2018, pp. 81–101.

Dietz 2012: Bettina Dietz, "Contribution and Co-production. The Collaborative Culture of Linnaean Botany", *Annals of Science* 69/4 (2012), pp. 551–569.

Dietz 2017: Bettina Dietz, *Das System der Natur. Die kollaborative Wissenskultur der Botanik im 18. Jahrhundert*. Cologne and Vienna: Böhlau, 2017.

Dietz 2019: Bettina Dietz, "Networked Names. Synonyms in Eighteenth-Century Botany", *History and Philosophy of the Life Sciences* 41 (2019), article 46 (online resource).

Dietz 2022: Bettina Dietz, "Iterative Books. Posthumous Publishing in Eighteenth-Century Botany", *History of Science* 60/2 (2022), pp. 166–182.

Dilg 1999: Peter Dilg, "The *Liber aggregatus in medicinis simplicibus* of Pseudo-Serapion: An Influential Work of Medical Arabism", in Charles Burnett, and Anna Contadini (eds), *Islam and the Italian Renaissance*. London: Warburg Institute (London University Press), 1999, pp. 221–231.

Draelants 2011: Isabelle Draelants, "Expérience et autorité dans la philosophie naturelle d'Albert le Grand", in Thomas Bénatouïl, and Isabelle Draelants (eds), *Expertus sum. L'expérience par les sens dans la philosophie naturelle médiévale* (Micrologus Library 40). Florence: SISMEL–Edizioni del Galluzzo, 2011, pp. 89–122.

Duccini 2008: Hélène Duccini, "Traditions iconographiques et partage des modèles en Europe (première moitié du XVIIe siècle)", *Le Temps des médias* 11/2 (2008), pp. 10–24.

Duistermaat 2020: Leni H. Duistermaat, *H. Heukels' Flora van Nederland*. Groningen/Utrecht: Noordhoff - Naturalis Biodiversity Center, 2020.

Dunlop 1967: Douglas Morton Dunlop, *The History of the Jewish Khazars* (Princeton Oriental Studies 16). Princeton, NJ: Princeton University Press, 1954 (2nd edition, New York, NY: Schocken Books, 1967).

Dupèbe 2017: Jean Dupèbe (ed.), *Michel Servet – Apologie contre Leonhart Fuchs* (Cahiers d'Humanisme et Renaissance 143). Geneva: Droz, 2017.

Durand 2016: Gilbert Durand, *Les structures anthropologiques de l'imaginaire. Introduction à l'archétypologie générale*. Paris: Bordas, 1969 (12th edition, Paris: Dunod, 2016).

Eamon 1994: William Eamon, *Science and the Secrets of Nature, Books of Secrets in Medieval and Early Modern Culture*. Princeton, NJ: Princeton University Press, 1994.

Eamon 2003: William Eamon, "Pharmaceutical Self-Fashioning or How to Get Rich and Famous in the Renaissance Medical Marketplace", *Pharmacy in History* 45/3 (2003), pp. 123–129.

Eamon 2014: William Eamon, "The Difference that Made Spain, the Difference that Spain Made", in Slater, López Terrada and Pardo Tomás 2014, pp. 231–243.

Eamon 2018: William Eamon, "Corn, cochineal, and quina: The "Zilsel thesis" in a Colonial Iberian Setting", *Centaurus* 60/3 (2018), pp. 141–158.

Earles 1961: Melvin P. Earles, "Early theories of the mode of action of drugs and poisons", *Annals of Science* 17/2 (1961), pp. 97–110.

Earles 1963: Melvin P. Earles, "Experiments with drugs and poisons in the seventeenth and eighteenth centuries", *Annals of Science* 19/4 (1963), pp. 241–254.

Egmond 2007a: Florike Egmond, "Clusius and Friends: Cultures of Exchange in the Circles of European Naturalists", in Florike Egmond, Paul Hoftijzer, and Robert Visser (eds), *Carolus Clusius: Towards a Cultural History of Renaissance Naturalists*. Amsterdam: Koninklijke Nederlandse Akademie van Wetenschappen, 2007, pp. 9–48.

Egmond 2007b: Florike Egmond, "Correspondence and Natural History in the Sixteenth-century: Cultures of exchange in the circle of Carolus Clusius", in Francisco Bethencourt, and Florike Egmond (eds), *Cultural Exchange in Early Modern Europe* (Cultural Exchange in Early Modern Europe 3). Cambridge: Cambridge University Press, 2007, pp. 104–142.

Egmond 2008: Florike Egmond, "Apothecaries as Experts and Brokers in the Sixteenth-Century Network of Carolus Clusius", *History of Universities* 23 (2008), pp. 59–91.

Egmond 2012: Florike Egmond, "Names of Naturalia in the Early Modern Period: Between the Vernacular and Latin, Identification and Classification", in Harold J. Cook, and Sven Dupré (eds), *Translating Knowledge in the Early Modern Low Countries* (Low Countries Studies on the Circulation of Natural Knowledge 3). Vienna: Lit, 2012, pp. 131–161.

Egmond 2017: Florike Egmond, "Experimenting with Living Nature: Documented Practices of Sixteenth-Century Naturalists and Naturalia Collectors", *Journal of Early Modern Studies* 6/1 (2017), pp. 21–46.

Egmond 2021: Florike Egmond, "Sixteenth-Century University Gardens in a Medical and Botanical Context", in Fabrizio Baldassarri, and Fabio Zampieri (eds), *Scientiae in the History of Medicine* (Studies in the History of Medicine 4). Rome and Bristol, CT: L'Erma di Bretschneider, 2021, pp. 89–120.

Egmond 2022: Florike Egmond, "Plants and Medicine", in Andres Dalby, and Annette Giesecke (eds), *A Cultural History of Plants in the Early Modern Era* (Annette Giesecke, and David J. Mabberley [eds], *A Cultural History of Plants*, vol. 3) (The Cultural Histories Series). London and New York, NY: Bloomsbury, 2022, pp. 117–136.

Ek 2011: Renske Ek, *George Clifford Herbarium (1685–1760)* [see Clifford, *Herbarium*].

Ekholm 2010: Karin J. Ekholm, "Fabricius's and Harvey's representations of animal generation", *Annals of Science* 67/3 (2010), pp. 329–352.

Elliott 1970: John. H Elliott, *The Old World and the New (1492–1650)*. Cambridge: Cambridge University Press, 1970.

Endtz 1972: Lambertus J. Endtz, *De Hage-professoren: geschiedenis van een chirurgische school*. Amstelveen: Specia, 1972.

Estes 1995: J. Worth Estes, "The European Reception of the First Drugs from the New World", *Pharmacy in History* 37/1 (1995), pp. 3–23.

Estrella 1992: Eduardo Estrella, "Los sistemas médicos precolombinos", in López Piñero 1992a, pp. 13–33.

Faivre 1995: Daniel Faivre, "Les représentations primitives du monde des morts chez les Hébreux", *Dialogues d'histoire ancienne* 21/1 (1995), pp. 59–80.

Fausti 2004: Daniela Fausti (ed.), *La complessa scienza dei semplici. Atti delle celebrazioni per il V centenario della nascita di Pietro Andrea Mattioli*. Siena: Accademia dei Fisiocritici, 2004.

Favaretti Camposampiero 2022: Matteo Favaretti Camposampiero, "Galenism in Early Modern Philosophy and Medicine", in Dana Jalobeanu, and Charles T. Wolfe (eds), *Encyclopedia of Early Modern Philosophy and the Sciences*. Cham: Springer, 2022, pp. 697–704.

Favaro 1911-1912: Antonio Favaro (ed.), *Atti della nazione germanica artista nello studio di Padova*. Venice: A spese della società, 1911–1912.

Fend 2019: Mechthild Fend, "Drawing the Cadaver *ad vivum*: Gérard de Lairesse's Illustrations for Govard Bidloo's *Anatomia Humani Corporis*", in Thomas Blfe, Joanna Woodall, and Claus Zittel (eds), *Ad vivum? Visual Materials and the Vocabulary of Life-Likeness in Europe before 1800* (Intersections 61). Leiden and Boston, MA: Brill, 2019, pp. 294–327.

Ferraces Rodriguez 1999: Arsenio Ferraces Rodriguez (ed.), *Estudio sobre textos latinos de fitoterapia entre la Antigüedad y la Alta Edad Media*. A Coruña: Universidade de Coruña, 1999.

Ferraces Rodriguez 2007: Arsenio Ferraces Rodriguez (ed.), *Tradición griega y textos médicos latinos en el periodo presalernitano*. A Coruña: Universidade de Coruña, 2007.

Ferraces Rodriguez 2009: Arsenio Ferraces Rodriguez (ed.), *Fito-zooterapia Antigua y altomedieval: textos y doctrinas*. A Coruña: Universidade de Coruña, 2009.

Ferri 1997: Sara Ferri (ed.), *Pietro Andrea Mattioli, Siena 1501–Trento 1578. La vita, le opere, con l'identificazione delle piante*. Perugia: Quattroemme, 1997.

Findlen 1994: Paula Findlen, *Possessing Nature: Museums, Collecting and Scientific Culture in Early Modern Italy*. Berkeley, CA: University of California Press, 1994.

Findlen 2006: Paula Findlen, "Natural History", in Katherine Park, and Lorraine Daston (eds), *The Cambridge History of Science*. Volume 3: *Early Modern Science*. Cambridge and New York, NY: Cambridge University Press, 2006, pp. 435–468.

Findlen 2017: Paula Findlen, "The Death of a Naturalist: Knowledge and Community in Late Renaissance Italy", in Gideon Manning, and Cynthia Klestinec (eds), *Professors, Physicians and Practices in the History of Medicine* (Archimedes: New Studies in History and Philosophy of Science and Technology 50). Cham: Springer 2017, pp. 155–196.

Fleischer 2017: Alette Fleischer, "Leaves on the Loose: The Changing of Nature of Archiving Plants and Botanical Knowledge", *Journal of Early Modern Studies* 6/1 (2017), pp. 117–135.

Fontes da Costa 2012: Palmira Fontes da Costa, "Geographical Expansion and the Reconfiguration of Medical Authority. Garcia de Orta's *Colloquies on the simples and drugs of India* (1563)", *Studies in History and Philosophy of Science* 43/1 (2012), pp. 74–81.

Fontes da Costa 2015: Palmira Fontes da Costa (ed.), *Medicine, Trade and Empire: Garcia de Orta's Colloquies on the Simples and Drugs of India (1563) in Context*. Farnham, and Burlington, VT: Ashgate, 2015.

Fortuna 2007: Stefania Fortuna, "The Prefaces to the First Humanist Medical Translations", *Traditio* 62/3 (2007), pp. 17–35.

Foucault 1976: Michel Foucault, *Histoire de la sexualité, I La volonté de savoir*. Paris: Gallimard, 1976.

Francis and Ramandi 2020: Sally Francis, and Maria Teresa Ramandi (eds), *Crocologia. A Detailed Study of Saffron, the King of Plants* (Brill's Texts and Sources in Intellectual History 22). Leiden and Boston, MA: Brill, 2020.

Fresquet Febrer 1992: José Luís Fresquet Febrer, "Terapéutica y materia médica americana de la obra de Andrés Laguna (1555)", *Asclepio* 44/2 (1992), pp. 53–82.

Funk 2016: Holger Funk, "The First Printed Latin Editions of Dioscorides's *De materia medica* (1478, 1512): An Inventory-Based Re-Evaluation", *Archives of Natural History* 43/2 (2016), pp. 237–254.

Garbari, Tongiorgi Tomasi and Tosi 1991: Fabio Garbari, Lucia Tongiorgi Tomasi, and Alessandro Tosi, *Il giardino dei semplici – L'orto botanico di Pisa dal XVI al XX secolo*. Pisa: Pacini, 1991.

Garosci 1934: Aldo Garosci, Jean Bodin. *Politica e diritto nel Rinascimento francese*. Milan: Corticelli, 1934.

Garrett 2003: Brian Garrett, "Vitalism and Teleology in the Natural Philosophy of Nehemiah Grew (1641–1712)", *The British Journal for the History of Science* 36/1 (2003), pp. 63–81.

Gentilcore 2003a: David Gentilcore, "Introduction to *The World of the Italian Apothecary*: Apothecaries, 'Charlatans' and the Medical Marketplace in Italy, 1400–1750", *Pharmacy in History* 45/3 (2003a), pp. 91–94.

Gentilcore 2003b: David Gentilcore, "'For the Protection of Those Who Have Both Shop and Home in this City': Relations Between Italian Charlatans and Apothecaries", *Pharmacy in History* 45/3 (2003b), pp. 108–122.

Gerber 1927: C. Gerber, "Une controverse: Les progrès de la botanique au XVIIe siècle", *Bulletin de la Société d'histoire de la pharmacie* 56 (1927), pp. 477–491.

Gibson 2015: Susannah Gibson, *Animal, Vegetable, Mineral? How Eighteenth-Century Science Disrupted the Natural Order*. Oxford: Oxford University Press, 2015.

Giglioni 2021: Guido Giglioni, "Plantanimal Imagination: Life and Perception in Early Modern Discussions of Vegetative Power", in Baldassarri and Blank 2021, pp. 325–345.

Giglioni 2022: Guido Giglioni, "Glisson, Francis, and the Irritable Life of Nature", in Dana Jalobeanu, and Charles T. Wolfe (eds), *Encyclopedia of Early Modern Philosophy and the Sciences*. Cham: Springer, 2022, pp. 759–770.

Giglioni and Ferrini 2020: Guido Giglioni, and Maria Fernanda Ferrini (eds), Περὶ φυτῶν. *Greek Botanical Treatises in the West and the East* (The Aristotelian tradition: text and contexts, Technical and scientific treatises 2). Macerata: EUM, 2020.

Gillett 2018: Alexander James Gillett, "Invention through bricolage: epistemic engineering in scientific communities", *RT: A Journal on Research Policy & Evaluation* 6/1 (2018), pp. 1–17.

Givens 2006: Jean A. Givens, "Reading and Writing the Illustrated *Tractatus de herbis*", in Jean A. Givens, Karen M. Reeds, and Alain Touwaide (eds), *Visualizing Medieval Medicine and Natural History, 1200–1500* (AVISTA Studies in the History of Medieval Technology, Science and Art 5). Aldershot, and Burlington, VT: Ashgate, 2006, pp. 115–145.

Golb 1997/2013: Norman Golb (ed.), *Judaeo-Arabic Studies: Proceedings of the Founding Conference of the Society for Judaeo Arabic Studies*. Amsterdam: Harwood Academic Press, 1997 (reprint: New York, NY: Routledge, 2013).

Gómez López 2005: Susana Gómez López, "Natural Collections in the Spanish Renaissance", in Marco Beretta (ed.), *From Private to Public: Natural Collections and Museums*. Sagamore Beach, MA: Science History Publications, 2005, pp. 13–41.

González Bueno 2007: Antonio González Bueno, "El descubrimiento de la naturaleza del Nuevo Mundo: Las Plantas Americanas en la Europa del siglo XVI", *Circumscribere* 2 (2007), pp. 10–25.

Grafton 2004: Anthony Grafton, "Conflict and Harmony in the *Collegium Gellianum*", in Leofranc Holford-Strevens, and Amiel D. Vardi (eds), *The Worlds of Aulus Gellius*. Oxford: Oxford University Press, 2004, pp. 318–342.

Grafton, Shelford, and Siraisi 1992: Anthony Grafton, April G. Shelford, and Nancy Siraisi, *New Worlds, Ancient texts: The Power of Tradition and the Shock of Discovery*. Cambridge, MA, and London: Belknap Press of Harvard University Press, 1992.

Grämiger 2016: Gregory Grämiger, "Reconstructing Order: the Spatial Arrangements of Plants in the Hortus Botanicus of Leiden University in Its First Years", in Hubertus Fischer, Volker R. Remmert, and Joachim Wolschke-Bulmahn (eds), *Gardens, Knowledge and the Sciences in the Early Modern Period* (Trends in the History of Science). Basel and Cham: Birkhäuser and Springer, 2016, pp. 235–251.

Grendler 2002: Paul F. Grendler, *The Universities of the Italian Renaissance*. Baltimore, MD: Johns Hopkins University Press, 2002.

Grieco 1993: Allen J. Grieco, "Les plantes, les régimes végétariens et la mélancolie à la fin du Moyen Age et au début de la Renaissance italienne", in Allen J. Grieco, Odile Redon, and Lucia Tongiorgi Tomasi (eds), *Le Monde végétal (XIIe–XVIIe siècles): Savoirs et usages sociaux*. Saint-Denis: Presses Universitaires de Vincennes, 1993, pp. 11–29.

Griffin 2020: Clare Griffin, "Disentangling Commodity Histories: *pauame* and sassafras in the early modern global world", *Journal of Global History* 15/1 (2020), pp. 1–18.

Groom 1981: Nigel Groom, *Frankincense and Myrrh: A Study of the Arabian Incense Trade*. London and New York, NY: Longman, 1981.

Grove 1996: Richard Grove, "Indigenous Knowledge and the Significance of South-West India for Portuguese and Dutch Constructions of Tropical Nature", *Modern Asian Studies* 30 (1996), pp. 121–143.

Guerra 1977a: Francisco Guerra, "The Introduction of Cinchona in the Treatment of Malaria. Part I", *Journal of Tropical Medicine and Hygiene* 80/6 (1977a), pp. 112–118.

Guerra 1977b: Francisco Guerra, "The Introduction of Cinchona in the Treatment of Malaria. Part II", *Journal of Tropical Medicine and Hygiene* 80/7 (1977b), pp. 135–140.

Guthrie 1961: Douglas Guthrie, "Plants as Remedies: The Debt of Medicine to Botany", *Transactions of the Botanical Society of Edinburgh* 39/2 (1961), pp. 184–195.

Haggis 1941a: Alex W. Haggis, "Fundamental Errors in the Early History of Cinchona: Part II", *Bulletin of the History of Medicine* 10/3 (1941a), pp. 417–459.

Haggis 1941b: Alex W. Haggis, "Fundamental Errors in the Early History of Cinchona: Part II", *Bulletin of the History of Medicine* 10/4 (1941b), pp. 568–592.

Halikowski Smith 2008: Stefan Halikowski Smith, "Meanings behind Myths: the Multiple Manifestations of the Tree of the Virgin at Mataria", *Mediterranean Historical Review* 23/2 (2008), pp. 101–128.

Hall and Hall 1965-1986: Alfred Rupert Hall, and Marie Boas Hall (eds), *The Correspondence of Henry Oldenburg*, 13 vols. Madison, WI: University of Wisconsin Press, and London: Taylor & Francis, 1965–1986.

Hall 1974: Alfred Rupert Hall, "Medicine in the Early Royal Society", in Allen G. Debus (ed.), *Medicine in Seventeenth Century England*. Berkeley, CA: University of California Press, 1974, pp. 421–452.

Hanbury and Ince 1876: Daniel Hanbury, and Joseph Ince, *Science Papers: Chiefly Pharmacological and Botanical*. London: Macmillan, 1876.

Hardy and Totelin 2016: Gavin Hardy, and Laurence Totelin, *Ancient Botany* (Sciences of Antiquity). London and New York, NY: Routledge, 2016.

Harris 2018: Stephen A. Harris, "Seventeenth-century plant lists and herbarium collections: a case study from Oxford Physic Garden", *Journal of the History of Collections* 30/1 (2018), pp. 1–23.

Harvey-Gibson 1919: Robert John Harvey-Gibson, *Outlines of the History of Botany*. London: A&C Black, 1919 (reprint: New York, NY: Arno Press, 1981).

Healy 2001: Margaret Healy, *Fictions of Disease in Early Modern England. Bodies, Plagues and Politics*. Basingstoke: Palgrave Macmillan, 2001.

Henniger 1986: J. Henniger, *Hendrik Adriaan van Reede tot Drakenstein (1636–1691) and Hortus Malabaricus. A Contribution to the History of Dutch Colonial Botany*. Rotterdam and Boston, MA: A.A. Balkema, 1986.

Heyd and Raynaud 1923: Wilhelm Heyd, and Furcy Raynaud, *Histoire du commerce du levant au moyen âge*, éd. française refondue et considérablement augmentée, 2 vols. Leipzig: Harrassowitz, 1923.

Hickman 2019: Clare Hickman, "'The want of a proper Gardiner': late Georgian Scottish botanic gardeners as intermediaries of medical and scientific knowledge", *The British Journal for the History of Science* 52 (2019), pp. 543–567.

Hirai 2014: Hiro Hirai, "Images, Talisman and Medicine in Jacques Gaffarel's Unheard-of Curiosities", in Hiro Hirai (ed.), *Jacques Gaffarel between Magic and Science* (Bruniana&Campanelliana 39). Rome: Serra, 2014, pp. 73–84.

Holford-Strevens 2014: Leofranc Holford-Strevens, "Aulus Gellius", in Greti Dinkova-Bruun (ed.), *Catalogus Translationum et Commentariorum. Mediaeval and Renaissance Latin Translations and Commentaries: Annotated Lists and Guides* 10. Toronto: Pontifical Institute of Mediaeval Studies, 2014, pp. 273–330.

The Holy Bible: Alfred Rahlfs, and Robert Hanhart (eds),*Septuaginta: id est Vetus Testamentum graece iuxta LXX interpretes*. Stuttgart: Württembergische Bibelanstalt, 2006.

Holzman 1998: Robert S. Holzman, "The Legacy of Atropos, the Fate Who Cut the Thread of Life", *Anesthesiology* 89/1 (1998), pp. 241–249.

Howley 2018: Joseph Howley, *Aulus Gellius and Roman Reading Culture*. Cambridge: Cambridge University Press, 2018.

Huguet-Termes 2001: Teresa Huguet-Termes, "New World Materia Medica in Spanish Renaissance Medicine: From Scholarly Reception to Practical Impact", *Medical History* 45/3 (2001), pp. 359–376.

Iluz et al. 2010: David Iluz, Miri Hoffman, Nechama Gilboa-Garber, and Zohar Amar, "Medicinal properties of Commiphora gileadensis", *African Journal of Pharmacy and Pharmacology* 4/8 (2010), pp. 516–520.

Ineichen 1959: Gustav Ineichen, "Bemerkungen zu den pharmakognostischen Studien im Spätmittelalter im Bereiche von Venedig", *Zeitschrift für romanische Philologie* 75 (1959), pp. 438–466.

IJpelaar and Chavannes-Mazel 2016: Linda IJpelaar, and Claudine A. Chavannes-Mazel (eds), *De Groene Middeleeuwen: Duizend Jaar Gebruik van Planten (600–1600)*. Eindhoven: Lecturis, 2016.

Irissou 1949: Louis Irissou, "La mandragore: Louis Tercinet, Vue d'ensemble sur une Solanacée délaissée par la Pharmacopée moderne", *Revue d'histoire de la pharmacie* 124 (1949), pp. 469–470.

Israel 1995/1996: Jonathan Israel, *The Dutch Republic. Its Rise, Greatness, and Fall, 1477–1806*, 2 vols. Oxford: Oxford University Press, 1995 (Dutch translation: *De Republiek, 1477–1806*, trans. by Bert Smilde, 2 vols. Franeker: Van Wijnen, 1996).

Jacquart 1996: Danielle Jacquart, "The Influence of Arabic Medicine in the Medieval West", in Roshdi Rashed (ed.), *Encyclopedia of the History of Arabic Science*, 3 vols. London and New York, NY: Routledge, 1996, vol. 3, pp. 963–984.

Jacquart and Micheau 1990: Danielle Jacquart, and Françoise Micheau, *La médecine arabe et l'Occident médiéval* (Islam et Occident 7). Paris: Maisonneuve et Larose, 1990.

Jäger 2017: Eckehart J. Jäger (ed.),*Rothmaler: Exkursionsflora von Deutschland. Gefäßpflanzen: Grundband 21., durchgesehene Auflage*. Berlin: Springer, 2017.

Jalobeanu and Matei 2020: Dana Jalobeanu, and Oana Matei, "Treating Plants as Laboratories: A Chemical Natural History of Vegetation in 17th-Century England", *Centaurus* 62/3 (2020), pp. 542–561.

Jalobeanu and Matei 2022: Dana Jalobeanu, and Oana Matei, "Spiritual Technologies: Cider-Making and Natural Philosophy in Early Modern England", *Nuncius* 37/2 (2022), pp. 315–345.

Jarcho 1993: Saul Jarcho, *Quinine's predecessor: Francesco Torti and the Early History of Cinchona* (The Henry E. Sigerist Series in the History of Medicine). Baltimore, MD: Johns Hopkins University Press, 1993.

Jarvis 2007: Charlie Jarvis, *Order out of Chaos. Linnaean plant names and their types*. London: Linnean Society of London in association with the Natural History Museum, 2007.

Jarvis 2016: Charlie Jarvis, *Dataset: Clifford Herbarium. Natural History Museum Data Portal* (data.nhm.ac.uk) (online resource).

Jensen 2000: Kristian Jensen, "Description, Division, Definition – Cæsalpinus and the Study of Plants as an Independent Discipline", in Marianne Pade (ed.), *Renaissance Readings of the* Corpus aristotelicum*: Proceedings of the Conference held in Copenhagen 23–25 April 1998*. Copenhagen: Museum Tusculanum Press, 2000, pp. 185–206.

Johnson 2007: Brian Johnson, "The Changing Face of the Botanic Garden", in Nadine Käthe Nomen (ed.), *Botanic Gardens: A Living History*. London: Black Dog, 2007, pp. 64–81.

Joly 2007: Bernard Joly, "À propos d'une prétendue distinction entre la chimie et l'alchimie au XVIIe siècle: Questions d'histoire et de méthode", *Revue d'histoire des sciences* 60 (2007), pp. 167–184.

Jouanna 1999: Jacques Jouanna, *Hippocrates*. New York, NY: John Hopkins University Press, 1999 (English translation of Jouanna, *Hippocrate*. Paris: Fayard, 1992).

Jullien 1904: Michel Jullien, *L'arbre de La Vierge à Matarieh: Souvenirs du séjour de la Sainte Famille en Égypte*. Cairo: Imprimerie Nationale, 1904.

Kahn 2007: Didier Kahn, *Alchimie et paracelsisme en France à la fin de la Renaissance (1567–1625)* (Cahiers d'humanisme et Renaissance 80). Geneva: Droz, 2007.

Karstens and Kleibrink 1982: Willem Karstens, and Herman Kleibrink, *De Leidse Hortus: een botanische erfenis*. Zwolle: Waanders, 1892.

Keeble 1997: T. W. Keeble, "A Cure for the Ague: The Contribution of Robert Talbor (1642–81)", *Journal of the Royal Society of Medicine* 90/5 (1997), pp. 285–290.

Keller 2014: Vera Keller, "Nero and the Last Stalk of *Silphion*: Collecting Extinct Nature in Early Modern Europe", *Early Science and Medicine* 19 (2014), pp. 424–447.

Kerényi 1976/2020: Carl Kerényi, *Dionysos: Archetypal Image of Indestructible Life*, translated by Ralph Manheim. Princeton, NJ: Princeton University Press, 1976 (reprint: 2020).

Kikuchihara and Hirai 2022: Yohei Kikuchihara, and Hiro Hirai, "Signatura rerum theory", in Marco Sgarbi (ed.), *Encyclopedia of Renaissance Philosophy*. Cham: Springer, 2022, pp. 3016–3022.

Klein 2003: Ursula Klein, "Experimental History and Herman Boerhaave's Chemistry of Plants", *Studies in History and Philosophy of Biological and Biomedical Sciences* 34 (2003), pp. 533–567.

Klein 2005: Ursula Klein, "Shifting Ontologies, Changing Classifications: Plant Materials from 1700 to 1830", *Studies in History and Philosophy of Science* 36 (2005), pp. 261–329.

Klein 2008: Ursula Klein, "The Laboratory Challenge. Some Revisions of the Standard View of Early Modern Experimentation", *Isis* 99 (2008), pp. 769–782.

Klein and Pieters 2016: Wouter Klein, and Toine Pieters, "The Hidden History of a Famous Drug: Tracing the Medical and Public Acculturation of Peruvian Bark in Early Modern Western Europe (c. 1650–1720)", *Journal of the History of Medicine and Allied Sciences* 71/4 (2016), pp. 400–421.

Klerk 2014: Saskia Klerk, "The Trouble with Opium. Taste, Reason and Experience in Late Galenic Pharmacology with Special Regard to the University of Leiden (1575–1625)", *Early Science and Medicine* 19 (2014), pp. 287–316.

Klerk 2015: Saskia Klerk, *Galen Reconsidered. Studying drug properties and the foundations of medicine in the Dutch Republic ca. 1550–1700*. Unpublished PhD thesis. Utrecht: Utrecht University, 2015.

Klotzsch 1860: Friedrich Klotzsch, "Linné's natürliche Pflanzenklasse *Tricoccae* des Berliner Herbarium's im Algemeinen und die natürliche Ordnung *Euphorbiaceae* insbesondere", *Aus den Abhandlungen der Königlichen Akademie der Wissenschaften zu Berlin 1859*. Berlin: Ferdinand Dümmler, 1860.

Kostylo 2015: Joanna Kostylo, "Pharmacy as a Centre for Protestant Reform in Renaissance Venice", *Renaissance Studies* 30/2 (2015), pp. 236–253.

Kremers and Urdang 1951: Edward Kremers, and Georg Urdang, *History of Pharmacy. A Guide and a Survey*. Philadelphia: Lippincott, 1951 (first edition, 1940).

Kristeller 1945: Paul Oskar Kristeller, "The School of Salerno: its Development and its Contribution to the History of Learning", *Bulletin of the History of Medicine* 17/2 (1945), pp. 138–194.

Kristeller 1963–1997: Paul Oskar Kristeller, *Iter Italicum: A Finding List of Uncatalogued or Incompletely Catalogued Humanistic MSS of the Renaissance in Italian and Other Libraries*, 6 vols. and index vol. London: Warburg Institute, and Leiden: Brill, 1963–1997.

Krul 1891: R. Krul, *Haagsche doctoren, chirurgen en apothekers in den ouden tijd*. The Hague: W.P. van Stockum & zoon, 1891.

Kuijlen et al. 1983: J. Kuijlen, Carla Oldenburger-Ebbers, and Dirk Onno Wijnands, *Paradisus Batavus: Bibliografie van plantencatalogi van onderwijstuinen, particuliere tuinen en kwekerscollecties in de Noordelijke en Zuidelijke Nederlanden (1550–1839)*. Wageningen: Pudoc, 1983.

Kuntz 1975: Marion Leathers Daniels Kuntz, "Introduction", in Jean Bodin, *Colloquium of the Seven about Secrets of the Sublime: Colloquium Heptaplomeres de Rerum Sublimium Arcanis Abditis*, translation with introduction, annotations, and critical readings. Princeton, NJ: Princeton University Press, 1975.

Kusukawa 2009: Sachiko Kusukawa, "Image, Text and *Observatio*: The *Codex Kentmanus*", *Early Science and Medicine* 14/4 (2009), pp. 445–475.

Kusukawa 2012: Sachiko Kusukawa, *Picturing the Book of Nature: Image, Text, and Argument in Sixteenth-Century Human Anatomy and Medical Botany*. Chicago, IL: University of Chicago Press, 2012.

Kyle 2017: Sarah Kyle, *Medicine and Humanism in Late Medieval Italy: The Carrara Herbal in Padua* (Medicine in the Medieval Mediterranean 8). London and New York, NY: Routledge, 2017.

Kyle 2022: Sarah R.Kyle, "Representation of Plants: Mediators of Body and Soul", in Alain Touwaide (ed.), *A Cultural History of Plants in the Post-Classical Era* (Annette Giesecke, and David J. Mabberley [eds], *A Cultural History of Plants*, vol. 2) (The Cultural Histories Series). London and New York, NY: Bloomsbury, 2022, pp. 167–186.

Landsberg 1996: Sylvia Landsberg, *The Medieval Garden*. New York, NY: Thames and Hudson, 1996.

Le Fanu 1990: William Le Fanu, *Nehemiah Grew M.D., F.R.S. A Study and Bibliography of his Writings*. Winchester-Detroit, MI: St. Paul's Bibliographies-Omnigraphics Inc., 1990.

Lehoux 2017: Daryn Lehoux, "Observation Claims and Epistemic Confidence in Aristotle's Biology", *Isis* 108/2 (2017), pp. 241–258.

Lenoble 1969: Robert Lenoble, *Esquisse d'une histoire de l'idée de Nature*. Paris: Albin Michel 1969.

Leong and Rankin 2017: Elaine Leong, and Alisha Rankin, "Testing Drugs and Trying Cures: Experiment and Medicine in Medieval and Early Modern Europe", *Bulletin of the History of Medicine* 91/2 (2017), pp. 157–182.

Lévi-Strauss 2009: Claude Lévi-Strauss, *Mythologiques T.2; Du miel aux cendres*. Paris: Plon, 2009.

Leyel 1959: Hilda W. Leyel, "Editor's introduction", in Hilda W. Leyel (ed.), *Maud Grieve, A Modern Herbal: The Medicinal, Culinary, Cosmetic and Economic Properties, Cultivation and Folklore of Herbs, Grasses, Fungi, Shrubs and Trees with all their Modern Scientific Uses*. New York, NY: Hafner, 1959 (first edition: San Diego, CA: Harcourt, Brace & Company, 1931).

Lindeboom 1979: Gerrit A. Lindeboom, *Haller in Holland: het dagboek van Albrecht von Haller van zijn verblijf in Holland (1725–1727)*. Amsterdam: Rodopi, 1979.

Lindeboom 1968/2007: Gerrit A. Lindeboom, *Herman Boerhaave; the man and his work*. Rotterdam: Erasmus Publishing, 2007 (first edition: London: Methuen & Co, 1968).

Lines 2006: David A. Lines, "Humanism and the Italian Universities", in Christopher S. Celenza, and Kenneth Gouwens (eds), *Humanism and Creativity in the Renaissance. Essays in Honor of Ronald G. Witt* (Brill's Studies in Intellectual History 136). Leiden and Boston, MA: Brill, 2006, pp. 327–346.

López Piñero 1992a: José María López Piñero (ed.), *Viejo y Nuevo Continente: la medicina en el encuentro de dos mundos*. Madrid: Saned, 1992a.

López Piñero 1992b: José María López Piñero, "Los primeros estudios científicos sobre la materia médica americanas: La *Historia medicinal* de Nicolás Monardes y la Expedición de Francisco Hernández a Nueva España", in López Piñero 1992a, pp. 221–279.

López Piñero and Calero 1992: José María López Piñero, and Francisco Calero, *De Pulvere Febrifugo Occidentalis Indiae (1663) de Gaspar Caldera de Heredia y la Introducción de la Quina en Europa* (Cuadernos Valencianos de Historia de la Medicina y de la Ciencia 39). Valencia: Instituto de Estudios Documentales e Histórico sobre la Ciencia, Universitat de València, 1992.

López Piñero and López Terrada 1997: José María López Piñero, and María Luz López Terrada, *La influencia Española en la introducción en Europa de las plantas americanas (1493–1623)* (Cuadernos Valencianos de Historia de la Medicina y de la Ciencia 53). Valencia: Instituto de Estudios Documentales e Históricos sobre la Ciencia, Universitat de València, 1997.

López Piñero and Pardo Tomás 1994: José María López Piñero, and José Pardo Tomás, *Nuevos materiales y noticias sobre la Historia de las plantas de Nueva España, de Francisco Hernández* (Cuadernos Valencianos de Historia de la Medicina y de la Ciencia 44). Valencia: Instituto de Estudios Documentales e Históricos sobre la Ciencia, Universitat de Valencia, 1994.

López Piñero and Pardo Tomás 1996: José María López Piñero, and José Pardo Tomás, *La influencia de Francisco Hernández (1515–1587) en la constitución de la botánica y la materia médica modernas*, (Cuadernos Valencianos de Historia de la Medicina y de la Ciencia 51). Valencia: Instituto de Estudios Documentales e Históricos sobre la Ciencia, Universitat de Valencia, 1996.

López Piñero and Pardo Tomás 2000: José María López Piñero, and José Pardo Tomás, "The Contribution of Hernández to European Botany and Materia Medica", in Varey, Chabrán and Weiner 2000, pp. 122–137.

Lucchetta 1997: Francesca Lucchetta. "Il medico del bailaggio di Costantinopoli: fra potere e politica (secc. XV–XVI)", *Quaderni di Studi Arabi* 15 (1997), pp. 5–50.

Lucchetta and Lucchetta 1986: Francesca Lucchetta, and Giuliano Lucchetta, "Un medico veneto in Siria nel Cinquecento: Cornelio Bianchi", *Quaderni di Studi Arabi* 4 (1986), pp. 1–56.

MacLeod 2000: Roy M. MacLeod (ed.), *Nature and Empire: Science and the Colonial Enterprise*. Chicago, IL: University of Chicago Press, 2000.

Maehle 1998: Andreas-Holger Maehle, "Pharmacological experimentation with opium in the eighteenth century", in Roy Porter, and Mikuláš Teich (eds), *Drugs and Narcotics in History*. Cambridge: Cambridge University Press, 1998, pp. 52–76.

Maehle 1999: Andreas-Holger Maehle, *Drugs on Trial: Experimental Pharmacology and Therapeutic Innovation in the Eighteenth Century* (Clio Medica 53. Wellcome Institute Series in the History of Medicine). Amsterdam: Rodopi, 1999.

Maire 2014: Brigitte Maire (ed.), *'Greek' and 'Roman' in Latin Medical Texts. Studies in Cultural Change and Exchange in Ancient Medicine* (Studies in Ancient Medicine 42). Leiden and Boston, MA: Brill, 2014.

Manning 2008: John Manning, *Field guide to Fynbos*. Cape Town: Struik, 2008.

Mandaville 2011: James P. Mandaville, *Bedouin Ethnobotany: Plant Concepts and Uses in a Desert Pastoral World*. Tucson, AZ: University of Arizona Press, 2011.

Marcon 1988: Susy Marcon, "Nicolò Roccabonella. *Liber de simplicibus*", in Rolando Bussi (ed.), *Di sana pianta: erbari e taccuini di sanità. Le radici storiche della nuova farmacologia*. Modena: Panini, 1988, pp. 154–155.

Marcon 2002: Susy Marcon, "Nicolò Roccabonella. *Liber de simplicibus*", in Silvia Fogliati, and Davide Dutto (eds), *Il giardino di Polifilo: ricostruzione virtuale dalla* Hypnerotomachia Poliphili *di Francesco Colonna stampata a Venezia nel 1499 da Aldo Manuzio*. Milan: Franco Maria Ricci, 2002, pp. 113–116.

Marcon 2003: Susy Marcon, "Codici: effetto natura. L'Erbario di Nicolò Roccabonella", *Alumina* 1 (2003), pp. 4–13.

Mariani Canova 1986: Giordana Mariani Canova, "Amadio, Andrea", in Günter Meißner (ed.), *Allgemeines Künstler-Lexikon: die bildenden Künstler aller Zeiten und Völker*, 2 vols. Leipzig: Seemann, 1986, vol. 2, p. 514.

Mariani Canova 1988: Giordana Mariani Canova, "La traduzione europea degli erbari miniati e la scuola veneta", in Rolando Bussi (ed.), *Di sana pianta: erbari e taccuini di sanità. Le radici storiche della nuova farmacologia*. Modena: Panini, 1988, pp. 21–28.

Margócsy 2010: Dániel Margócsy, "'Refer to folio and number': Encyclopedias, the Exchange of Curiosities, and Practices of Identification before Linnaeus", *Journal of the History of Ideas* 71/1 (2010), pp. 63–89.

Marroquín Arredondo 2019: Jaime Marroquín Arredondo, "The method of Francisco Hernández: Early Modern Science and the Translation of Mesoamerica's Natural History", in Jaime Marroquín Arredondo, and Ralph Bauer (eds), *Translating Nature: Cross-Cultural Histories of Early Modern Science*. Philadelphia, PA: University of Philadelphia Press, 2019, pp. 45–69.

Mateos 1956: Francisco Mateos (ed.), *Obras del P. Bernabé Cobo de la compañía de Jesús*, 2 vols (Biblioteca de autores españoles 91–92). Madrid: Atlas, 1956.

Matthew 2002: Louisa C. Matthew, "'Vendecolori a Venezia': The Reconstruction of a Profession", *The Burlington Magazine* 144 (2002), pp. 680–686.

Maxson 2014: Brian Maxson, *The Humanist World of Renaissance Florence, 1400–1480*. Cambridge: Cambridge University Press, 2014.

McClive and Pellegrin 2010: Cathy McClive, and Nicole Pellegrin (eds), *Femmes en fleurs, femmes en corps, sang, santé, sexualités, du Moyen Âge aux Lumières* (L'école du genre. Série Nouvelles recherches). Saint-Etienne: Publications de l'Université de Saint-Etienne, 2010.

McVaugh 1975/2008: Michael McVaugh, "Rufinus", in Charles Coulston Gillispie (ed.), *Dictionary of Scientific Biography*, 16 vols. New York, NY: Charles Scribner's Sons, 1970–1980, vol. 11, 1975, p. 601 (reprint in Charles Coulston Gillispie, Frederic Lawrence and Noretta Koertge [eds], *Complete Dictionary of Scientific Biography*, 27 vols. Detroit: Charles Scribner's Sons, 2008, vol. 11, p. 601).

Messer 1987: Ellen Messer, "The hot and cold in Mesoamerican indigenous and hispanicized thought", *Social Science and Medicine* 25/4 (1987), pp. 339–346.

Milwright 2001: Marcus Milwright, "Balsam in the Mediaeval Mediterranean: A Case Study of Information and Commodity Exchange", *Journal of Mediterranean Archaeology* 14/1 (2001), pp. 3–23.

Milwright 2003: Marcus Milwright, "The Balsam of Maṭariyya: An Exploration of a Medieval Panacea", *Bulletin of the School of Oriental and African Studies* 66/2 (2003), pp. 193–209.

Minelli 1995: Alessandro Minelli (ed.), *The Botanical Garden of Padua, 1545–1995*. Venice: Marsilio 1995.

Minio 1952–1953: Michelangelo Minio, "Il Quattrocento Codice 'Rinio'. Integralmente rivendicato al medico Nicolò Roccabonella", *Atti dell'Istituto Veneto di Scienze, Lettere ed Arti* 111 (1952–1953), pp. 49–64.

Minuzzi 2016: Sabrina Minuzzi, *Sul filo dei segreti. Farmacopea, libri e pratiche terapeutiche a Venezia in età moderna*. Milan: Unicopli, 2016.

Molhuysen 1913: Philipp C. Molhuysen (ed.), *Bronnen tot de geschiedenis der Leidsche universiteit*, 4 vols. 's-Gravenhage: Martinus Nijhoff, 1913.

Moran 2017: Bruce T. Moran, "Preserving the Cutting edge: Traveling Woodblocks, Material Networks, and Visualizing Plants in Early Modern Europe", in Matteo Valleriani (ed.), *The Structures of Practical Knowledge*. Cham: Springer, 2017, pp. 393–419.

Morelli 1802: Jacopo Morelli, *Bibliothecae Regiae divi Marci Venetiarum Custodis bibliotheca manuscripta graeca et latina tomus primus*. Bassano: Ex typographia remondiniana, 1802.

Moroni 1840: Gaetano Moroni, *Dizionario di erudizione storico-ecclesiastica da S. Pietro sino ai nostri giorni specialmente intorno ai principali santi . . .* 103 vols. Venice: Tipografia Emiliana, 1840–1861.

Morton 1981: Allan G. Morton, *History of Botanical Sciences. An Account of the Development of Botany from Ancient Times to the Present Day*. London, New York, NY, Toronto, Sydney, San Francisco, CA: Academic Press, 1981.

Moss 1996: Ann Moss, *Printed Commonplace-Books and the Structuring of Renaissance Thought*. Oxford: Clarendon, 1996.

Moulinier 1993: Laurence Moulinier, "Deux jalons de la construction d'un savoir botanique en Allemagne au XIIe–XIIIe siècles: Hildegarde de Bingen et Albert le Grand", in Allen J. Grieco, Odile Redon, and Lucia Tongiorgi Tomasi (eds), *Le Monde végétal (XIIe–XVIIe siècles): Savoirs et usages sociaux*. Saint-Denis: Presses Universitaires de Vincennes, 1993, pp. 89–105.

Munger 1949: Robert S. Munger, "Guaiacum, the Holy Wood from the New World", *Journal of the History of Medicine and Allied Sciences* 4/2 (1949), pp. 196–229.

Napolitano Valditara 2007: Linda N. Napolitano Valditara, *Platone e le 'ragioni' dell'immagine*. Milan: Vita e Pensiero, 2007.

Nardo et al. 2009: Luigi Nardo, Riccardo Naccari, Giorgio Boscolo, and Emmanuele Bello, "Pioppo", in Luigi Nardo (ed.), *Dizionario italiano-veneto: a sercar parole*. Padua: Programma, 2009, p. 536.

Naturalis Bioportal: Bioportal Naturalis biodiversity center (online resource).

Nieto Olarte 2000: Mauricio Nieto Olarte, *Remedios para el Imperio: Historia natural y la apropiación del Nuevo Mundo*. Bogotá: Instituto Colombiano de Antropología e Historia, 2000.

Nissen 2007: Dorothy Nissen MFA, "The Earth Is an Alembic, an 'imaginal' reading of the images in Athanasius Kircher's *Mundus subterraneus* [Atlas of the Underworld]", *Jung Journal: culture & psyche* 1/4 (2007), pp. 18–31.

Norton 2006: Marcy Norton, "Tasting Empire: Chocolate and the European Internalization of Mesoamerican Aesthetics", *The American Historical Review* 111/3 (2006), pp. 660–691.

Nutton 1985a: Vivian Nutton, "Humanistic Surgery", in Andrew Wear, Roger K. French, and Iain M. Lonie (eds), *The Medical Renaissance of the Sixteenth Century*. Cambridge: Cambridge University Press, 1985a, pp. 75–99.

Nutton 1985b: Vivian Nutton, "John Caius and the Eton Galen: Medical Philology in the Renaissance", *Medizinhistorisches Journal* 20 (1985b), pp. 227–252.

Nutton 1988: Vivian Nutton, "*Prisci Dissectionum Professores*: Greek Texts and Renaissance Anatomists", in Anna Carlotta Dionisotti, Anthony Grafton, and Jill Kraye (eds), *The Uses of Greek and Latin: Historical Essays*. London: Warburg Institute, 1988, pp. 111–126.

Nutton 1995: Vivian Nutton, "The Changing Language of Medicine, 1450–1550", in Olga Weijers (ed.), *Vocabulary of Teaching and Research between Middle Ages and Renaissance: Proceedings of the Colloquium, London, Warburg Institute, 11–12 March 1994* (Études sur le vocabulaire intellectuel du Moyen Âge 8). Turnhout: Brepols, 1995, pp. 184–198.

Nutton 1997: Vivian Nutton, "The Rise of Medical Humanism: Ferrara, 1464–1555", *Renaissance Studies* 11 (1997), pp. 2–19.

Nutton 2004: Vivian Nutton, *Ancient Medicine*. London and New York, NY: Routledge, 2004.

O'Malley Pierson 2000: Peter O'Malley Pierson, "Philip II. Imperial Obligations and Scientific Vision", in Varey, Chabrán and Weiner 2000, pp. 11–18.

Offerhaus 2020: Aleida Offerhaus, "Het Zierikzee herbarium: Onderzoek naar Inhoud en Oorsprong van een raadselachtig Herbarium", *Kroniek van het land van de meermin (Schouwen-Duiveland)* 45 (2020), pp. 29–47.

Offerhaus et al. 2021: Aleida Offerhaus, Emma de Haas, Henk Porck, Adriaan Kardinaal, Renske Ek, Omar Pokorni, and Tinde van Andel, "The Zierikzee Herbarium: contents and origins of an enigmatic 18th Herbarium", *Blumea* 66 (2021), pp. 1–52.

Offerhaus et al. 2022: Aleida Offerhaus, Anastasia Stefanaki, and Tinde van Andel, "The true 'Boerhaave Herbarium': an Analysis of the Specimens of Herman Boerhaave contained in the van Royen Collection at Naturalis", *Botany Letters* 169/4 (2022), online publication: doi.org/10.1080/23818107.2022.2114545

Ogilvie 1997: Brian W. Ogilvie, *Observation and Experience in Early Modern Natural History*. Unpublished PhD thesis. Chicago, IL: University of Chicago, 1997.

Ogilvie 2003: Brian W. Ogilvie, "The Many Books of Nature: Renaissance Naturalists and Information Overload", *Journal of the History of Ideas* 64/1 (2003), pp. 29–40.

Ogilvie 2006: Brian W. Ogilvie, *The Science of Describing: Natural History in Renaissance Europe*. Chicago, IL: University of Chicago Press, 2006.

Olariu 2014: Dominic Olariu, "The Misfortune of Philippus de Lignamine's Herbal, or New Research Perspectives in Herbal Illustrations from an Iconological Point of View", in Stefan Kiedroń, Anna-Maria Rimm, and Patrycja Poniatowsha (eds), *Early Modern Print Culture in Central Europe. Proceedings of the Young Scholars section of the Wrocław Seminars, September 2013*. Wrocław: Wydawnictwo UNiwesytetu Wrocławskiego, 2014, pp. 39–62.

Olariu 2018: Dominic Olariu, "Die Kunst, die Natur zu kennen. Produktion, Fixierung und Transfer von Naturwissen in der Frühen Neuzeit", in Wolfgang Meighörner (ed.), *Cranach natürlich. Hieronymus in der Wildnis*. Innsbruck: Tiroler Landesmuseum Ferdinandeum, 2018, pp. 153–161.

Olmi 1991: Giuseppe Olmi, "Molti amici in vari luoghi: studio della natura e rapporti epistolari nel XVI secolo", *Nuncius* 6 (1991), pp. 3–31.

Olmi 1992: Giuseppe Olmi, *L'inventario del mondo. Catalogazione della natura e luoghi del sapere nella prima età moderna*. Bologna: Il Mulino, 1992.

Olmi 2009: Giuseppe Olmi, "Sfogliare le pagine del *codex naturae*: il viaggio scientifico nella prima età moderna", *Antologia Vieusseux* 45 (2009), pp. 5–32.

Ongaro 1961-1963: Giuseppe Ongaro, "Contributi alla biografia di Prospero Alpini", *Acta medicae historiae Patavina* 8–9 (1961–62/1962–63), pp. 79–168.

Ongaro 2009: Giuseppe Ongaro, "Introduzione", in Elsa Mariella Cappelletti, Luciano Cremonini, and Giuseppe Ongaro (eds), *Prospero Alpini. Le piante dell'Egitto – Il balsamo (Venezia 1592)* (Centro studi Prospero Alpini-Marostica 1). Treviso: Antilia, 2009, pp. 5–16.

Ongaro 2011a: Giuseppe Ongaro (ed.), *Alpiniana Studi e Testi 1*. Treviso: Antilia, 2011a.

Ongaro 2011b: Giuseppe Ongaro, "Ottaviano Rovereti e Prospero Alpini", in Ongaro 2011a,. Treviso: Antilia, 2011b, pp. 273–311.

Ongaro 2016: Giuseppe Ongaro (ed.), *Alpiniana Studi e Testi 2*. Treviso: Antilia, 2016.

Ongaro 2018: Giuseppe Ongaro (ed.), *Alpiniana Studi e Testi 3*. Treviso: Antilia, 2018.

Ortiz Crespo 1995: Fernando Ortiz Crespo, "Fragoso, Monardes and pre-Chinchonian Knowledge of Cinchona", *Archives of Natural History* 22/2 (1995), pp. 169–181.

Ottaviani 2002: Alessandro Ottaviani, "Scuola Galileiana e cartesianesimo nella polemica fra Marcello Malpighi e Giovan Battista Trionfetti sulla generazione delle piante", in Maria T. Marcialis, and Francesca M. Crasta (eds), *Descartes e l'eredità cartesiana nell'Europa Sei-Settecentesca*, Atti del Convegno "Cartesiana 2000", Cagliari, 30 novembre-2 dicembre 2000. Lecce: Conte, 2002, pp. 261–276.

Pächt 1950: Otto Pächt, "Early Italian Nature Studies and the Early Calendar Landscape", *Journal of the Warburg and Courtauld Institutes* 13/1–2 (1950), pp. 13–47.
Pagallo 2011: Giulio F. Pagallo, "Della inedita traduzione di Prospero Alpini del *De animalibus* di Averroè: note introduttive e trascrizione del testo", in Ongaro 2011a, pp. 161–230.
Paganelli and Cappelletti 1996: Francesco Paganelli, and Elsa M. Cappelletti, "Il codice erbario Roccabonella (sec. XV) e suo contributo alla storia della farmacia", *Atti e memorie dell'Accademia Patavina di scienze, lettere ed arti* 13 (1996), pp. 111–116.
Palmer 1984: Richard Palmer, "The Influence of Botanical Research on Pharmacists in Sixteenth-Century Venice", *NTM –Schriftenreihe Geschichte der Naturwissenschaften und Technik* 21/2 (1984), pp. 69–80.
Palmer 1985a: Richard Palmer, "Pharmacy in the Republic of Venice in the Sixteenth Century", in Andrew Wear, Roger K. French, and Iain M. Lonie (eds), *The Medical Renaissance of the Sixteenth Century*. Cambridge: Cambridge University Press, 1985a, pp. 100–117.
Palmer 1985b: Richard Palmer, "Medical Botany in Northern Italy in the Renaissance", *Journal of the Royal Society of Medicine* 78/2 (1985b), pp. 149–157.
Panarelli 2020: Marilena Panarelli, "Albert the Great's *De Vegetabilibus* and its unique position among the medieval commentaries on *De plantis*", in Giglioni and Ferrini 2020, pp. 137–162.
Panarelli 2021: Marilena Panarelli, "Can Plants Desire? Aspects of the Debate on *desiderium naturale*", in Baldassarri and Blank 2021, pp. 91–104.
Panese 2003: Francesco Panese, "The doctrine of signatures and graphical technologies at the dawn of modernity", *Gesnerus* 60/1–2 (2003), pp. 6–24.
Paravicini Bagliani 2009: Agostino Paravicini Bagliani (ed.), *Le monde végétal. Médecine, botanique, symbolique* (Micrologus Library 30). Florence: SISMEL-Edizioni del Galluzzo, 2009.
Pardo Tomás 1991: José Pardo Tomás, "Obras españolas sobre historia natural y materia médica americanas en la Italia del siglo XVI", *Asclepio* 43/1 (1991), pp. 51–94.
Pardo Tomás 2002: José Pardo Tomás, *Oviedo, Monardes, Hernández. El Tesoro natural de América: colonialismo y ciencia en el siglo XVI*. Madrid: Nivola, 2002.
Pardo Tomás 2007: José Pardo Tomás, "Two glimpses of America from a distance: Carolus Clusius and Nicólas Monardes", in Florike Egmond, Paul Hoftijzer, and Robert Visser (eds), *Carolus Clusius: Towards a Cultural History of a Renaissance Naturalist*. Amsterdam: Koninklijke Nederlandse Akademie van Wetenschappen, 2007, pp. 173–193.
Pardo Tomás 2015: José Pardo Tomás, "East Indies, West Indies: Garcia de Orta and the Spanish Treatises on Exotic Materia Medica", in Fontes da Costa 2015, pp. 195–212.
Pardo Tomás and López Terrada 1992: José Pardo Tomás, and María Luz López Terrada, "Alimentos, Drogas y medicamentos en las primeras relaciones y crónica de Indias", in López Piñero 1992a, pp. 195–219.
Pavord 2005: Anna Pavord, *The Naming of Names: The Search for Order in the World of Plants*. New York, NY: Bloomsbury, 2005.
Pelayo 2003: Francisco Pelayo, "Las expediciones científicas francesas y su influencia en la España del siglo XVIII", in Felix Muñoz Garmendia (ed.), *La Botánica al servicio de la Corona. La expedición de Ruiz, Pavón y Dombey al Virreinato del Perú (1777–1831)*. Barcelona: Lunwerg, 2003, pp. 14–49.
Pelusi 2011: Simonetta Pelusi, "'Quel libro . . . che vale un tesoro'. La circolazione dei manoscritti slavi a Venezia: dalle biblioteche religiose alla Pubblica Libreria", in Dejan Ajdačić, and Persida Lazarević Di Đakomo (eds), *Venecija i Slovenske Književnosti. Zbornik Radova*. Belgrade: SlovoSlavia, 2011, pp. 113–149.
Pennuto 2008: Concetta Pennuto (ed.), *Girolamo Fracastoro De sympathia et antipathia rerum liber unus*. Rome: Storia e letteratura, 2008.
Perez 2003: Stanis Perez, "Louis XIV et le quinquina", *Vesalius* 9/2 (2003), pp. 25–30.
Pesenti 1984: Tiziana Pesenti, *Professori e promotori di medicina nello studio di Padova dal 1405 al 1509*. Trieste: Lint, 1984.

Pesenti 1999: Tiziana Pesenti, "Studio dei farmaci e produzione di commenti nell'Università di arti e medicina di Padova nel primo ventennio del Trecento", *Annali di storia delle università italiane* 3 (1999), pp. 61–78.

Peterson 1993: Kristen E. Peterson, *Translatio libri Avicennae De Viribus Cordis et Medicinis Cordialibus de Arnaldi de Villanova*. Unpublished PhD thesis. Cambridge, MA: Harvard University, 1993.

Piechocki 2019: Katharina N. Piechocki, *Cartographic Humanism: The Making of Early Modern Europe*. Chicago, IL: University of Chicago Press, 2019.

Pierini 2021: Edoardo Pierini "Different Peoples, Different Inebriations: The Recognition of Different Cultures of Intoxication in Early Modern English Medicine", *Canadian Bulletin of Medical History* 38/S1 (2021), pp. S72–S92.

Piomelli and Pollio 1994: Daniele Piomelli, and Antonino Pollio, "*In upupa o strige*. A Study in Renaissance Psychotropic Plant Ointments", *History and Philosophy of the Life Sciences* 16/2 (1994), pp. 241–273.

Pitacco 2002: Francesca Pitacco, "Un prestito mai rifuso: la vicenda del *Liber de simplicibus* di Benedetto Rinio", in Linda Borean, and Susy Marcon (eds), *Figure di collezionisti a Venezia tra Cinque e Seicento*. Udine: Forum, 2002, pp. 11–23.

Pomata and Siraisi 2005: Gianna Pomata, and Nancy G. Siraisi, *Historia: Empiricism and Erudition in Early Modern Europe*. Cambridge, MA: MIT Press, 2005.

Porter 1989: Roy Porter, "The Early Royal Society and the Spread of Medical Knowledge", in Roger French and Andrew Wear (eds), *The Medical Revolution of the Seventeenth Century*. Cambridge: Cambridge University Press, 1989, pp. 272–293.

***POWO* 2019:** *Plants of the World Online*. Facilitated by the Royal Botanic Gardens, Kew. Published on the Internet, 2019 (online resource).

Pratt 1992: Mary Louise Pratt, *Imperial Eyes: Travel Writing and Transculturation*. London and New York, NY: Routledge, 1992.

Prigioniero et al. 2020: Antonello Prigioniero, Pierpaolo Scarano, Valentino Ruggieri, Mario Marziano, Maria Tartaglia, Rosaria Sciarrillo, and Carmine Guarino, "Plants named 'Lotus' in Antiquity: Historiography, Biogeography, and Ethnobotany", *Harvard Papers in Botany* 25/1 (2020), pp. 59–71.

Pugliano 2017: Valentina Pugliano, "Pharmacy, Testing, and the Language of Truth in Renaissance Italy", *Bulletin of the History of Medicine* 91/2 (2017), pp. 233–273.

Pugliano 2018: Valentina Pugliano, "Natural history in the apothecary's shop", in Helen-Anne Curry, Nicholas Jardine, James A. Secord, and Emma C. Spary (eds), *Worlds of Natural History*. Cambridge: Cambridge University Press, 2018, pp. 44–60.

Ragland 2017a: Evan R. Ragland, "'Making Trials' in Sixteenth- and Early Seventeenth-Century European Academic Medicine", *Isis* 108/3 (2017a), pp. 503–528.

Ragland 2017b: Evan R. Ragland, "Experimental Clinical Medicine and Drug Action in Mid-Seventeenth-Century Leiden", *Bulletin of the History of Medicine* 91/2 (2017b), pp. 331–361.

Raj 2007: Kapil Raj, *Relocating Modern Science: Circulation and the Construction of Knowledge in South Asia and Europe, 1650–1900*. New York, NY: Palgrave Macmillan, 2007.

Rankin 2017: Alisha Rankin, "On Anecdote and Antidotes: Poison Trials in Sixteenth-Century Europe", *Bulletin of the History of Medicine* 91/2 (2017), pp. 274–302.

Rauschert 1970: Stephan Rauschert, "Das Herbarium von Paul Hermann (1646–1695) in der Forschungsbibliothek Gotha", *Hercynia* 7 (1970), pp. 301–328.

Reeds 1976: Karen Meier Reeds, "Medical Humanism and Botany", *Annals of Science* 33 (1976), pp. 519–542.

Reeds 1991: Karen Meier Reeds, *Botany in Medieval and Renaissance Universities* (Harvard Dissertations in the History of Science). New York, NY, and London: Garland Publishing, Inc., 1991.

Reeds 2012: Karen Meier Reeds, "Saint John's Wort (*Hypericum perforatum* L.) in the Age of Paracelsus and the Great Herbals: Assessing the Historical Claims for a Traditional Remedy", in Anne van Arsdall, and Timothy Graham (eds), *Herbs and Healers from the Ancient Mediterranean through the Medieval West*.

Essays in Honor of John M. Riddle (Medicine in the medieval Mediterranean). Farnham, and Burlington, VT: Ashgate, 2012, pp. 265–306.

Repici 2000: Luciana Repici, *Uomini capovolti. Le piante nel pensiero dei Greci*. Rome: Laterza, 2000.

Repici 2003: Luciana Repici, "Teodoro Gaza traduttore e interprete di Teofrasto: la ricezione della botanica antica tra Quattro e Cinquecento", *Rinascimento* 43 (2003), pp. 417–505.

Repici 2009: Luciana Repici, "Il *De plantis* pseudo-aristotelico nella tradizione antica e medievale", in Paravicini Bagliani 2009, pp. 77–94.

Repici 2010: Luciana Repici, "Le jardin des *Parva Naturalia*: les plantes chez Aristote et après lui", in Christophe Grellard, and Pierre-Marie Morel (eds), *Les Parva naturalia d'Aristote. Fortune antique et médiévale*. Paris: Publications de la Sorbonne, 2010, pp. 31–45.

Repici 2012: Luciana Repici, "Platone e il buon uso delle immagini", in Andrea Balbo, Federica Bessone, and Ermanno Malaspina (eds), *Tanti affetti in tal momento: studi in onore di Giovanna Garbarino*. Alessandria: Edizioni dell'Orso, 2012, pp. 755–764.

Reveal 1992: James L. Reveal, *Gentle Conquest: The Botanical Discovery in North America with Illustrations from the Library of Congress*. Washington, D.C.: Starwood Publishing, 1992.

Reveal 1996: James L. Reveal, "What's in a Name: Identifying Plants in Pre-Linnaean Botanical Literature", in Bart K. Holland (ed.), *Prospecting for Drugs in Ancient and Medieval European Texts. A Scientific Approach*. Amsterdam: Harword Academic, 1996, pp. 57–90.

Riddle 1971/2008: John M. Riddle, "Dioscorides", in Charles Coulston Gillispie (ed.), *Dictionary of Scientific Biography*, 16 vols. New York, NY: Charles Scribner's Sons, 1970–1980; vol. 4, 1970, pp. 119–123 (reprint in Charles Coulston Gillispie, Frederic Lawrence and Noretta Koertge [eds], *Complete Dictionary of Scientific Biography*, 27 vols. Detroit: Charles Scribner's Sons, 2008, vol. 4, pp. 119–123).

Riddle 1981: John M. Riddle, "Pseudo-Dioscorides' *Ex herbis femininis* and Early Medieval Medical Botany", *Journal of the History of Biology* 14/1 (1981), pp. 43–82.

Riddle 1985: John M. Riddle, *Dioscorides on Pharmacy and Medicine* (History of Sciences Series 3). Austin, TX: University of Texas Press, 1985.

Riddle 2010: John Riddle, John, *Goddesses, Elixirs, and Witches: Plants and Sexuality Throughout Human History*. New York, NY: Palgrave McMillan, 2010.

Rieder and Zanetti 2018: Philip Rieder, and François Zanetti (eds), *Materia Medica: Savoirs et usages des médicaments aux époques médiévales et modernes* (Rayon d'histoire). Geneva: Droz, 2018.

Rinaldi 2011: Massimo Rinaldi, "Sulla fortuna Settecentesca di Prospero Alpini", in Ongaro 2011a, pp. 231–259.

Rinaldi 2016: Massimo Rinaldi, "'*Non domini nostri, sed duces*'. Sulla circolazione del *De balsamo* di Prospero Alpini tra Cinque- e Seicento", in Ongaro 2016, pp. 299–328.

Rioux 2004: Jean-Antoine Rioux, *Le Jardin des plantes de Montpellier: les "leçons de l'histoire"*. Montpellier: Sauramps Medical, 2004.

Risse 1984: Guenter B. Risse, "Transcending Cultural Barriers: The European Reception of Medicinal Plants from the America", in Wolfgang-Hagen Hein (ed.), *Botanical Drugs of the Americas in the Old and New Worlds. Invitational Symposium at the Washington Congress, 1983*. Stuttgart: Wissenschaftliche Verlagsgellschaft, 1984, pp. 31–42.

Risse 1987: Guenter B. Risse, "Medicine in New Spain", in Ronald L. Numbers (ed.), *Medicine in the New World: New Spain, New France, and New England*. Knoxville, TN: University of Tennessee Press, 1987, pp. 12–63.

Roos 2007: Anna Marie Roos, *The Salt of the Earth: Natural Philosophy, Medicine, and Chymistry in England, 1650–1750* (History of Science and Medicine Library 3). Leiden and Boston, MA: Brill, 2007.

Roos 2011: Anna Marie Roos, *Web of Nature: Martin Lister (1639–1712), the First Arachnologist* (History of Science and Medicine Library 22). Leiden and Boston, MA: Brill, 2011.

Rose 1977: P. L. Rose, "Review of Jean Bodin, *Colloquium of the Seven about Secrets of the Sublime (Colloquium Heptaplomeres de Rerum Sublimium Arcanis Abditis)*, ed. by Marion Leathers and Daniel

Kuntz, Princeton, NJ: Princeton University Press, 1975", *Bibliothèque d'Humanisme et Renaissance* 39/2 (1977), pp. 409–410.

Ross 2016: Sarah Gwenyth Ross, *Everyday Renaissances: The Quest for Cultural Legitimacy in Venice* (I Tatti Studies in Italian Renaissance History). Cambridge, MA, and London: Harvard University Press, 2016.

Rotelli 2018: Federica Rotelli, "Exotic Plants in Italian Pharmacopoeia (16th–17th centuries)", *Medicina nei secoli*
30/3 (2018), pp. 827–880.

Rusu 2017: Doina-Cristina Rusu, "Rethinking *Sylva Sylvarum*: Bacon's use of Della Porta's *Magia Naturalis*", *Perspectives on Science* 25/1 (2017), pp. 1–35.

Rusu 2020: Doina-Cristina Rusu, "Using instruments in the study of animate beings: Della Porta's and Bacon's experiments with plants", *Centaurus* 62/3 (2020), pp. 393–405.

Safier 2010: Neil Safier, "Global knowledge on the Move. Itineraries, Amerindian narratives, and Deep Histories of Science", *Isis* 101/1 (2010), pp. 133–145.

Salomon-Bayet 1972: Claire Salomon-Bayet, "Opiologia, Imposture Et Célébration De L'opium", *Revue d'histoire des Sciences*
25/2 (1972), pp. 125–150.

Santamaría Hernández 2012: Maria Teresa Santamaría Hernández (ed.), *Textos médicos grecolatinos antiguos y medievales: estudios sobre composición y fuentes* (Colección humanidades 123). Cuenca: Universidad de Castilla-La-Mancha, 2012.

Scarborough and Nutton 1982: John Scarborough, and Vivian Nutton, "The Preface to Dioscorides' *Materia medica*: introduction, translation, and commentary", *Transactions and Studies of the College of Physicians of Philadelphia*
5/4 (1982), pp. 187–227.

Schiebinger 2004: Londa Schiebinger, *Plants and Empire. Colonial Bioprospecting in the Atlantic World*. Cambridge, MA: Harvard University Press, 2004.

Schiebinger 2005: Londa Schiebinger, "Prospecting for Drugs: European Naturalists in the West Indies", in Schiebinger and Swan 2005, pp. 119–133.

Schiebinger and Swan 2005: Londa Schiebinger, and Claudia Swan (eds), *Colonial Botany: Science, Commerce, and Politics in the Early Modern World*. Philadelphia, PA: University of Pennsylvania Press, 2005.

Shackelford 2004: Jole Shackelford, *A Philosophical Path for Paracelsian Medicine: The Ideas, Intellectual Context, and Influence of Petrus Severinus (1540–1602)*. Copenhagen: Museum Tusculanum Press, 2004.

Shapin 1989: Steven Shapin, "The invisible technician", *American Scientist* 77/6 (1989), pp. 554–563.

Siegel and Poynter 1962: Rudolph E. Siegel, and Frederick Noël Lawrence Poynter, "Robert Talbor, Charles II, and Cinchona: A Contemporary Document", *Medical History* 6/1 (1962), pp. 82–85.

Sigerist 1935: Henry E. Sigerist, "A Doctor's Family in the Fifteenth Century", *Bulletin of the Institute of the History of Medicine* 3/2 (1935), pp. 159–162.

Simon 2019: Anne Simon, "Creuser la terre, creuser la langue. Zoopoétique de la vermine", *Communications* 105/2 (2019), pp. 221–234.

Siraisi 1981: Nancy G. Siraisi, *Taddeo Alderotti and His Pupils*. Princeton, NJ: Princeton University Press, 1981.

Siraisi 1987a: Nancy G. Siraisi, *Avicenna in Renaissance Italy: The Canon and Medical Teaching in Italian Universities after 1500*. Princeton, NJ: Princeton University Press, 1987.

Siraisi 1987b: Nancy G. Siraisi, "The Physician's Task: Medical Reputations in Humanist Collective Biographies", in Nancy G. Siraisi, and Alistair Cameron Crombie (eds), *The Rational Arts of Living: Ruth and Clarence Kennedy Conference in the Renaissance 1982* (Smith College Studies in History 50). Northampton, MA: Department of History, Smith College, 1987, pp. 105–133 (reprint in Siraisi 2001, pp. 157–183).

Siraisi 1990: Nancy G. Siraisi, *Medieval and Early Renaissance Medicine*. Chicago, IL: University of Chicago Press, 1990.

Siraisi 1994: Nancy G. Siraisi, "Il Canone di Avicenna e l'insegnamento della medicina pratica in Europa", in Francesca Vannozzi (ed.), *L'insegnamento della medicina in Europa (secoli XIV–XIX). Atti del Convegno tenutosi a Siena in occasione delle celebrazioni del 750 anni dalla fondazione dell'Università di Siena* (Monografie di Quaderni Internazionali di Storia della Medicina e della Sanità 6). Siena: Tipografia Senese, 1994, pp. 9–24.
English translation: "Avicenna and the Teaching of Practical Medicine", in Siraisi 2011, pp. 63–78.

Siraisi 1997: Nancy G. Siraisi, *The Clock and the Mirror: Girolamo Cardano and Renaissance Medicine*. Princeton, NJ: Princeton University Press, 1997.

Siraisi 2001: Nancy G. Siraisi, *Medicine and the Italian Universities 1250–1600* (Education and Society in the Middle Ages and Renaissance 12). Leiden, Boston, MA, and Koln: Brill, 2001.

Siraisi 2007: Nancy G. Siraisi, *History, Medicine, and the Traditions of Renaissance Learning*. Ann Arbor, MI: University of Michigan Press, 2007.

Slater, López Terrada and Pardo Tomás 2014: John Slater, María Luz López Terrada, and José Pardo Tomás (eds), *Medical Cultures of the Early Modern Spanish Empire*. Farnham, and Burlington, VT: Ashgate, 2014.

Slater, Pardo Tomás and López Terrada 2014: John Slater, José Pardo Tomás, and María Luz López Terrada, "Introduction", in Slater, López Terrada and Pardo Tomás 2014, pp. 1–17.

Smith 2004: Pamela H. Smith, *The Body of the Artisan: Art and Experience in the Scientific Revolution*. Chicago, IL, and London: Chicago University Press, 2004.

Smith 2008: Pamela H. Smith, "Artisanal Knowledge and the Representation of Nature in Sixteenth-Century Germany", in Therese O'Malley, Amy R.W. Meyers (eds), *The Art of Natural History: Illustrated Treatises and Botanical Paintings, 1400–1850* (Studies in the History of Art 69). New Haven, CT, and London: Yale University Press, 2008, pp. 14–31.

Smith and Findlen 2002: Pamela H. Smith, and Paula Findlen (eds), *Merchants & Marvels: Commerce, Science, and Art in Early Modern Europe*. New York, NY: Routledge, 2002.

Sozinskey 1891: Thomas S. Sozinskey, *Medical Symbolism in Connection with Historical Studies in the Arts of Healing and Hygiene*. London: F.A. Davis, 1891.

Spary 2022: Emma C. Spary, "Opium, Experimentation, and Alterity in France", *The Historical Journal* 65/1 (2022), pp. 49–67.

Sprague 1939: T.A. Sprague, "Boerhaave as a botanist", in Committee for the commemoration of Boerhaave (ed.), *Memorialia Herman Boerhaave optimi medici*. Haarlem: Erven F. Bohn, 1939, pp. 91–94.

Stacey 2007: Peter Stacey, *Roman Monarchy and the Renaissance Prince*. Cambridge and New York, NY: Cambridge University Press, 2007.

Stannard 1969/1999a: Jerry Stannard, "The Herbal as a Medical Document", *Bulletin of the History of Medicine* 43 (1969), pp. 212–220 *(reprint in* Katherine E. Stannard, and Richard Kay [eds], *Herbs and Herbalism in the Middle Ages and Renaissance* [Variorum Collected Studies Series CS650]. Aldershot: Ashgate, 1999a, chap. II).

Stannard 1974/1999b: Jerry Stannard, "Medieval Herbals and their Development", *Clio Medica* 9 (1974), pp. 23–33 (reprint in Katherine E. Stannard, and Richard Kay [eds], *Herbs and Herbalism* [Variorum Collected Studies Series CS650]. Aldershot: Ashgate, 1999b, chap. III).

Stannard 1980/1999c: Jerry Stannard, "Albertus Magnus and Medieval Herbalism", in James A. Weisheipl (ed.), *Albertus Magnus and the Sciences*. Toronto: Pontifical Institute of Medieval Studies, 1980, pp. 355–377 (reprint in Katherine E. Stannard, and Richard Kay [eds], *Pristina Medicamenta. Ancient and Medieval Botany* [Variorum Collected Studies Series CS646]. Aldershot: Ashgate, 1999c, chap. XVIII).

Stearn 1962: William T. Stearn, "The Influence of Leyden on Botany in the Seventeenth and Eighteenth Centuries", *The British Journal for the History of Science* 1/2 (1962), pp. 137–158.

Stefanaki et al. 2018: Anastasia Stefanaki, Gerard Thijsse, Gerda van Uffelen, Marcel Eurlings, and Tinde van Andel, "The En Tibi herbarium, a 16th century Italian treasure", *Botanical Journal of the Linnean Society* 187/3 (2018), pp. 397–427.

Stefanaki et al. 2019: Anastasia Stefanaki, Henk Porck, Ilaria Maria Grimaldi, Nikolaus Thurn, Valentina Pugliano, Adriaan Kardinaal, Jochem Salemink, Gerard Thijsse, Claudine Chavannes-Mazel, Erik Kwakkel, and Tinde van Andel, "Breaking the silence of the 500-year-old smiling garden of everlasting flowers", *PLoS ONE* 14/6 (2019).

Stobart and Francia 2014: Anne Stobart, and Susan Francia, "The Fragmentation of Herbal History: The Way Forward", in Susan Francia, and Anna Stobart (eds), *Critical Approaches to the History of Western Herbal Medicine. From Classical Antiquity to the Early Modern Period*. London, New Delhi, New York, NY, and Sydney: Bloomsbury, 2014, pp. 1–20.

Stolberg 2013: Michael Stolberg, "Medizinische *Loci communes*. Formen und Funktionen einer ärztlichen Aufzeichnungspraxis im 16. und 17. Jahrhundert", *NTM – Zeitschrift für Geschichte der Wissenschaften, Technik und Medizin* 21 (2013), pp. 37–60.

Stolberg 2014: Michael Stolberg, "John Locke's 'New Method of Making Common-Place-Books': Tradition, Innovation and Epistemic Effects", *Early Science and Medicine* 19 (2014), pp. 448–470.

Stolberg 2016: Michael Stolberg, "Medical Note-Taking in the Sixteenth and Seventeenth Centuries", in Cevolini 2016, pp. 243–264.

Suggi 2005: Andrea Suggi, *Sovranità e armonia: la tolleranza religiosa nel* Colloquium Heptaplomeres *di Jean Bodin*. Rome: Storia e letteratura, 2005.

Swan 2006: Claudia Swan, "The Uses of Realism in Early Modern Illustrated Botany", in Jean A. Giverns, Karen M. Reeds, and Alain Touwaide (eds), *Visualizing Medieval Medicine and Natural History, 1200–1550* (AVISTA Studies in the History of Medieval Technology, Science and Art 5). Aldershot: Ashgate, 2006, pp. 239–250.

Swan 2008: Claudia Swan, "The Uses of Botanical Treatises in the Netherlands, c. 1600", in Therese O'Malley, Amy R.W. Meyers (eds), *The Art of Natural History: Illustrated Treatises and Botanical Paintings, 1400–1850* (Studies in the History of Art 69). New Haven, CT, and London: Yale University Press, 2008, pp. 63–81.

Swart et al. 2019: Ingeborg Swart, Mieke Beumer, Wouter Klein, and Tinde van Andel, "Bodies of the plant and Animal Kingdom: An illustrated manuscript on materia medica in the Netherlands (ca. 1800)", *Journal of Ethnopharmacology* 237 (2019), pp. 236–244.

Tafur 1874: Pero Tafur, *Andanças é viajes de Pero Tafur por diversas partes del mundo avidos (1435–1439)*. Madrid: Impr. de M. Ginesta, 1874.

Talbot 1976: Charles H. Talbot, "America and the European Drug Trade", in Fredi Chiappelli (ed.), *First Images of America: The Impact of the New World on the Old*, 2 vols. Berkeley, CA, and London: University of California Press, 1976, vol. 2, pp. 833–844.

Tannier 2018: Bernard Tannier, "L'hermétisme à la Renaissance", in François Laroque (ed.), *Histoire et secret à la Renaissance*. Paris: Presses Sorbonne Nouvelle, 2018, pp. 85–97.

Teigen 1987: Philip M. Teigen, "Taste and Quality in 15th- and 16th- Century Galenic Pharmacology", *Pharmacy in History* 29/2 (1987), pp. 60–68.

Teza 1897–1898: Emilio Teza, "Il 'De Simplicibus' di B. Rinio nel codice Marciano", *Atti del Regio Istituto Veneto di Scienze, Lettere ed Arti* 9 (1897–1898), pp. 18–29.

Thijsse 2018: Gerard Thijsse, "A Contribution to the History of the Herbaria of George Clifford III (1685–1760)", *Archives of Natural History* 45/1 (2018), pp. 134–148.

Thijsse and Veldkamp 2003: Gerard Thijsse, and Jan-Frits Veldkamp, *Guide: Van Royen Herbarium*. Leiden and Boston, MA: Brill, 2003.

Thijsse and Wesseling 2021: Gerard Thijsse, and Margreet Wesseling, "Herman Boerhaave: A search for his herbarium collections", *Taxon* 70/1 (2021), pp. 170–181.

Thomas 1955: H. Hamshaw Thomas, "Presidential Address: Experimental Plant Biology in Pre-Linnaean Times", *Bulletin of the British Society for the History of Science* 2/12 (1955), pp. 15–22.

Thorogood 2016: Chris Thorogood, *Field guide to the wild flowers of the Western Mediterranean*. Richmond, Surrey: Royal Botanic Gardens, Kew, 2016.

Thorogood 2019: Chris Thorogood, *Field guide to the wild flowers of the Eastern Mediterranean*. Richmond, Surrey: Royal Botanic Gardens, Kew, 2019.

Thorndike 1923–1958: Lynn Thorndike, "The Attack on Pliny", in Lynn Thorndike, *A History of Magic and Experimental Science*, 8 vols. New York, NY: Columbia University Press, 1923–1958, vol. 4, 1934, pp. 593–610.

Thorndike 1932: Lynn Thorndike, "Rufinus: A forgotten botanist of the thirteenth century", *Isis* 18/1 (1932), pp. 63–76.

Thorndike 1946: Lynn Thorndike, *The Herbal of Rufinus* (Corpus of mediaeval scientific texts 1). Chicago, IL: University of Chicago Press, 1949.

Tjon Sie Fat 1991: Leslie Tjon Sie Fat, "Clusius' garden: a reconstruction" in Leslie Tjon Sie Fat and Erik de Jong (eds),
The Authentic Garden: A Symposium on Gardens. Leiden: Clusius Foundation, 1991, pp. 3–12.

Tomasello 1999: Michael Tomasello, *The Cultural Origins of Human Cognition*. Cambridge, MA, and London: Harvard University Press, 1999.

Tomlinson 2012: Rowan Cerys Tomlinson, "'Plusieurs choses qu'il n'avoit veuës': Antoine Du Pinet's Translation of Pliny the Elder (1562)", *Translation and Literature* 21/2 (2012), pp. 145–161.

Tongiorgi Tomasi 1997: Lucia Tongiorgi Tomasi, *An Oak Spring Flora: Flower Illustration from the Fifteenth Century to the Present Time; a Selection of the Rare Books, Manuscripts and Works of Art in the Collection of Rachel Lambert Mellon*. New Haven, CT: Yale University Press, 1997.

Tongiorgi Tomasi and Willis 2009: Lucia Tongiorgi Tomasi, and Tony Willis, *An Oak Spring Flora: Herbs and Herbals from the Fourteenth to the Nineteenth Century; a Selection of the Rare Books, Manuscripts and Works of Art in the Collection of Rachel Lambert Mellon*. Upperville Virginia, VA: Oak Spring Garden Library, 2009.

Totelin 2004: Laurence M. V. Totelin, "Mithradates' Antidote: A Pharmacological Ghost", *Early Science and Medicine* 9/1 (2004), pp. 1–19.

Totelin 2008: Laurence M. V. Totelin, *Hippocratic Recipes- Oral and Written Transmission of Pharmacological Knowledge in Fifth- and Fourth-Century Greece* (Studies in Ancient Medicine 34). Leiden and Boston, MA: Brill, 2008.

Totelin 2018: Laurence M. V. Totelin, "Therapeutics", in Peter E. Pormann (ed.), *The Cambridge Companion to Hippocrates*. Cambridge: Cambridge University Press, 2018, pp. 200–216.

Touwaide 1996: Alain Touwaide, "The Aristotelian School and the Birth of Theoretical Pharmacology in Ancient Greece", in Regine Pötzsch (ed.), *The Pharmacy: Windows on History*. Basel: Editiones Roche, 1996, pp. 11–22.

Touwaide 1997: Alain Touwaide, "La thérapeutique médicamenteuse de Dioscoride à Galien: du *pharmaco-centrisme* au *médico-centrisme*", in Armelle Debru (ed.), *Galen on Pharmacology. Philosophy, History and Medicine* (Studies in Ancient Medicine 16). Leiden, New York, NY, and Köln: Brill, 1997, pp. 255–282.

Touwaide 2000: Alan Touwaide, "Loquantur ipsi ut velint . . . modo quis serpens sit tirus . . . non ignorent: Leoniceno's contribution to Renaissance epistemological approach to scientific lexicography", in Wouter Bracke, and Herwig Deumens (eds), *Medical Latin from the Late Middle Ages to the Eighteenth Century. Proceedings of the European Science Foundation Exploratory Workshop in the Humanities, Organized Under the Supervision of Albert Derolez in Brussels on 3 and 4 September 1999* (Academia regia Belgica medicinae. Dissertationes. Series historica 8). Brussels: Koninklijke Academie voor Geneeskunde van België, 2000, pp. 151–173.

Touwaide 2007: Alain Touwaide, "Art and Sciences: Private Gardens and Botany in the Roman Empire", in Michael Conan, and W. John Kress (eds), *Botanical Progress, Horticultural Innovation and Cultural Changes* (Dumbarton Oaks Colloquium on the History of Landscape Architecture 28). Washington, DC: Dumbarton Oaks, 2007, pp. 37–50.

Touwaide 2008a: Alain Touwaide, "Leoniceno, Nicolò", in Noretta Koertge (ed.), *New Dictionary of Scientific Biography*. Detroit, MI: Charles Scribner's Sons, 2008a, vol. 4. pp. 264–267.

Touwaide 2008b: Alain Touwaide, "Botany and Humanism in the Renaissance: Background, Interaction, Contradictions", in Therese O'Malley, Amy R.W. Meyers (eds), *The Art of Natural History: Illustrated Treatises and Botanical Paintings, 1400-1850* (Studies in the History of Art 69). New Haven, CT, and London: Yale University Press, 2008b, pp. 33–61.

Touwaide 2012: Alain Touwaide, "*Quid pro Quo*: Revisiting the Practice of Substitution in Ancient Pharmacy", in John M. Riddle, Anne Van Arsdall, and Timothy Graham (eds), *Herbs and Healers from the Ancient Mediterranean through the Medieval West. Essays in Honor of John M. Riddle* (Medicine in the Medieval Mediterranean). Farnham, and Burlington, VT: Ashgate, 2012, pp. 19–62.

Touwaide 2013a: Alain Touwaide, "Foreword: In Defense of Medical Tradition", in Hakima Amri, Mones Abu-Asab, and Mark S. Micozzi (eds), *Avicenna's Medicine: A New Translation of the 11th-Century Canon with Practical Applications for Integrative Health Care*. Rochester, VT, and Toronto: Healing Arts Press, 2013a, pp. ix–xiii.

Touwaide 2013b: Alain Touwaide, *Tractatus de Herbis. Sloane Ms. 4016*. Barcelona: Moleiro, 2013b.

Touwaide 2022a: Alain Touwaide, "Mattioli, Pietro Andrea", in Marco Sgarbi (ed.), *Encyclopedia of Renaissance philosophy*. Cham: Springer, 2022a, pp. 2112–2114.

Touwaide 2022b: Alain Touwaide, "Leoniceno, Nicolò", in Marco Sgarbi (ed.), *Encyclopedia of Renaissance philosophy*. Cham: Springer, 2022b, pp. 1895–1899.

Touwaide and Dendle 2008: Alain Touwaide, and Peter Dendle, "Introduction", in Peter Dendle, and Alain Touwaide (eds), *Health and Healing from the Medieval Garden*. Woodbridge, Suffolk and Rochester, NY: Boydell Press, 2008, pp. 1–14.

Trimen 1887: Henry Trimen, "Hermann's Ceylon Herbarium and Linnaeus' Flora Zeylanica", *Journal of the Linnean Society (Botany)* 24 (1887), pp. 129–155.

Truitt 2009: Elly R. Truitt, "The Virtues of Balm in Late Medieval Literature", *Early Science and Medicine* 14/6 (2009), pp. 711–736.

Turchetti 2015: Mario Turchetti, "Jean Bodin", in Edward N. Zalta (ed.), *The Stanford Encyclopedia of Philosophy*, Spring 2015 (online resource).

Ubrizsy Savoia 1996: Andrea Ubrizsy Savoia, "The influence of New World Species on the Botany of the 16th Century", *Asclepio* 48/2 (1996), pp. 163–172.

Ubrizsy Savoia 2014: Andrea Ubrizsy Savoia, "500 anni fa iniziava l'insegnamento della botanica s.l. all'Università 'La Sapienza' di Roma", *Annali di storia delle Università italiane* 18 (2014), pp. 341–354.

Valentinelli 1868–1873: Joseph Valentinelli, *Bibliotheca manuscripta ad S. Marci venetiarum, digessit et commentarium addidit Joseph Valentinelli praefectus. Codices mss. Latini*, 6 vols. Venice: ex Typographia Commercii, 1868–1873.

Valverde 1982: José Luís Valverde, "La experimentación farmacológica de Drogas americanas", *Ars Pharmaceutica* 23/2 (1982), pp. 151–192.

Valverde 2010: José Luís Valverde, *Evaluation of Latin American Materia Medica and its Influence on Therapeutics*. Granada: International Academy of History of Pharmacy, 2010.

Van Andel 2017: Tinde van Andel, *Open the Treasure Room and Decolonize the Museum*. Leiden: Universiteit Leiden, 2017 (Inaugural lecture delivered on the occasion of the acceptance of the position of Special professor of the Clusius chair of History of Botany and Gardens, 6 January 2017) (online resource).

Van Andel and Barth 2018: Tinde van Andel, and Nadine Barth, "Paul Hermann's Ceylon Herbarium (1672 1679) at Leiden, the Netherlands", *Taxon* 67/5 (2018), pp. 977–988.

Van Andel et al. 2018: Tinde van Andel, Jagannath Mazumdar, Nadine Barth, and Jan-Frits Veldkamp, "Possible Rumphius Specimens Detected in Paul Hermann's Ceylon Herbarium (1672–1679) in Leiden, the Netherlands", *Blumea* 63/1 (2018), pp. 11–19.

Van Arsdall 2014: Anne Van Arsdall, "Evaluating the Content of Medieval Herbals", in Susan Francia, and Anne Stobart (eds), *Critical Approaches to the History of Western Herbal Medicine. From Classical Antiquity to the Early Modern Period*. London, New Delhi, New York, NY, and Sydney: Bloomsbury, 2014, pp. 47–65.

van der Eijk 1997: Philip J. van der Eijk, "Galen's use of the concept of 'qualified experience' in his dietetic and pharmacological works";, in Armelle Debru (ed.), *Galen on Pharmacology* (Studies in Ancient Medicine 16). Leiden, New York, NY, and Koln: Brill, 1997, pp. 35–57.

Van der Ham 2006: Raymond van der Ham, "Een Haagse simpliciakast met inhoud", *Foliolum* 29/3 (2006), pp. 27–32.

Van Leerdam 2021: Andrea van Leerdam, "Popularising and Personalising an Illustrated Herbal in Dutch", *Nuncius* 36/2 (2021), pp. 356–393.

Van Oostroom 1937: S. J. van Oostroom, "Hermann's Collection of Ceylon Plants in the Rijksherbarium (National Herbarium) at Leiden", *Blumea* 29 (1937), pp. 193–209.

Varey, Chabrán and Weiner 2000: Simon Varey, Rafael Chabrán, and Dora B. Weiner (eds), *Searching for the Secrets of Nature: The Life and Works of Dr. Francisco Hernández*. Stanford, CA: Stanford University Press, 2000.

Veendorp and Baas Becking 1938/1990: Hesso Veendorp, and Lourens Baas Becking, *Hortus Academicus Lugduno-Batavus 1587-1937*. Leiden: Rijksherbarium/Hortus Botanicus, 1938 (reprint: Leiden: Rijksherbarium, Hortus Botanicus, 1990).

Veldman 2012: Sarina Veldman, "Prins der botanici. De reizen, verzamelingen en studies van Paul Hermann", in Esther van Gelder (ed.), *Bloeiende Botanie*. Hilversum: Verloren, 2012, pp. 146–157.

Ventura 2009: Iolanda Ventura, "Introduction", in Iolanda Ventura (ed.), *Tractatus de herbis: MS London, British Library, Egerton 747* (Edizione nazionale "La Scuola Medica Salernitana" 5). Florence: SISMEL-Edizioni del Galluzzo, 2009, pp. 1–188.

Ventura 2013: Iolanda Ventura, "Changing Representations of Botany in Encyclopaedias from the Middle Ages to the Renaissance", in Anja-Silvia Goeing, Anthony T. Grafton, and Paul Michel (eds), *Collectors' Knowledge: What is Kept, What is Discarded/Aufbewahren oder wegwerfen – Wie Sammeler entscheiden*. Leiden and Boston, MA: Brill, 2013, pp. 97–143.

Ventura 2016: Iolanda Ventura, "Botany, Dietetics, and Pharmacy in Pietro D'Abano's *Expositio Problematum*: on Sections XX, XXI, and XXII", in Pieter De Leemans, and Maarten J.F.M. Hoenen (eds), *Between Text and Tradition: Pietro D'Abano and the Reception of Pseudo-Aristotle's* Problemata Physica *in the Middle Ages*. Leuven: Leuven University Press, 2016, pp. 163–200.

Ventura 2017: Iolanda Ventura, "Classification System and Pharmacological Theory in Medieval Collections of Materia Medica: A Short History from the Antiquity to the End of the 12th Century", in Tanja Pommerening, and Walter Bisang (eds), *Classification from Antiquity to Modern Times: Sources, Methods, and Theories from an Interdisciplinary Perspective*. Berlin and Boston, MA: De Gruyter, 2017, pp. 101–166.

Ventura 2018: Iolanda Ventura, "Medicina e farmacologia «scolastica» nei commenti dell'*Antidotarium Nicolai*", in Cecilia Panti, and Nicola Polloni (eds), *Vedere nell'ombra. Studi su natura, spiritualità e scienze operative offerti a Michela Pereira* (Micrologus Library 90). Florence: SISMEL-Edizioni del Galluzzo, 2018, pp. 277–297.

Verwaal 2020: Ruben E. Verwaal, *Bodily Fluids, Chemistry and Medicine in the Eighteenth-century Boerhaave School*. Cham and London: Springer, Palgrave MacMillan, 2020.

Vine 2019: Angus E. Vine, *Miscellaneous Order: Manuscript Culture and the Early Modern Organization of Knowledge*. Oxford: Oxford University Press, 2019.

Waller 1938: François Gerard Waller, *Biographisch woordenboek van Noord Nederlandsche graveurs*. The Hague: Martinus Nijhoff, 1938.
Wallis 2012: Patrick Wallis, "Exotic Drugs and English Medicine: England's Drug Trade, c. 1550–c. 1800", *Social History of Medicine* 25/1 (2012), pp. 20–46.
Wear 1999: Andrew Wear, "The Early modern debate about foreign drugs: localism versus universalism in medicine", *The Lancet* 354 (1999), pp. 149–151.
Webb 2011: Jennifer D. Webb, "All is not fun and games: conversation, play, and surveillance at the Montefeltro court in Urbino", *Renaissance Studies* 26/3 (2011), pp. 417–440.
Webster 1966: Charles Webster, "The Recognition of Plant Sensitivity by English Botanists in the Seventeenth Century", *Isis* 57/1 (1966), pp. 5–23.
Weidenbaum 2009: Shira Chaya Weidenbaum, *Patterns of persuasion: Religious literary dialogue in Renaissance France*. Unpublished PhD thesis. New Haven, CT: Yale University, 2009.
Welch 2008: Evelyn Welch, "Space and spectacle in the Renaissance pharmacy", *Medicina & Storia* I5 (2008), pp. 127–158.
Wijnands 1988: D. Onno Wijnands, "Hortus auriaci: The Gardens of Orange and their Place in late 17th-Century Botany and Horticulture", *Journal of Garden History* 8/2 (1988), pp. 61–86.
Wijnands and Heniger 1991: D. Onno Wijnands, and Johannes Heniger, "The origins of Clifford's herbarium", *Botanical Journal of the Linnean Society* 106 (1991), pp. 129–146.
Wijnands et al. 1994: D. Onno Wijnands, Erik Zevenhuizen, and Johannes Heniger, *Een Sieraad voor de Stad: De Amsterdamse Hortus Botanicus 1638–1993*. Amsterdam: Amsterdam University Press, 1994.
Wilson 1995: Catherine Wilson, *The Invisible World: Early Modern Philosophy and the Invention of the Microscope*. Princeton, NJ: Princeton University Press, 1995.
Winterbottom 2014: Anna Winterbottom, "Of the China Root: A Case Study of the Early Modern Circulation of Materia Medica", *Social History of Medicine* 28/1 (2014), pp. 22–44.
Witt 2009: Ronald G. Witt, "The Rebirth of the Romans as Models of Character. De viris illustribus", in Victoria Kirkham, and Armando Maggi (eds), *Petrarch: A Critical Guide to the Complete Works*. Chicago, IL: The University of Chicago Press, 2009, pp. 103–111.
Wittop Koning 1954: Dirk Arnold Wittop Koning, "De geschiedenis van de groothandel in geneesmiddelen", *Pharmaceutisch Tijdschrift België* 7 (1954), pp. 1–5.
Woolley 2004: Benjamin Woolley, *The Herbalist: Nicholas Culpeper and the Fight for Medical Freedom*. London: HarperCollins, 2004.
Yates 1966: Frances A. Yates, *The Art of Memory*. London: Routledge, 1966.
Yates 1976: Frances A. Yates, *The Mystery of Jean Bodin*, "The New York Review" October 14, 1976.
Yeo 2014: Richard Yeo, *Notebooks, English Virtuosi, and Early Modern Science*. Chicago, IL, and London: University of Chicago Press, 2014.
Yoo 2018: Genie Yoo, "Wars and Wonders. The Inter-Island Information Networks of Georg Everhard Rumphius", *The British Journal of the History of Science* 51/4 (2018), pp. 559–584.
Zipser 2013: Barbara Zipser (ed.), *Simon of Genoa's Medical Lexicon*. London: Versita, and Warschau and Berlin: De Gruyter, 2013.
Zonta and Brotto 1922: Gasparo Zonta, and Augusto Giovanni Brotto (eds), *Acta graduum academicorum Gymnasii Patavini ab anno 1406 ad annum 1450*. Padua: Typis Seminarii, 1922.

Index

Abdachim 53, 61, 62, 64, 66, 68
Abdella 53, 61, 62, 64, 66, 68
Académie des Sciences XII, 158, 165, 188, 189, 193, 194, 211, 216, 218
Achillea millefolium L. X, 100, 101
Acmella 109
Aconitum lycoctonum L. 92
Aconitum napellus L. (monkshood) 92
Adam and Eve 72, 84
Adelmann, Howard B. 126, 135, 140, 143, 144, 225
Adversaria. See notebooks, types of
Aeneid. See Virgil
Aesculapius 42
Al-Adwīa al-qalbīya (De viribus cordis). See Avicenna
Al-Qānūn fi'l-ṭibb (Liber Canonis). See Avicenna
Al-Rāzī. See Rhazes
Albert the Great 122, 123, 229, 240, 242, 246
Alchemilla xanthochlora Rothm. (lady's mantle) 22
Alchemy 14, 74, 81
Aldrovandi, Ulysses 11, 19, 54, 67, 121
– *Dendrologiae* 121, 212
Alpago, Andrea 56
Alpini, Prospero 10, 59–70, 229, 231, 240–242, 244
– *De balsamo dialogus* 53–54, 56, 59, 60, 63–65, 68, 70, 212
– *De medicina Aegyptiorum* 53–54, 59, 62, 64, 66, 212
– *De plantis Aegypti* 10, 53–54, 56, 60–62, 64–68, 212, 223
Amadio, Andrea IX, 6, 19, 22, 24–31, 33–35, 44, 46–47, 51–52, 211, 237
Amber. See Ovid, *Metamorphoses*
American materia medica 170, 176–177, 179–182, 184, 195, 225, 249
American plants 58, 169–173, 175, 177, 181, 183–184, 191, 193
Anatome plantarum. See Marcello Malpighi
Anatomy of Plants. See Nehemiah Grew
Anemone hepatica L. (common hepatica) 22
Antidotes 54, 81, 146, 150–155, 158–159, 163, 201, 220, 243, 248
Apollo 42
Apothecary College of Naples 68
Apothecaries XIII, 6, 10, 31, 42, 46, 51, 54, 59, 66, 68–69, 94, 163–165, 174, 191, 207, 230, 233, 243
– errors made by 20–21

– goods sold by 58
– Jesuit apothecary 186
– Venice and 56, 60
– Zara (Zadar, Croatia) and 33–34, 50
Apuleius Platonicus 72, 101
– *De herbis masculinis* 37
– *Herbarium Apuleii Platonici* X, 72, 100–101, 211
Aqrish, Yitzhak Abraham 63
Arber, Agnes 4, 8, 10, 131, 224
Aristotle 17, 35, 66, 121–122, 124, 126, 180, 237
Arnaud de Villanova 36
Arnold, David 55, 225
Aromatari, Giuseppe degli 139
Ars medendi. See Copho
Asplenium scolopendrium L. 116
Astruct, Jean
– *De morbis venereis libri sex* 162, 212
Atropa belladonna L. (nightshade or belladonna) 22, 72, 95
Atropa mandragora L. See mandrake
Augustine 44, 76
Authority 17, 21, 54, 56, 60–62, 65–69, 151, 174, 178, 180–182, 187, 195, 233
Averroes (Ibn Rushd) IX, 49–50, 64, 66, 242
Avicenna (Ibn Sīnā) 35, 42–43, 61, 242, 246, 249
– *De viribus cordis (al-Adwīa al-qalbīya)* 36
– *Liber canonis (al-Qānūn fi 'l-ṭibb)* 36
– Latin translation of. See Gerard of Cremona
Ayn Shams 57

Bacon, Francis XII, 14–15, 194, 212, 245
Bado, Sebastiano 187
– *Anastasis cortices peruviae* 189
Balm of Gilead 57, 224–225
Balsam or *Commiphora gileadensis* 53, 225, 235
Balsam from America. See American materia medica
Bakhtīshū', Jibra'il ibn (Nestorian)
– *Kitab na't al-hayawā* 47
Bartolomeo of Salerno IX, 49
– *Practica* 50
Bartholomeus Anglicus 38
– *De proprietatibus rerum* 38
Basalla, George 35
Baucis. See Ovid, *Metamorphoses*
Bauhin, Caspar 15
– *Pinax* 106, 109, 204–205, 207, 212

Bauhin, Johann 109
Bedrunia 62
Belladonna, see *Atropa belladonna* L.
Bellorini, Cristina 4, 59, 225
Belon, Pierre 56, 163
– *Les Observations de plusieurs singularitez* 158, 212
Bély, Lucien 87, 225
Bertoloni Meli, Domenico V, 6, 124, 128, 132, 135, 139, 141, 227
Besler, Basilius 109
Bey, Haly 64
Bey, Horrem 64
Bidens Pilosa L. 109
Bidloo, Govard 141, 212, 232
Blegny, Nicolas de
– *Le Remede anglois pour la Guerison des fievres* 191, 213
Blood 92, 122, 124, 126, 130–132, 141, 159, 219
– animal injection 154–155, 157
– circulation of 126–128, 130–131, 146, 153, 155, 168
– experiment 130, 160, 177, 190, 193
Boaistuau, Pierre (also known as Seigneur de Launay) 87
– *Histoires prodigieuses* X, 93, 211
Boccaccio, Giovanni 85
Bodin, Jean 63, 231, 235, 242, 244, 246, 248
Boerhaave, Herman X, XIII, 68, 105–114, 117, 119–120, 125, 141–143, 211, 213, 220, 236, 240, 243, 246–247, 250
– *Elementa Chemiae* 142, 213
– *Index alter plantarum* 106, 113–114, 116, 118, 213
– *Index plantarum* 106, 110, 116, 213
– *Index Seminum Satorum* 119, 211
– *Institutiones et experimenta chemiae* 142, 213
Bologna V, XIII, 19, 36, 54, 67, 102, 211
– Botanical garden 11
Botanical Garden, Bologna. See Bologna
Botanical Garden, Leiden. See Hortus Botanicus
Botanical Garden, Padua. See Padua
Boyle, Robert 153–155, 158–159, 161, 162, 168, 213, 222, 229
Brancher, Dominique 72–74, 76, 79, 81, 84, 97, 121, 123, 228
Brasavola, Antonio Musa 8, 10
Bravo de Sobremonte, Gaspar 186
– *Disputatio apologetic pro Dogmatica Medicinae praestantia* 186, 213
Breviarium. See Stephen of Antioch
Breyne, Jakob 198, 206, 210

Breyne, Johannes 198
Browne, Thomas 15
Brunfels, Otto 6
– *Herbarum vivae eicones* IX, 7, 213
Bruno, Giordano 65, 74
Burke, Peter 62, 227
Burman, Johannes 199, 202–203, 209
– *Flora indica* 210
– *Thesaurus Zeylanicus* XII, 198–199, 203–204, 207, 213

Cairo 53, 57, 59, 61, 63–65
Calancha, Antonio de la
– *Coronica moralizada del Orden de San Augustin en el Peru* 185, 213
Caldera de Heredia, Gaspar 186, 238
– *Tribunalis Medici illustrationes et observationes practicae* 186, 214
Camerarius, Rudolf Jakob 141
Camphora officinarum Nees (camphor) 206
Cañizares-Esguerra, Jorge 55, 227
Cardano, Girolamo 12, 246
Carrara, Francesco II "il Novello" 26, 27
Carrara Herbal (London, British Library, Egerton 2020) IX, XII, 5, 22–23, 26–28, 47, 237
– illustrative cycle in 26
– textual content of 33
– *Tractatus de herbis* (Paris, Bibliothèque nationale de France, *Lat.* 6823) and 47
Carreto, Giovanna 34
Cassia grandis L. (canafistula) 171
– sive *Canella* 204
Cassini, Giovanni 131, 213
Catelan, Laurent 72, 94
– *Rare et curieux discours* 94, 213
Celsus, Aulus Cornelius 145, 150, 214
Cerberus 92–93
Cesalpino, Andrea 11–12, 226
Ceylon XII, 197–199, 201, 203–207, 210
– Herbarium VII, 197, 217, 249–250
Charas, Moyse 146, 163–166, 168
– *Histoire naturelle des . . . dans la composition de la thériaque d'Andromachus* 163, 214
– *Pharmacopée royale galénique et chymique* 163–165, 214
– *Nouvelles observations sur l'Opium par Charas* 165
Charlemagne 99
Chinese root 91
Cibo, Gherardo 90

Cichorium intybus L. (chicory)
- illustration in *Roccabonella Herbal* IX, 28, 30, 32
- sun and 28–29
Cieza de Léon, Pedro 172
- *Parte primera dela Chronica del Peru* 172, 214
Cinnamon 150, 204, 206, 207
Cinamomum sive Canella Zeylanica (*Cinnamomum zeylanicum* Blume) XII, 204–205, 207
- *Cinnamomi cortex* 207
Cinchona bark (*Cinchona officinalis* L.) VII, XII, 16, 118, 169–171, 184–185, 188, 225, 230, 234, 236, 241, 245
- Classification 187, 189–190
- experiment with 186–187, 191–193
- medical debates on 187, 191–193, 195
- remedy 118, 178, 185–186, 191–193
Circa instans. *See* Platearius, Matthaeus
Circe (Circae) 85
Cirsium lanceolatum L. (bull thistle) 34
Citrullus vulgaris Schrad. (watermelon) 23
Clifford, George 106, 120, 245
- *Hortus Cliffortianus* (Clifford Herbarium) X, 105–106, 108, 110, 113–116, 120, 214, 218, 230, 236, 251
Clusius, Carolus XIV, 11, 109, 158, 231–232, 238, 242, 248–249
- *Exoticorum libri decem* 158, 212, 214
Clutia pulchella L. X, 114–115
Clymene 28
Cluyt, Auger 114
Cluyt, Dirck Outgaerts 103, 112
Cobo y Peralta, Bernabé (or Bartolomé) 188, 239
- *Historia del Nuevo mundo* 188, 211
Coldness 147–148, 150, 152, 168
Colin, Antoine 68, 212
Collaboration 180
- Literal 22, 31, 46, 50–51
- Metaphorical 22, 50–51
- virtual VII, 19, 21, 23–25, 27, 29, 31, 33, 35, 37, 39, 41, 43, 45, 47, 49, 51
Collegium Romanum 186–187
Collenuccio, Pandolfo 42
Commelin, Caspar 107, 214
Commelin, Jan 106, 114, 116, 203, 224
Commonplace books. *See also* notebooks, types of
Commonplacing 39–40, 35
Condemned criminals
- poison experiment 146–148, 150–151, 153, 162, 167–168

Conegliano 19
Confirmation (Christian rite of) 57
Consilia 35–36, 41, 224
Constantine "the African" 37, 47, 50
Conversation 2, 45, 47, 52, 61–62, 69–70, 250
- cross-referencing as 35, 46
- portraits and 50–51
Convolvulus mechoacan Vandelli (mechoacan root) 171
Cook, Harold 55, 62, 102, 147, 158, 169, 179, 184, 192, 229–230, 232
Cooper, Alex 55, 230
Copenhagen 207, 236, 245
Copho 37
- *Ars medendi* 37
Cordus, Euridicius 8
Cordus, Valerius 8
Cornaro, Francesco 113
Costeo, Giovanni 12, 124
Courten, William 159, 168, 214
Crescenzi, Piero de 37
- *Liber ruralium commodorum* 37
Crete (Greece) 33
Crisciani, Chiara 8, 36, 41, 122, 224, 228–229
Cross-cultural production VII, 53–57, 59, 61–63, 65, 67, 69, 178, 238
Cross-culturality 64
Croton sp. 171
Cuba, Johannes von 78
- *Herbarius* X, 86, 214
Cucumis flexuosus L. (serpent melon) 23
Cucumis melo L. (melon) 23
Cucumis sativus L. (garden cucumber) 23
Cucurbita Lagenaria L. (bottle gourd) 23
Cucurbitaceae family (plant) 23
Culpeper, Nicholas 91, 248
Curran, Brian 65, 229

D'Oignies, Simon Joseph 108, 225
- *Herbarium* X, XIII, 105, 108–111, 113–117, 120
Dalmatia. *See* Venice, territories of
Dante, Alighieri 88
da Orta, Garcia 91, 242
- *Colóquios dos simples* 63, 91, 215, 231
De Bry, Théodore X, 77, 83, 216, 221
De dietis universalibus et particularibus. *See* Isaac Judaeus
De dinamidia. *See* Pseudo-Galen
De fracturis. *See* Hippocrates

De herbis masculinis. See Apuleius Platonicus
De La Brosse, Guy 14
De La Bruyère, Jean 87–88
– *Caractères* 87, 218
De La Fontaine, Jean 85
Delaporte, François 128, 141, 229
Della Porta, Giovanni Battista 14, 73, 76, 245
– *Magiae naturalis* 73, 215, 245
de l'Obel, Matthias (or Mathias de Lobel) 15, 109
De materia medica (Περὶ ὕλης ἰατρικῆς). See also Dioscorides
– books in 10, 47, 107, 172, 203, 206–209, 247
– curriculum and lectures 4, 9, 12, 206
– materia medica VII–VIII, 15–17, 23, 37–38, 52–55, 57–59, 61, 63, 65, 67, 69, 175–176, 180–184, 195, 197, 199, 205, 216, 219, 226–228, 230, 233, 235, 238, 244, 248, 249–251
– organisation of 118, 170, 172
De medicina Aegyptiorum. See Prospero Alpini
De plantis Aegypti. See Prospero Alpini
De plantis. See Pseudo-Aristotle
De Plinii et plurium aliorum medicorum in medicina erroribus. See Leoniceno, Niccolò
De politiae litterariae. See Decembrio, Angelo
De proprietatibus rerum. See Bartholomeus Anglicus
De simplicium medicamentorum temperamentis ac facultatibus. See Galen
De Tyard, Pontus 87
de Villasante, Antonio 58
De viribus cordis (al-Adwīa al-qalbīya). See Avicenna
De virtutibus herbarum. See Rufinus
De virtutibus herbarum et aromatum. See Macer Floridus; Odo de Meung
Demeter 72, 82
de' Medici, Cosimo I 59, 101
de' Medici, Francesco I 59
Diodatus 64, 66
Dioscorides IX, XI, 5, 7, 9, 17, 23, 26, 32, 35–37, 42, 61, 67, 72, 80–81, 85, 93, 99, 101, 145–149, 153, 168, 172–173, 204, 212, 215, 219, 221, 233, 244, 248
– *De materia medica* (Περὶ ὕλης ἰατρικῆς) 5, 8, 23, 36, 47, 87, 99, 145, 148, 172–173, 215, 233, 245
Doctrine of Signatures 11, 14, 72–73, 76, 85, 90, 123, 133, 227, 236, 242
Dodoens, Rembert 109
– *Stirpium historiae* 146
Donzelli, Giuseppe 68, 214–215

Dorn, Gérard
– *Congeries Paracelsicae chemiae* 83, 215
Du Laurens, André 122
Dutch East India Companies (VOC) 102, 197
Dutch West India Companies (WIC) 102

East Indies 16, 91, 228, 240
Egmond, Florike V, 4, 10, 14, 21, 31, 35, 124, 184, 225, 231–232, 242
Elaphrium tecomaca (D.C.) Standl. (tacamahac) 171
Eleusinian mysteries 82
Elsholtz, Johann Sigismund XI, 155–157, 159, 168, 215
En Tibi Herbarium 102, 247
Erastus, Thomas 123
Eryngium maritimum L. (sea holly) 91
Este, Isabella d' (Marchesa of Mantua) 50, 227
Evelyn, John 15
Ex herbis feminis. See Pseudo-Dioscorides
Exoticorum libri decem. See Carolus Clusius
Experimentation

Fabaceae family (plant) 26
Fabricius ab Aquapendente, Hieronimus XI, 122, 124–126, 144, 215, 232
Falloppio, Gabriele 146, 148, 150, 152, 167
– *De tumoribus* 146, 148, 150
Fernel, Jean 79, 122
Fertility 72, 90–92, 94, 96
Fevers 54, 118, 148, 150, 179, 185–186, 189–193, 206–207
Ficino, Marsilio 74
Filago germanica L. (common cottonrose) 22
Findlen, Paula 4, 11, 42, 54–55, 101–102, 123, 226, 232–233, 246
Flavius Josephus 61
Florilegia 38
Fludd, Robert 80–81
– *Utriusque cosmi maioris* X, 80–81, 215
Fracastoro, Girolamo 72, 79, 83–84, 242
– *Syphilis, sive Morbi Gallici* 84, 96, 216
Fragaria viridis Weston (strawberries) 34
Fragoso, Juan 185, 239
– *Discursos de las cosas Aromaticas* 185, 216
French Revolution 57
Fuchs, Leonhart 6–8, 11–12, 79, 81, 232
– *De historia stirpium* 6, 81–82, 123, 216
– *L'hystoire des plantes* 82, 216

Galen IX, 17, 35–36, 42–43, 48–49, 56, 61, 78, 84, 88, 121, 124, 126, 128, 144–148, 153, 163, 173, 211, 226
- *De Antidotis* 150, 215
- *De simplicium medicamentorum temperamentis ac facultatibus* 5, 36, 145, 147–148, 173, 215
- *De Theriaca ad Pisonem* 153, 216
- medical school curriculum and Galenic medicine VII, 10, 16, 48, 56, 150–152, 167, 169, 175, 177–178, 190, 232, 240
- humoral theory 16, 73, 70, 170, 176, 192, 195
- pharmacology and Galenic pharmacy 16, 58, 146, 152, 162–163, 168, 177. 185, 190–193, 215, 237, 247–248, 250
- theriaca 159
Galilei, Galileo 65
Garcin, Laurent 209
Gart des Gesundheit. *See Hortus sanitatis*
Gellius, Aulus 44–46, 51
- *Noctes Atticae* 44–46, 50, 216
Gentiana acaulis L. 110
Gentiana major L. 110
Gerard of Cremona 36–37
- translation of Avicenna's *Liber canonis* 36
Gerard of Sabloneta 36
Gerard, John 91–92
- *The Herball* 92
Gessner, Conrad 8
Ghini, Luca 11, 101, 103
Giglioni, Guido 122, 127, 226, 233, 242
Gillenia trifoliata L. 109
Ginzburg, Carlo 82
Gleditsch, Johannes Gottlieb 94, 216
Glisson, Francis 125, 127–128, 131–132, 143, 233
- *Anatomia hepatis* 128, 216
Gloriosa superba L. X, 116–117
Glossaries 23
Gotha 198, 214
Grasshoff, Johann (Latinized in Johannes Grassaeus or Crassaeus) 76
- *Dyas chymica* X, 76, 77, 83, 216, 219
Grew, Nehemiah 121, 125, 127–128, 131–135, 143–144, 224, 227, 233, 237
- *Anatomy of Plants* XI, 15, 132–135, 217
- *Musaeum Regalis Societatis* 133, 217
Guaiacum officinale L. 171
Guarini, Guarino 44
Guilandino, Melchir 60, 67
Günther, August 207, 219

Haak, Cornelis
- *Catalogus* 111, 216
Hades 82
Hales, Stephen 125, 141–143
- *Vegetable Staticks* 142–143, 217, 225
Haller, Albrecht 113, 236
Harvey, William VII, 121, 125–127, 143–144
- *De motu cordis* 126, 217
- *Exercitationes de Generatione Animalium* 126–127, 217
Harvey-Gibson, Robert John 3–4, 235
Healy, Margaret 71, 73–74, 96, 235
Hebrew Bible 57
Heliades. *See* Ovid, *Metamorphoses*
Heliopolis 57
Herbals
- botanical knowledge 16, 81, 87, 92, 101, 241, 247, 249
- herbaria and 1, 13, 16, 72
- herbalist 2, 55, 90, 250
- illustrated VII, IX, 6, 8, 22, 26, 28, 47, 229, 240, 249
- materia medica and medicine 17, 90, 95, 103, 109, 236, 244
- natural history 10, 81
- printed 6
- production of 5
- *see also* Carrara Herbal
- *see also* Roccabonella Herbal
Herbarium Apuleii Platonici. *See* Apuleius Platonicus
Herbarius in Dyetsch 6
Herbarius latinus 6
Herbarum vivae eicones. *See* Otto Brunfels
Heredia, Pedro Miguel de 186, 217
Hermann, Paul 106, 109, 112–113, 116, 197–198, 206, 210–211, 219, 222
- *Ceylon-Herbaria* VII, 197–198, 203, 206, 208–209, 217, 249–250
- *Cynosura materiae medicae* 206, 217
- *Florae Lugduno-Batavae* 217
- *Herbarium* XII, 241, 246–247, 199, 202, 204–205, 207, 242
- *Horti academici* 106, 112, 116, 197, 203, 217
- *Lapis materiae medicae* 204, 206
- *Musaeum Zeylanicum* 198–199, 201, 203, 205, 207, 217
Hermes. *See* Ovid, *Metamorphoses*
Hermes Trismegistus
- *Tabula smaragdina* 83, 218, 223

Hernández, Francisco 182–183, 238–239, 242, 245, 250
- *Historia de las plantas de Nueva España* 183, 238
Hertodt von Todtenfeld, Johann Ferdinand 123
- *Crocologia* 123, 233
Hippocrates 35, 42–43, 47–48, 50, 121, 173, 190, 236, 248
- *Aphorismi* 35, 49
- *De fracturis* 35
- humoral theory, medical practice, pharmacology 73, 79, 84, 145, 225, 247
- medical school curriculum and IX, 50
- portraits of 48
- *Prognostica* 35
Historia medicinal. See Nicolas Monardes
Hoffmann, Friedrich 193
Holy Bible 74, 78, 89, 218
- Plants 226
Homo arbor inversa 74, 84, 241
Hooke, Robert 128–130, 132, 143
- *Micrographia* XI, 128–130, 218
Horace 28
- *Odes* 28
Hordeum distichum L. (two-rowed barley) 23
Hordeum hexastichum L. (common barley) 23
Hortus botanicus, Amsterdam 114, 251
Hortus botanicus, Utrecht 15
Hortus botanicus or *Hortus academicus*, Leiden 11, 14, 99, 103, 234, 250
Hortus Malabaricus 102, 202–203, 207–208, 224, 235
Hortus sanitatis 6
Humanism 42–43, 236, 240, 249
- medicine and 235, 238, 243
Humoral medicine. See Galen
Humulus lupulus L. (common hops) 34

Iatrochemistry 146, 160, 192
Iatromechanics 192
Iberian colonial world 179–180, 231
Ibn Rushd. See Averroes
Ibn Sarabī. See Serapion the Younger
Ibn Sīnā. See Avicenna
Ilceus 84
Imitation. See also copying 28, 31, 42
Imperato, Ferrante 54
Ipecacuana 109
Isagoge in Artem parvam Galeni. See Johannitius (Ḥunayn b. Isḥāḳ)

Israeli ben Solomon, Isaac (or, in the full form of his name, Abu Ya'qub Ishaq ibn Suleiman al-Isra'ili; Latinized in Isaac Judaeus) 37

Jacquart, Danielle 37, 234
Japan 102, 226
Jardin des plantes
- Jardin royal des plantes in Paris 14, 165
- in Montpellier 242
Jesuit missionaries 185
Jode, Gérard de 74, 76
Johannitius (Ḥunayn b. Isḥāḳ) IX, 48, 50
- *Isagoge in Artem parvam Galeni* 48
Jones, John 166–168
- *Mysteries of opium* 166–167, 218
Jungius, Joachim 15
Jupiter (god). See Ovid, *Metamorphoses*

Kaempfer, Engelbert 206, 208
- *Amoenitatum exoticarum fasciculi* 208
Khunrath, Heinrich
- *Amphitheatrum sapientae aeternae* 83, 218
Kircher, Athanasius
- *Mundus subterraneus* 79, 83, 218, 240
- *Ars magna sciendi* 79, 218
Kitab na't al-hayawā. See Bakhtīshū', Jibra'il ibn
Klerk, Saskia 16, 146, 150, 192, 237
Knowledge-building
- process of 24
Kusukawa, Sachiko 6, 90, 237

La Condamine, Charles Marie de XII, 187–189, 218
Laguna, Andrés 10, 173, 233
- *Acerca de la materia medicinal* 8, 172, 219
Laurus XII, 205, 207–209
Leibniz, Gottfried Wilhelm 123, 227
Lemaistre De Sacy, Louis-Isaac 89, 219
Lemnius, Levinus 151–152, 167
- *Les Occultes Merveilles* 151–152, 219
Leoniceno, Niccolò 42, 249
- *De Plinii et plurium aliorum medicorum in medicina erroribus* 42
Lepidium sativum L. (nasturtium) 34
Lettore dei semplici 67
Levens, Peter 88–89, 218
Lévi-Strauss, Claude 71, 238
Liber canonis (al-Qānūn fī 'l-ṭibb). See Avicenna
Liber ruralium commodorum. See Crescenzi, Piero de'

Liber Serapionis aggregatus in medicinis simplicibus.
  See also Carrara Herbal, textual content of;
  Roccabonella Herbal, sources cited; Serapion
  the Younger; Shēm-Tōb, Abrāhām ben; Simon
  of Genoa
Ligtvoet, Jakob 111, 120, 216
Linnaeus, Carolus 2, 104–105, 109, 114, 116, 120,
  141, 199, 206–207, 210, 239
- *Flora Zeylanica* XII, 198, 204, 219, 207–208, 249
- *Hortus Cliffortianus* 105, 108, 116, 207, 219
- *Materia medica* XII, 207–209, 219
- *Species plantarum* 1, 105, 114, 116, 187, 220
Lipare 84
Liquidambar styraciflua L. (liquidambar) 171
Lister, Martin 130–131, 220, 244
locus communis 39
López de Gómara, Francisco 172
- *La Istoria de las Indias* 172, 220
López Piñero, José María 172, 174–175, 183, 186,
  227, 232, 238, 242
Lotus corniculatus L. (bird's foot trefoil) IX, 25–27
Lugo, Juan de 187

Macer Floridus. *See* Odo de Meung
Machiavelli, Niccolò 85, 91
- *La mandragora* 220
Macrobius 44
Maeier, Michael 77
Major, Johann Daniel 130
malaria 170, 185–186, 190, 192–193, 195, 227, 233
Malpighi, Marcello VII, 15, 121, 125, 131–132,
  135–144, 221, 226–227, 241
- *Anatome plantarum* XI, 135–139, 221
- *Opera posthuma* XI, 140, 143–144, 221
Mandrake X, 10–11, 71–73, 78, 82–88, 90, 92–97,
  219, 228
- *Atropa mandrake* L. 95
- *Mandragora* 72, 85, 95, 101
- *Mandragora officinarum* L. 95
Manfredus de Monte Imperiale 47–50
Masci, Girolamo. *See* Nicolas IV (pontificate
  1288–1292)
Maranta, Bartolomeo 58–59
- *Della theriaca* 58, 220
Marggraf, Georg (or Markgraf) 184, 201, 205, 222
Mariano, Paolo 62, 64
Martyr d'Anghiera, Peter 58
Mataria garden of 57–58, 60, 62, 64, 234

Materia medica. *See also* American; Dioscorides;
  medical school, curriculum; pharmacopeia
Mattioli, Pietro Andrea 8, 10, 12, 58–59, 85,
  147–148, 151, 153, 229–230, 232, 249
- *Commentarii* IX, 8–9, 221
- *Di Pedacio Dioscoride Anazarbeo Libri cinque* 8,
  146, 148, 172, 221
- *Discorsi* XI, 58, 148–149, 221
Medical botany 1, 3–6, 10–13, 15, 17, 36, 84, 236,
  242, 244
Medical school 35, 40, 43, 226
- curriculum in 35–37
- liberal arts and 43
- medical College of Venice 60
- medical theory 1, 36, 54, 65, 71, 73, 81, 95,
  147, 148
Melilotus neapolitanus Ten. (neapolitan clover) 26
Melilotus officinalis [L.] Pall. (sweet clover) 26
Mercuriale, Girolamo 123
Merrett, Christopher 130
Messinor 62
Mesue (Yuḥannā Ibn Māsawayh) IX, 49–50
Messir 62
Mexía, Pedro 87, 94, 220
milk 67, 133
Minio, Michelangelo 19, 33–34, 237
Minot, Jacques 193
- *De la nature, et des causes de la fièvre* 193, 220
Miscellanies (miscellanea) 24, 38–39, 44–45, 94,
  217, 250
Mithridatium 59
Momordica balsamina L. (balsam apple) 22
Monardes, Nicolás 58, 68, 91, 171, 182–185,
  241–242
- *Dos libros* 58–59, 221, XI
- *Historia medicinal* 58, 68, 171, 177, 222, 238
- *Las cosas* 183, 185, 222
Mondina da Cividale, Giovanni 36
Montefeltro, Federico da (Duke of Urbino) 50, 251
Morison, Robert 109, 114
- *Plantarum historiae* 114, 222
Morosini, Andrea 65
Morosini, Giovanni 65
Morpheus 82
Morton, Allan G. 4, 132, 240
Mount Sinai 57
Münster, Sebastian 87
Myroxylon balsamum L. 171

Natural History Museum in London X, XII, 105, 114–115, 198, 205, 214, 217, 234
Natural knowledge 54–56, 62, 65, 69–70, 226, 230
Natural philosophy XIII, 5, 15, 24, 36, 184, 213, 225–227, 233, 236, 244
Naturalis historia. *See* Pliny
Naturalis biodiversity Centre X, XIV, 105, 114–115, 117, 198, 231, 240
New World 133, 227
– assimilation and influences on the Old 58, 169–171, 173, 175, 177–181, 183, 185, 187, 189, 191, 193, 195, 246–248
– balsam 58–59
– materia medica, medicine and pharmacology, and botany VII, XI, 176, 195, 226–227, 230, 233, 240, 244
– *Nueva España* 183, 238
– Spanish exploration of 182
Nietzel, Dirck 120
Nicolas IV (pontificate 1288–1292) 31
Nicholas of Cusa 44
Nicotiana tabacum L. (tobacco) 171
Noctes Atticae. See also Gellius, Aulus
Notebooks
– commonplace books 24, 39–41, 45, 122, 240
– humanists' 24
– indexing of 41
– physicians' 38, 41
Note-taking 38–41, 44, 51, 226, 246

Oderisio, Roberto d' IX, 48–49
Odo de Meung
– *De virtutibus herbarum et aromatum* 37
Ogilvie, Brian W. 4, 10, 12–13, 31, 42, 54, 241
Old World balsam 58–59, 169
Oldenburg, Henry 121, 132, 159, 221, 233
Ongaro, Giuseppe 59–60, 64–65, 212, 231, 241, 244
Opium VII, XI, 16, 82, 138, 139, 145–168, 191, 217, 237, 238, 245–246
Opobalsam 177
Organisational systems 23–24, 44 *See also* florilegia; miscellanies; notebooks; note-taking
– alphabetical 23–24, 32–33, 40, 218
– bibliographical 37, 39, 46
– compiling and compilations 20, 26, 35, 38, 41, 73, 87
– cross-referencing 38, 44, 46
– mental mapping 51

– morphological 23, 183
– synonymic 1, 22–23, 28, 31, 33–34, 38, 46, 51, 198–199, 201–203, 205, 207, 212, 219, 227, 229
– taxonomical 1, 13, 31, 104, 174, 207, 210, 224
Ottoman Empire 56, 66, 102, 145, 222
Ovid 28
– *Metamorphoses* 28
Oviedo, Gonzalo Fernández de 172, 181, 222, 242

Pächt, Otto 19, 26, 242
Padua
– Botanical garden 11, 14, 65, 67, 113, 124, 239
– lords (*signori*) of. *See also* Carrara, Francesco II "il Novello" 28,
– physicians and apothecaries 69
– University of Padua and medical School 19–21, 36, 53, 59–60, 68, 125, 198
Palsgrave, John 88–89, 222
Paracelsus, Philippus Aureolus Theophrastus Bombastus von Hohenheim 78, 81, 152, 191, 222, 243
Pardo Tomás, José 172–173, 175, 177, 180–181, 183 232, 238, 242, 246
Pausanias 61
Peony 91–92, 94, 103
Persephone 72, 82
Petrarch 42, 44, 251
Petrollini, Francesco 102
Phaethon. *See* Ovid, *Metamorphoses*
Pharmacopeia or pharmacopoeia 16, 23, 36–37, 72–73, 85, 153, 170–172, 175–178, 181–182, 184, 186, 192–193, 195, 207, 222, 245
Pharmacopoea Hagana 106, 109, 117, 222
Philemon. *See* Ovid, *Metamorphoses*
Philosophical Transactions 130–131, 139, 145, 158–159, 193–194, 209, 213–214, 218, 220–221
Pinax. See Caspar Bauhin
Piso, Willem 184, 201, 205, 209, 222
Plain notebooks. *See* notebooks, types of
plants. *See individual plant names*
Platearius, Matthaeus
– *Circa instans* 37
Plato 76, 121–122, 222, 238, 241
– *Protagoras* 78
Pliniana defensio. See Collenuccio, Pandolfo
Pliny the Elder 8, 28, 35–36, 67, 81, 101, 180, 228, 248
– *Naturalis historia* 29, 42, 99, 201, 223

Plukenet, Leonard 113, 203, 223
Poaceae family (plant) 23
Poison 81, 90, 146, 153, 159, 162, 190, 231, 243
- poisonus plants 72, 87, 90, 92, 96, 148, 150–154, 158–159, 201, 214
Polydorus, Chrysogonus
- *De alchemia* 217, 223
Poppy XI 35, 81–82, 96, 147–149, 151, 154, 157, 163, 168
Populus nigra L. (black poplar) IX, 28–29, 34
Porphyry IX, 49–50
Practica. See Bartolomeo of Salerno
Praga, Giuseppe 19
Pre-Hispanic medical system 176, 180
Priuli, Francesco 64
Pseudo-Apuleius. See Apuleius platonicus
Pseudo-Aristotle 122, 226, 244
- *De plantis* 122, 226
Pseudo-Dioscorides
- *Ex herbis feminism* 37, 244
Pseudo-Galen
- *De dinamidia* 36, 38

Qualities
- system of medicine 6, 10, 12, 123, 147, 152, 177
- logic of opposition and combination 123, 147
- therapeutic 54, 123, 190, 193, 195, 199, 210

Raj, Kapil 55, 241
Rauwolf, Leonhard 56
Ray, John 17
Recipes (medical or pharmacological) 3, 20, 35, 42, 73, 85, 101, 103, 118, 226, 245
Reeds, Karen M. V, VII, 1–3, 5, 42, 173, 204, 234, 243, 247
Reformers of the Studio of Padua 60
Regius, Henricus 15
Rhetorical strategies 39, 54, 60, 65, 71, 76, 94
Reinhardt, Giovanni di Ermanno 34
Renardi, Iohannes filius Hermanni 34
Repici, Luciana 8, 72, 76, 121–122, 244
Rhazes (al-Rāzī) 35
Rinio, Benedetto 19, 230, 239, 243, 247
Riolan the Younger, Jean 122
Roccabonella Herbal (*Liber de simplicibus*, Venice, Biblioteca Nazionale Marciana, *Lat*. VI, 59 [coll. 2548]) VII, IX, 2, 6, 19–35, 37, 39, 41, 43, 45, 47, 49, 51–52

Roccabonella, Jacopo 19, 21
Roccabonella, Lodovico 33–34
Roccabonella, Lodovicus 21–22
Roccabonella, Nicolò IX, 2, 6, 19, 32, 239
- translations by 211
Rovereti, Ottaviano 59, 241
Royal Society 15, 125, 127–128, 130, 132, 143, 145–146, 158, 160, 168, 193–194, 217, 227, 235–236, 242–243
Ruscelli, Girolamo 73, 223
Ruel, Jean 8
Rufinus 37, 239, 248
- *De virtutibus herbarum* 37
Rycaut, Paul
- *The Present State of the Ottoman Empire* 145, 223

Salerno IX, 37, 47, 49, 100–101, 227, 237
Sarsaparilla (*Smilax officinalis* H.B.K.) 171
Salvia canariensis L. X, 113–114
Salvia rosmarinus Spenn. (rosemary) 34
Sarpi, Paolo 65
Sassafran albidum L. (sassafras root) 171
Sbaraglia, Girolamo 143–144
Scaliger, Julius Caesar 12
Schegk, Jacob 123
Schondorff, Johann Balthasar 193, 223
Schreber, Johann Christian Daniel 208–209, 219
Schwencke, Martinus Gill 106, 118–119, 223
Scribonius Largus 145, 223
Serapion the Younger (Ibn Sarabī) 33, 35–36, 61, 67
- *Liber Serapionis aggregatus in medicinis simplicibus* 22, 33, 36, 211
- Pseudo-Serapion 231
Sesamum orientale L. (sesame) 22
Seville 170, 182, 184, 186, 228
Shakespeare, William 85
Shēm-Tōb, Abrāhām ben 33
Sherard, William 198, 201, 203, 223
Symeon Seth 61
Simon of Genoa
- medical lexicon 251
- *Synonyma medicinae seu Clavis sanationis* 31, 38, 228
Simplices (*simplicia*) or simples VII, X, 4–6, 8, 11, 21, 63, 67, 91, 99–119, 153, 212, 214–215, 233
Siraisi, Nancy 12, 36, 41–42, 54, 56, 64, 66, 81, 175, 204, 234, 243, 245–246

Sloane, Hans 159, 214, 249
Smet, Heinrich 68
Smilax aspera L. 109
Smilax china L. 118
Smilax officinalis H.B.K. 171
Spanish chroniclers of the Indies 172
Spanish crown 58, 170, 173, 178, 181–182, 195
Species plantarum. *See* Carolus Linnaeus
Spigelius, Adrianus 14, 146
- *Isagoges in rem herbariam* 14, 146
Spon, Jacob 137, 139–140, 220
Stannard, Jerry XIII, 5–6, 122, 246
Stephan of Antioch 35
- *Breviarium* 31, 35
- Simon of Genoa and 33
Stobart, Anna 3, 247, 250
Strabo 61
Studiolo 50, 227
Swan, Claudia 3, 6, 8, 55, 227, 229, 245–247
Swinden, Tobias 90, 224
Syen, Arnold 198, 206, 224
Synonyma medicinae seu Clavis sanationis.
    See Simon of Genoa
Syphilis 72, 83–84, 91, 95–96, 174, 177, 216, 226

Tafur, Pero 58, 245
Talbor, Robert 191–192, 236, 245
Taste 43, 147–148, 152, 163, 167, 177, 185, 190–191, 207, 237, 247
Taxonomy. *See* organisational systems
Textbooks 27
Theophrastus 61, 67, 80–81, 87–88, 101, 124, 145, 152
- *Historia plantarum* 145, 152, 224
Theriaca. *See* Galen and Maranta
Thesaurus Zeylanicus. *See* Johannes Burman
Tongiorgi Tomasi, Lucia 6, 10–11, 233–234, 240, 248
Tournefort, Joseph Pitton de 17, 106, 109, 223–224
Touwaide, Alain II, V, 4–5, 8, 23, 31, 36, 42, 50, 58, 121, 145, 169, 225, 234, 237, 247–249
Tractatus de herbis et plantis (Herbal of Manfredus de Monte Imperiale) IX, 47–49
Tragopogon pratensis L. (goat's beard) 22
Translation and transmission 23, 42
Trial 5, 82, 146–147, 157–158, 238, 243
Triticum sativum Lam. (wheat) 23
Turner, William
- *A new herbal* 91, 224

Underworld VII, X, 71–97, 240
Urinaria Indica erecta vulgaris 199
- *Pithawakka* 199, 201
- *Jathagembula* 201

Van Beverningh, Hieronymus 198
Van der Mij, Hieronymus 111
Van der Spijk, Johannes 111
Van Haecht Goidtsenhoven, Laurent (Latinized in Laurentius Haechtanus) X, 74
- Μικρόκοσμος: *Parvus mundus* 74–75, 224
Van Helmont, Jean Baptiste 152–153, 160
- *Ortus Medicinae* 152–153, 224
Van Reede tot Drakenstein, Hendrik 235
- *Hortus Indicus malabaricus*. *See also* Hortus malabaricus 102, 203
van Royen, Adriaan XII, 104–105, 107, 109, 112, 114, 116, 241, 247
- *Florae Leydensis* 116, 118, 224
Vanni, Lippo IX, 48–49
Vegetation 15, 121, 122, 124, 126, 130–132, 139, 141, 143, 217, 236
- sap 28, 54, 128, 130–133, 136, 138, 141–142, 217, 220
- vegetative XIII, 88, 122, 126–127, 143, 226, 229, 233
Venice V, IX, XIII, 19–20, 25, 29–30, 32, 34, 36, 43–44, 53, 56, 59–60, 65, 67, 110, 121, 211, 226, 231, 237, 242, 245
- Church of San Giovanni Crisostomo 20
- Church of San Giovanni Nuovo 20
- Church of San Lio 20
- Church of Santa Maria Formosa 20
- Church of San Salvatore 20
- neighbourhoods of (*sestieri*) 44
- territories of (Dalmatia) 34
Ventura, Iolanda V, 4–5, 12, 37, 122, 250
Vesling, Johann 68
- *De plantis aegyptis* 68, 224
Virgil
- *Aeneid* 46
- *Georgics* 28

West Indies 58, 169, 234, 242, 245
Weiditz, Hans 6
Weissmann, Johann Friedrich
- *Balsamum* 68, 224
Wier, Jean 87
Wilkins, John 154

Willis, Thomas 160–162, 167–168, 225
Witchcraft 74, 82, 84
Wren, Christopher 154–155, 158, 161

Xilobalsam 60, 212
Ximenes, Francisco
– *Quatro libros de la naturaleza y virtudes de las plantas y animals* 201

Zadar (Croatia). *See* Zara (Zadar, Croatia)
Zakynthos 66
Zalužanský y Zaluzian, Adam 12
– *Methodi Herbariae libri tres* IX, 12–13, 225
Zara (Zadar, Croatia) 33–34
Zeus 82
Zoophytes 126, 130